Topology and Geometry of Biopolymers

CONTEMPORARY MATHEMATICS

746

Topology and Geometry of Biopolymers

AMS Special Session
Topology of Biopolymers
April 21–22, 2018
Northeastern University, Boston, Massachusetts

Erica Flapan
Helen Wong
Editors

AMERICAN MATHEMATICAL SOCIETY

2010 *Mathematics Subject Classification.* Primary 57M25, 57M27, 05C10, 92C05, 92C40, 92D20, 92E10, 82D60, 82B41, 65C05.

Library of Congress Cataloging-in-Publication Data

Cataloging-in-Publication Data has been applied for by the AMS.
See http:www.loc.gov/publish/cip/.

∞ The paper used in this book is acid-free and falls within the guidelines
established to ensure permanence and durability.
Visit the AMS home page at `https://www.ams.org/`

10 9 8 7 6 5 4 3 2 1 25 24 23 22 21 20

Contents

Preface

Biopolymers are long flexible molecules which are produced by living organisms. The families of polynucleotides (including DNA and RNA) and polypeptides (including proteins) are especially important in basic cellular functions. Because of their flexibility, biopolymers can have complex topological conformations including knots, links, and other types of entanglements, which may affect how they interact with other molecules inside the cell. However, the exact role of such topological features and how they occur is not well understood. Techniques from topology, geometry, dynamics, probability, and statistics, in addition to simulations, and experimentation have all been used to identify and analyze topologically complex biopolymers. These approaches not only provide evidence that could help to explain how and why such features occur but might also pave the way to future uses of topologically complex biopolymers in treating disease or in developing new biodegradable products. By bringing together articles about the topology and geometry of DNA and proteins from many different perspectives, we hope to give readers a sense of the complexity and interconnections within this exciting and fast-moving field.

Part I. The Topology and Geometry of DNA

The first half of this volume is devoted to recent results about the topology and geometry of DNA and how they affect DNA behavior. We begin with an article by Jonathan M. Fogg and Lynn Zechiedrich reporting on experimental results about DNA supercoiling. In particular, they used electron cryo-tomography to explore the 3-dimensional structure of DNA minicircles each of whose length is only 336 base pairs. They also simulated how introducing sequences with a higher propensity to bend affects the 3-dimensional geometry of such minicircles. Their results show unexpected sequence-dependent and supercoiling-dependent structural changes in DNA which may play a role in the development of gene therapy for treating disease.

Next there is an article by Pengyu Liu, Ryan Polischuk, Yunan Diao, and Javier Arsuaga also presenting results about DNA minicircles. However, in this case, they study kinetoplast DNA which has the form of chainmail made up of thousands of minicircles linked together. The authors use numerical simulations to compare the density of minicircles in the kinetoplast DNA with that of a "random knotting" model that they develop. The paper concludes that minicircle density is the most important factor in network formation, with minicircle orientation restrictions playing a significant role as well. Their results may be useful in targeted drug delivery systems.

The third article, written by Alex Kasman and Brenton LeMesurier, studies the relationship between the geometry of a DNA molecule and the amino acid sequences it encodes. The authors use Monte Carlo and Gaussian sampling methods to approximate a multiple integral that represents both the amino acid sequences and the geometry of a DNA molecule. Their results imply the rather surprising result that the genetic code allows for less freedom in the geometry than what would be expected due to random chance. They offer interesting speculations as to what this result might say about the evolution of the genetic code.

This is followed by an article by Tetsuo Deguchi and Erica Uehara which uses simulations to evaluate the probability that a DNA chain of a given length contains a specific knot. In particular, they represent DNA molecules as self-avoiding polygons made up of rigid segments. They find that as the volume around one segment of the chain increases, the probability of obtaining a trefoil knot increases while that of obtaining more complex prime knots decreases. Furthermore, they observe that as the number of segments in a self-avoiding polygon goes to infinity, the probability that a knotted polygon contains a given prime knot as a summand approaches 1.

The article by Dorothy Buck and Danielle O'Donnol considers knotting of replication intermediates. They explain that while type II topoisomerases normally facilitate replication by performing crossing changes to unknot, unlink, or relieve torsional stress on a DNA molecule, topoisomerases can also accidentally cause knotted intermediates. In order to better understand how and why this occurs, the article models replication intermediates by θ-curves and uses the unknotting number together with six biologically justified restrictions to show that only ten knot types can be produced as intermediates in this way.

In the last article on DNA, Allison H. Moore and Mariel Vazquez use the topological operation of band surgery to model site-specific recombination. The article argues that if no such surgery is possible to get from one given knot to a second given knot, then site-specific recombination will not cause DNA containing the first knot to change into the second knot. The authors draw on techniques from 3-manifold topology to impose strong restrictions on when there is a band surgery between a given pair of knots. In addition to this topological approach, the article uses numerical simulation of knots in a cubic lattice to determine whether a band surgery between two knots is possible and, if so, what is the relative frequency of such a surgery.

Part II. The Topology and Geometry of Proteins

In contrast with DNA, biologists did not realize that proteins could contain knots until recently. As more and more knotted proteins are identified, understanding the possible function of knots in proteins and how such proteins fold becomes increasingly important. We begin the second half of this volume with a survey article written by Sophie Jackson on why proteins might knot. In particular, since the folding rates of certain deeply knotted proteins are slow compared with unknotted proteins, some scientists have hypothesized that such proteins must have an evolutionary advantage over faster folding unknotted structures. In this review, the evidence for and against this theory is discussed, including how a knot within a protein may affect the thermodynamic, kinetic, mechanical, and cellular stability of the protein.

The second article in Part II of the volume, written by Ana Nunes and Patrícia Faísca, presents a survey of lattice Gō models that have been used to better understand how proteins fold into a knotted native state. The article includes a discussion of how the results of such models fit together with experimental, theoretical, and other computational results on the folding of knotted proteins. Furthermore, the authors explain some folding properties of knotted proteins and present possible functional advantages of knots in proteins.

The next article, by Dimos Goundaroulis, Julien Dorier, and Andrzej Stasiak, explains how knotoids can be used to identify, analyze, and characterize knots in open protein chains. In particular, a knotoid is a diagram of an open knotted chain where the endpoints are fixed in the plane or sphere of projection. Using knotoids to represent proteins avoids the problem of having to decide which algorithm to use to close the endpoints of a knotted protein chain in order to characterize the knot. In addition to discussing how to use probability distributions of knotoids to measure entanglement of open chains, the paper explains how to incorporate intra-chain protein bonds in the analysis of entanglement. Finally, the article introduces their software package Knoto-ID which, so far, is the only computational tool that uses knotoids to analyze the topology of open curves.

In the fourth article in Part II of the volume, Kenneth C. Millett explains how to adapt Gauss linking numbers, which have played a role in analyzing the topological aspects of DNA molecules, to identify and quantify entanglements in proteins. In particular, the method introduced in this paper can be used to analyze local entanglements that resemble knotting or slipknotting. This makes it possible to compare the topological and geometric properties of entanglements in unknotted proteins to those in knotted proteins. Furthermore, the paper explains how this can be done with or without considering intra-chain bonds. Finally, the paper surveys the software that is currently available for analyzing linking and entanglement of protein structures.

The last paper in this volume, written by Eleni Panagiotou and Kevin W. Plaxco, explains how various topological and geometric properties of unknotted proteins can be used to predict folding rates. In particular, they consider the Gauss linking integral, the torsion, the average crossing number, and the number of sequence-distant contacts of certain proteins and show how all of these factors can cause folding rates to increase or decrease. Among their many results, they show that folding rates decrease as the writhe and torsion of the proteins becomes more negative.

This volume is based on an AMS special session on topology of biopolymers that we organized at Northeastern University in Boston, April 21–22, 2018. We very much enjoyed organizing the special session, listening to all of the talks, and putting together this collection of articles. We hope that readers will find the diversity of approaches to studying the topology and geometry of biopolymers in this volume as stimulating as we have. Finally, we want to thank the AMS for giving us the opportunity to organize such an interdisciplinary session and for inviting us to put together this volume.

Acknowledgments

Erica Flapan was supported in part by NSF grant DMS-1607744, and Helen Wong was supported in part by NSF grants DMS-1510453, DMS-1841221.

Part I

The Topology and Geometry of DNA

Contemporary Mathematics
Volume **746**, 2020
https://doi.org/10.1090/conm/746/14999

Beyond the static DNA model of Watson and Crick

Jonathan M. Fogg and Lynn Zechiedrich

ABSTRACT. DNA supercoiling, the coiling of the Watson-Crick double helix about itself, affects every aspect of how DNA is read, how it is replicated and transcribed, and its multiplicity of interactions with biological molecules. Despite its importance, much about supercoiled DNA (positively supercoiled DNA, in particular) remains unknown. We utilized electron cryo-tomography to investigate the 3D structures of individual and highly pure DNA minicircles, 336 bp in length, with specific and defined degrees of positive or negative supercoiling. Minicircles in each supercoiling state (topoisomer) adopt a wide distribution of 3D conformations unique to that specific supercoiling level. Probing for disruptions to base pairing revealed that localized distortions from the DNA double helix, evident as exposed DNA bases, allow the DNA to adopt conformations that would otherwise be energetically unfavorable. Counterintuitively, we also detected exposed bases in positively supercoiled minicircles beyond a sharp supercoiling threshold, probably as a consequence of extreme bending strain in the highly writhed minicircles. Our data support the "cooperative kinking model" of Lionberger, Stasiak, and colleagues, in which sharp bending at one location on the supercoiled minicircle induces kinking at a DNA site diametrically opposite. Modeling these DNA bending sites, we simulated how they modify the supercoiling-dependent 3D shapes of the minicircles. These experiments revealed unexpected DNA sequence- and supercoiling-dependent structural alterations in DNA and are a step toward creating gene therapy vectors with specific shapes for use in treating human diseases.

CONTENTS

1. Introduction

DNA is composed of two intertwined strands of biological polymers commonly referred to as the Watson and Crick strands. The number of times the strands wrap around each other is referred to as the linking number, Lk. In its most thermodynamically relaxed state, Lk_0, there are, on average, $\sim$10.5 base pairs per turn of the DNA helix (Bates and Maxwell, 2005). This value is called the helical repeat.

DNA is not found in its Lk_0 relaxed state in live cells. Instead, DNA is maintained in a slightly underwound state. This underwinding is called negative supercoiling. A more thorough review of DNA supercoiling and topology can be found in the following (Bates and Maxwell, 2005; Fogg *et al.*, 2012, 2009). When cells grow and divide, they rely on enzymes called polymerases to read and replicate the DNA. These processes require separation of the two strands to allow the polymerases access to the DNA genetic code. This strand separation is facilitated by negative supercoiling, which is likely one of the reasons why DNA is maintained in a negatively supercoiled state in the cell.

Overwound DNA accumulates in front of polymerases as they track along the double helix (Liu and Wang, 1987; Ma and Wang, 2016), and this overwound DNA is referred to as positive supercoiling. These positive supercoils are where many anticancer drugs and antibiotics exert their inhibitory effect on the enzymes that regulate DNA supercoiling, the DNA topoisomerases (Ashley *et al.*, 2017; McClendon *et al.*, 2005; McClendon and Osheroff, 2006). Although many aspects of exactly how DNA topoisomerases regulate supercoiling remain open questions, their roles as important anti-cancer and anti-microbial drug targets highlight the importance of understanding the structure of biologically relevant positively and negatively supercoiled DNA.

Topologically, linear DNA can only be transiently supercoiled (any unwinding or overwinding imparted upon the DNA double helix would quickly relax). To study DNA supercoiling, then, one must constrain the ends of DNA, which can be achieved by using closed circular DNA. A bacterial chromosome is millions of base-pairs (bp) in length and mammalian chromosomes are even longer. DNA of such great lengths is not amenable to biochemical or biophysical study.

A 336 bp closed circle is a highly tractable size for the study of DNA supercoiling. A relaxed DNA circle of this length has very close to 32 turns of the DNA helix (336 bp/10.5 bp/turn = 32 turns) thus there should be no additional puckering or strain from being out of "phase." The exact value of the helical repeat varies depending on solution conditions and sequence (Peck and Wang, 1981; Rybenkov *et al.*, 1997; Xu and Bremer, 1997), ranging for this 336 bp minicircle sequence from $\sim$10.42 to 10.57 bp/turn (Fogg and Zechiedrich, unpublished data). The value of Lk_0, however, remains very close to 32 ($\pm$0.25). Minicircles with this length are small enough to capture the conformational changes that are averaged out in larger molecules, yet large enough to yield a wide range of ten unique topoisomers. These different topoisomers were generated and purified as described (Fogg *et al.*, 2006; Irobalieva *et al.*, 2015). Briefly, supercoiled minicircles are first generated by λ-integrase mediated recombination on plasmids growing in *Escherichia coli* bacteria. The bacterial cells are subsequently lysed and the supercoiled minicircles isolated. To generate specific levels of supercoiling, the minicircles are nicked using an enzyme that cleaves just one of the two strands at a specific sequence. Addition of the intercalating agent ethidium bromide unwinds the DNA, introducing negative supercoiling. Different negatively supercoiled topoisomers are generated by varying the concentration of ethidium bromide added (Singleton and Wells, 1982). Positive supercoiling is generated by the addition of the protein HmfB which wraps the DNA in a positive writhe (LaMarr *et al.*, 1997). The nicked DNA is treated with ligase to seal the nick, thereby trapping the induced supercoiling, and the ethidium bromide or HmfB is subsequently removed, allowing the supercoiling to repartition between twist and writhe. Individual topoisomers are then isolated by polyacrylamide gel electrophoresis. As will be discussed below, each of these topoisomers can adopt a wide variety of conformations and these may not be accessible in smaller circles. The 336 bp minicircle has indeed proven useful for studying the structure of supercoiled DNA, how supercoiling affects DNA acting enzymes, and the drugs that affect them.

2. DNA supercoiling

The classic double-helical structure of B-DNA, postulated in 1953 by Watson and Crick (Watson and Crick, 1953), drawing from data collected by Franklin (Franklin and Gosling, 1953), remains one of the most iconic scientific structures. Sixty-five years of accumulated data on DNA structure subsequently, however, have shown that DNA often deviates from this iconic linear double helix.

Vinograd and colleagues discovered that DNA can be supercoiled (Vinograd *et al.*, 1965). To understand the basic principle of supercoiling, it is useful to model DNA as a simple elastic polymer, although, as we will describe below, this analogy is a gross oversimplification. The phenomenon of supercoiling can be modeled using a piece of rubber tubing. If a rubber tube is held with one end in each hand, and then gradually twisted, torsional strain in the tubing will increase until a critical amount of twist is added, at which point the tubing buckles and coils about itself. Similarly, underwinding or overwinding of DNA imparts torsional strain on the molecule. Beyond a critical threshold of winding, the DNA will buckle and coil about itself to relieve some of this torsional strain like a rubber tube. Vinograd turned to mathematician F. Brock Fuller to provide a means to quantify supercoiling (Fuller, 1978, 1971). Fuller introduced the now familiar terminology of twist (Tw), that describes the number of times the individual strands coil about the helical axis, and writhe (Wr), the coiling of the helical axis in three-dimensional space. These two properties together contribute to the linking number; thus:

$$Lk = Tw + Wr.$$

A topological change in linking number can be enumerated as:

$$\Delta Lk = \Delta Tw + \Delta Wr.$$

The relative partitioning of ΔLk into ΔTw and ΔWr has profound effects on the properties of DNA. Negative ΔTw increases the propensity of the individual strands to separate, thus facilitating access to the genetic code. Moderate positive ΔTw stabilizes the helix and makes it more resistant to denaturation. Writhe alters the global shape of the molecule, aiding in compaction of the DNA as well as facilitating the juxtaposition of distant sites on the molecule (Vologodskii and Cozzarelli, 1996). The relative portioning of twist and writhe is governed by both the torsional and bending rigidity of the DNA molecule. Twist imparts torsional strain whereas the introduction of writhe is inhibited by the intrinsic rigidity of the DNA helix. Electrostatic forces also exert a prominent effect (Feig and Pettitt, 1999; Manning, 1969a, 1969b; Rybenkov *et al.*, 1997). DNA writhing requires that negatively charged helices juxtapose. Positively charged ions in the solvent, as in live cells, help mitigate the repulsion between the DNA double helices (Fogg *et al.*, 2006; Randall *et al.*, 2006), thereby facilitating writhe.

Experimental exploration of DNA supercoiling is technically challenging. Properties of supercoiled DNA have been predicted on the basis of the results from experiments on short, linear DNA segments that are easier to study but are not supercoiled. For example, DNA flexibility is normally described by the persistence length, the length scale over which the final direction of the helix is correlated with the initial direction of the helix (average deflection = one radian). At lengths shorter than a persistence length (~ 150 bp, (Hagerman, 1988)), linear DNA behaves primarily as a rigid rod. When considering the persistence length of DNA, however, most studies relied upon how easily the two ends of a certain DNA sequence of a certain length would come close enough to be joined together by an enzyme called ligase (Horowitz and Wang, 1984; Shore *et al.*, 1981; Shore and Baldwin, 1983). Additional studies relied on altered mobilities of linear DNA as a measure of the degree of intrinsic bending (Marini *et al.*, 1982; Zimm and Levene, 1992). Both of these assays and many others utilize linear DNA, which is not supercoiled.

Theory and simulations incorporated these DNA bending measurements into elastic-rod or "worm-like chain" models, which were not only derived from non-supercoiled DNA but also assumed the elastic properties of DNA could be applied uniformly along the length of the DNA molecule. Most simulations used elastic-rod models to predict how DNA might behave when negatively supercoiled. A few considered positive supercoiling. The problem is that elastic-rod or worm-like chain models cannot account for internal structural changes, such as localized denaturation and loss of base-pairing, caused by torsional stress from supercoiling. The oversimplified models based on behavior of linear DNA also do not consider how the chirality of the DNA helix affects how two juxtaposed helices fit together. The groove-backbone interaction at the DNA crossovers have been proposed to differ between negatively and positively supercoiled DNA (Timsit *et al.*, 1999, 1998; Timsit and Moras, 1994).

When Strick, Bensimon, Croquette, Bustamante, Marko, and others developed techniques to constrain the linear ends of DNA, and stretch and pull it while undertwisting or overtwisting it, the measurements suggested new and completely unanticipated "phase" transitions (Bustamante *et al.*, 2003; Marko and Neukirch, 2013; Strick *et al.*, 2000, 1998). Having no way to directly visualize these changes, the researchers developed models to explain them. These new models for positively or negatively supercoiled DNA could not be reconciled with the elastic-rod models of DNA. Moreover, the predictions of the new models and the elastic-rod models diverged quickly, specifically at minimal levels of negative supercoiling and moderate levels of positive supercoiling.

With time, atomic-scale simulations of increasingly relevant lengths of DNA became possible and potential consequences of supercoiling on the local structure of DNA were revealed. Kinks (Harris *et al.*, 2008; Lankas *et al.*, 2006), base-flipping and the resulting denaturation from a series of base-flipping events (Randall *et al.*, 2009), and even Pauling-DNA (P-DNA) (Randall *et al.*, 2009), an inside-out

DNA conformation with the backbone on the outside and bases splayed outwards (Allemand *et al.*, 1998), have all been observed in these high-resolution simulations.

In our previous molecular dynamics simulations, we observed that these localized distortions act as a sink for the torsional strain thus allowing the remainder of the DNA to remain as regular B-form (Randall *et al.*, 2009). Short regions of DNA are sacrificed to preserve the overall integrity of the molecule. Instead of the twist being evenly distributed along the length of the DNA, as predicted by classical elastic rod models, the winding is instead biphasic, especially for negatively supercoiled DNA. Positively supercoiled DNA behaves much more like an elastic rod, with twist evenly distributed, up to the critical denaturation threshold where it too becomes biphasic.

3. How supercoiling affects DNA shape

Simulation does not always mirror real life. What about the actual biological molecule DNA? It is an ongoing quest, but we used 336 bp DNA minicircles to study how supercoiling affects the structure of supercoiled DNA in 3D by changing the Lk in steps of one (from $\Delta Lk = -6$ to $+3$ or $Lk = 26$ to 35). We used two main approaches to study these minicircles. We looked at them directly using cryo-electron tomography (cryo-ET) and we probed for un-paired bases using an enzymatic probe.

Cryo-ET is a sub-discipline of the increasingly powerful method cryo-electron microscopy (cryo-EM). This technique works by near-instantaneously freezing specimens in solution to capture a snapshot of the conformation at the instant of freezing. A more detailed description of cryo-EM can be found in the following excellent review, which is targeted at non-microscopists (Milne *et al.*, 2013).

In short, cryo-electron microscopy generates multiple 2D projections of a structure. To acquire 3D information, multiple projections of the structure in different orientations are used to computationally reconstruct the 3D structure by Fourier inversion. Probably the most commonly used cryo-EM method is single-particle reconstruction. In this approach data from very many particles, each randomly orientated in solution, are combined to generate a 3D reconstruction of the structure. The computational averaging of many thousands of images allows high-resolution structures to be obtained, even though each individual image has relatively low signal to noise. Successful single-particle reconstruction, however, requires that all particles are identical with very little heterogeneity (Cheng *et al.*, 2015), although sophisticated algorithms can allow multiple different conformations to be identified as long as the molecule is only present in a few different conformations (Ludtke, 2016).

Supercoiled DNA is very dynamic and conformationally heterogeneous, and therefore, was not amenable to single-particle reconstruction. Therefore, we turned to cryo-ET, which involves collecting a series of images of a sample with the sample rotated between each imaging. A 3D reconstruction of an individual DNA minicircle is obtained by computationally combining the data from the series of tilt images.

A brief mention of some of the caveats with this cryo-electron tomography technique is of use here. First, although efforts can be made to test whether a sample has reached equilibrium prior to freezing (in our case with minicircle DNA we pre-incubated the DNA at different temperatures and for different times, all with the same results (Irobalieva *et al.*, 2015)), we cannot be sure equilibrium has been reached. Second, in the $\sim$0.1 milliseconds it takes to freeze the sample, DNA may partially re-equilibrate with the changing temperature. Therefore, it is impossible to rule out an effect of the change in temperature over this short time scale on the 3D structure of DNA. Finally, the specimen holder rotates in each direction only $60 - 70°$ and cannot be completely turned. Thus, there is a missing bit of information referred to as the "missing wedge." This missing wedge is sometimes not as big a problem for continuous thin filaments like DNA molecules. In this case, two projections can be sufficient to reconstruct the object (Amzallag *et al.*, 2006; Bednar *et al.*, 1994; Demurtas *et al.*, 2009; Dustin *et al.*, 1991). The problems can arise, however, when there is noise in the dataset. Despite these caveats, the study provided unprecedented insight the structure and dynamics of supercoiled DNA.

In addition to cryo-electron tomography we also used Bal-31, an enzyme that can recognize DNA bases that are not base paired, to detect regions of distortion. Our 3D structures of 336 bp minicircles with defined ΔLk combined with the probing for un-paired bases with the enzyme Bal-31 revealed important details about "real" (not just simulated) closed circular supercoiled DNA:

(i) Supercoiled DNA exhibits remarkable structural diversity (Figure 1). Each topoisomer adopts a surprisingly wide variety of conformations. It is possible that each individual DNA molecule continuously fluctuates among the various conformations observed. Depending upon the currently undefined free energy of each conformation, the time spent in each potential meta-stable state may determine the spectrum of 3D structures seen. This model resembles that for protein folding (reviewed in Dill and MacCallum, 2012). Whether there is a single path each conformational change must follow (*e.g.*, a supercoiled minicircle with a specific ΔLk proceeds only from figure-8 to needle and back but never directly from figure-8 to rod-shaped) is unknown. The conformations we observed were determined by the state of each individual minicircle at the instant of freezing.

(ii) The 3D conformations adopted by minicircles with specific degree of supercoiling (ΔLk) are unique (from other topoisomers with different ΔLk) and reproducible (Figure 1). These conformations are rather like a "fingerprint" of a certain ΔLk for DNA circles of this particular length and sequence.

(iii) As negative or positive supercoiling increases, the 3D conformations tend to be more compact (decreased average radii of gyration), just as was simulated for DNA using the isotropic elastic rod or worm-like chain models (Vologodskii and Cozzarelli, 1994).

(iv) The greatest variability in conformations (ranging from open circle to highly writhed and every other 3D shape observed in the study) was seen for minicircle DNA at negative supercoiling levels near to that maintained by cells, leading to the thought provoking notion that proteins involved in the replication, transcription, and recombination of DNA, or any biomolecules involved in DNA metabolic processes may preferentially interact with different specific 3D conformations, all of which are provided at this critical level of supercoiling. If this notion is correct, then negative supercoiling does not stimulate biological activity because it induces a single conformational change of DNA, but rather it allows DNA to adopt a multiplicity of conformations that may be sampled by biomolecules that interact with DNA. Biomolecules furthermore may be able to bind to and stabilize otherwise transient conformations.

(v) DNA that is negatively supercoiled or extremely positively supercoiled is able to adopt conformations that would be energetically unfavorable, probably impossible, if the DNA were not allowed to deviate locally from isotropic elastic rod models. Using Bal-31, we found that negatively supercoiled DNA minicircles exhibited significant levels of base exposure (Irobalieva *et al.*, 2015). From our atomistic simulations we believe these to be base-flipping and denaturation (Randall *et al.*, 2009). When a base-pair is broken, the helix is no longer a rigid rod and, therefore, these disruptions act as hyperflexible sites to allow sharp bending of the double helix. Conversely, sharp bending, and the resultant strain, can disrupt base pairing.

We also observed exposed bases in positively supercoiled DNA, which is somewhat counterintuitive given that overwinding should stabilize the DNA helix to make it more resistant to denaturation. These exposed bases were not detected until a high threshold ΔLk was reached. The presence of exposed bases may implicate Pauling-DNA (Allemand *et al.*, 1998), or may result from bending-induced strain at the superhelical apices of highly writhed minicircles.

(vi) We discovered highly bent and contorted shapes (referred to as "other" in Irobalieva *et al.*, 2015). The appearance of these contorted shapes occurs in hypernegatively and hyperpositively supercoiled topoisomers and coincides with a high preponderance of exposed bases susceptible to Bal-31. Thus, these highly bent and contorted shapes may only be achievable because of the flexibility afforded by disruptions in base pairing.

(vii) There are striking structural differences between positive and negative supercoiling. These differences qualitatively agree with those observed in the stretching/twisting single molecule work reviewed in (Marko and Neukirch, 2013). Single-molecule experiments have shown that negatively supercoiled DNA deviates from idealized elastic behavior at fairly low levels of supercoiling, whereas positively supercoiled DNA behaves as an isotropic elastic polymer up to much higher degrees of superhelical density. Similarly, in our study, the conformational distributions observed with positively supercoiled DNA do not mirror those for the equivalent negatively supercoiled topoisomer. The differences can be explained by the much higher prevalence of base-flipping and denaturation in negatively supercoiled DNA and the concomitant appearance of hyperflexible sites, as revealed by our results with Bal-31.

(viii) We used the WEBSIDD algorithm (Bi and Benham, 2004) to predict where base disruption was most likely to occur on 336 bp minicircles. Surprisingly, however, we physically mapped Bal-31 cleavage to a site located diametrically opposite on the minicircle to the predicted site. We hypothesized that base-flipping in the more readily underwinding-induced base-flipped or denatured bending sequence resulted in a hyperflexible site and that, consequently, this sequence becomes localized to one of the superhelical apices. This site-specific flexibility results in the sequence

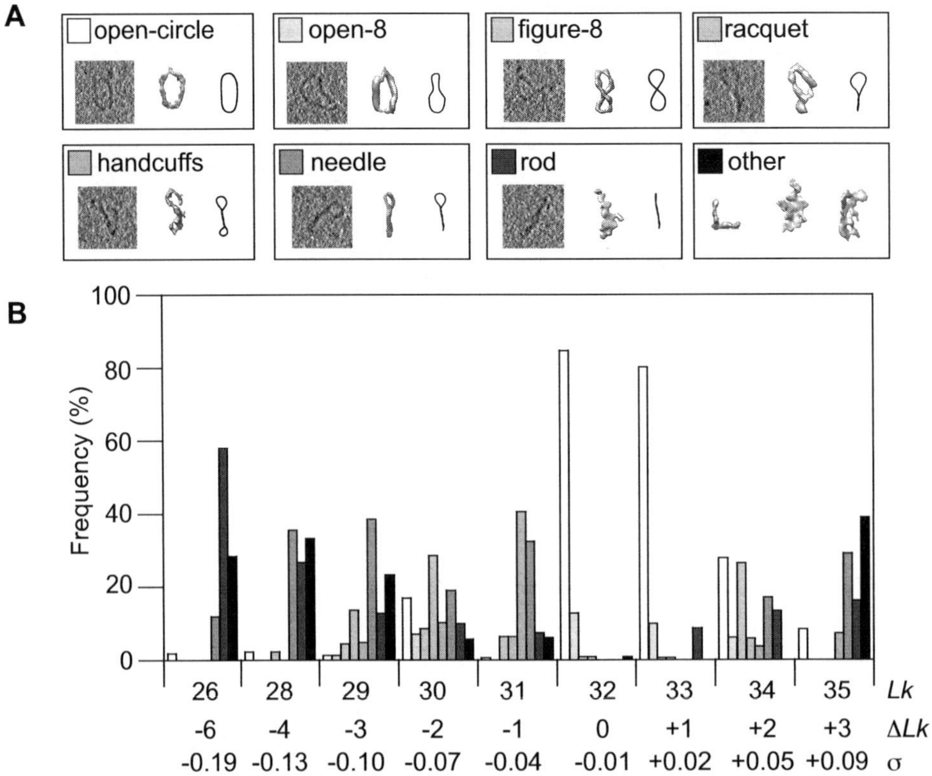

FIGURE 1. Minicircle geometries observed by cryo-ET. **A:** Observed minicircle geometries. For each shape an example of a 2-D projection, the derived 3-D tomogram obtained from multiple projections, and a drawing depicting the overall shape is shown. A few examples of tomograms are shown for the category of "other." **B:** Conformational frequency distribution for each topoisomer population. Data from Irobalieva *et al.*, (2015).

180° away along the circumference of the circle to be localized to the opposite superhelical apex, and the resultant bending-induced strain may result in breaking of base-pairs.

Lionberger, Stasiak, and colleagues had previously observed a similar cooperative effect whereby the formation of a kink at one site stimulated kinking at a site diametrically opposite in torsionally strained minicircles (Lionberger *et al.*, 2011). They proposed a cooperative kinking model to explain this phenomenon, which means that local alteration in supercoiled DNA can "communicate" over distance and strongly influence distant sites, a remarkable feature of DNA. A similar observation of co-operative kinking was subsequently observed in molecular dynamics simulations of ~100 bp DNA minicircles by Harris and co-workers (Sutthibutpong *et al.*, 2016).

The Lionberger *et al.*, study used DNA minicircles ~100 bp in size and measured their ellipticity by cryo-EM to detect kinking. Although torsionally strained, these very small minicircles were planar open circles with no writhe. We found that hyperflexible sites had an even more dramatic effect on the global structure of the DNA when it was able to writhe. This interplay between supercoiling-induced denaturation and writhe leads to sharp bending in many of the conformations observed.

4. Using DNA sequence and supercoiling to create new 3D shapes of DNA

An advanced coarse-grained model of supercoiled DNA that allows for DNA sequence-dependent base-pair opening has been recently developed (Wang *et al.*, 2015; Wang and Pettitt, 2016). Motivated by our findings described above, we collaborated with Pettitt's group to better predict the behavior of supercoiled DNA. Because base-pair opening makes the DNA locally flexible, we were able to use our cryo-electron tomograph datasets and the new coarse-grained model to align the DNA sequence register in the various cryo-electron micrograph images of minicircles to determine which sequences were found at sharp bends (Wang *et al.*, 2017). These sequences were in agreement with the sites identified by Bal-31 cleavage (Irobalieva *et al.*, 2015).

The coarse-grained modeling also allowed us to simulate the effect of adding additional sharp bending sites to our simulated supercoiled 336 bp minicircles, something that would take a long time to test biochemically or with cryo-EM. Indeed, our simulations show that mutating several base-pairs within a 13 bp segment to introduce an additional one or more DNA bending sites in the 336 bp minicircle sequence resulted in novel simulated minicircle shapes, indicating that we can use a combination of DNA sequence and degree of negative DNA supercoiling to design shapes of DNA minicircles. This finding has important implications for the potential clinical applications of minicircles.

Informed by the knowledge of how DNA sequence and supercoiling affect shape, our modeling suggests that we may be able to generate minicircles with specific shapes. We predict that these nanoparticle DNA 3D structures will have tissue specificity for gene therapy based upon empiric work with inert nanoparticles of various shapes (Wang *et al.*, 2011).

5. More work to be done

Although the work described above represents important steps toward understanding how supercoiling affects DNA geometry and, thus, biological activity, we are far from done. How do enzymes distinguish various 3D structures? How do the anticancer and antibiotic drugs that block the topoisomerases affect the observed 3D structures? How does DNA supercoiling affect these drugs? Do DNA minicircles form the 3D shapes we predicted computationally? These are just the obvious next questions. The impact beyond these immediate goals includes supercoiling-mediated structural changes on the formation or stability of non-B structures (Z-DNA motifs, four-stranded G quadruplex motifs, hairpin structures, cruciforms, three-stranded H-DNA, slipped structures), chromosome packing, and more.

6. Conclusion

The B-DNA helix of Watson and Crick is the heart of DNA structure yet it is how the double helix deviates from this lowest free energy state that drives DNA metabolism. Understanding these deviations is of fundamental importance and the foundational role of mathematics in studying this problem has been and will continue to be of paramount importance.

Acknowledgements

We thank Dr. Graham L. Randall and Dr. De Witt Sumners for critically evaluating this paper. We thank Dr. Steve Ludtke for advice.

Conflict of Interest

The authors hold numerous patents on the generation and purification of minicircle and minivector DNA, and the use of minivector DNA for gene therapy. Both authors have equity stake in Twister Biotech, Inc.

References Cited

Allemand, J.F., Bensimon, D., Lavery, R., Croquette, V., 1998. Stretched and overwound DNA forms a Pauling-like structure with exposed bases. Proc. Natl. Acad. Sci. U. S. A. 95, 14152–14157.

Amzallag, A., Vaillant, C., Jacob, M., Unser, M., Bednar, J., Kahn, J.D., Dubochet, J., Stasiak, A., Maddocks, J.H., 2006. 3D reconstruction and comparison of shapes of DNA minicircles observed by cryo-electron microscopy. Nucleic Acids Res. 34, e125.

Ashley, R.E., Dittmore, A., McPherson, S.A., Turnbough, C.L., Neuman, K.C., Osheroff, N., 2017. Activities of gyrase and topoisomerase IV on positively supercoiled DNA. Nucleic Acids Res. 45, 9611–9624.

Bates, A.D., Maxwell, A., 2005. DNA topology, 2nd ed. ed. Oxford University Press, Oxford ; New York.

Bednar, J., Furrer, P., Stasiak, A., Dubochet, J., Egelman, E.H., Bates, A.D., 1994. The twist, writhe and overall shape of supercoiled DNA change during counterion-induced transition from a loosely to a tightly interwound superhelix. Possible implications for DNA structure in vivo. J. Mol. Biol. 235, 825–847.

Bi, C., Benham, C.J., 2004. WebSIDD: server for predicting stress-induced duplex destabilized (SIDD) sites in superhelical DNA. Bioinforma. Oxf. Engl. 20, 1477–1479.

Bustamante, C., Bryant, Z., Smith, S.B., 2003. Ten years of tension: single-molecule DNA mechanics. Nature 421, 423–427.

Cheng, Y., Grigorieff, N., Penczek, P.A., Walz, T., 2015. A primer to single-particle cryo-electron microscopy. Cell 161, 438–449.

Demurtas, D., Amzallag, A., Rawdon, E.J., Maddocks, J.H., Dubochet, J., Stasiak, A., 2009. Bending modes of DNA directly addressed by cryo-electron microscopy of DNA minicircles. Nucleic Acids Res. 37, 2882–2893.

Dill, K.A., MacCallum, J.L., 2012. The protein-folding problem, 50 years on. Science 338, 1042–1046.

Dustin, I., Furrer, P., Stasiak, A., Dubochet, J., Langowski, J., Egelman, E., 1991. Spatial visualization of DNA in solution. J. Struct. Biol. 107, 15–21.

Feig, M., Pettitt, B.M., 1999. Sodium and chlorine ions as part of the DNA solvation shell. Biophys. J. 77, 1769–1781.

Fogg, J.M., Catanese Jr, D.J., Randall, G.L., Swick, M.C., Zechiedrich, L., 2009. Differences between positively and negatively supercoiled DNA that topoisomerases may distinguish, in: Mathematics of DNA Structure, Function and Interactions. Springer, pp. 73–121.

Fogg, J.M., Kolmakova, N., Rees, I., Magonov, S., Hansma, H., Perona, J.J., Zechiedrich, E.L., 2006. Exploring writhe in supercoiled minicircle DNA. J. Phys. Condens. Matter Inst. Phys. J. 18, S145–S159.

Fogg, J.M., Randall, G.L., Pettitt, B.M., Sumners, D.W.L., Harris, S.A., Zechiedrich, L., 2012. Bullied no more: when and how DNA shoves proteins around. Q. Rev. Biophys. 45, 257–299.

Franklin, R.E., Gosling, R.G., 1953. Molecular configuration in sodium thymonucleate. Nature 171, 740–741.

Fuller, F.B., 1978. Decomposition of the linking number of a closed ribbon: A problem from molecular biology. Proc. Natl. Acad. Sci. U. S. A. 75, 3557–3561.

Fuller, F.B., 1971. The writhing number of a space curve. Proc. Natl. Acad. Sci. U. S. A. 68, 815–819.

Hagerman, P.J., 1988. Flexibility of DNA. Annu. Rev. Biophys. Biophys. Chem. 17, 265–286.

Harris, S.A., Laughton, C.A., Liverpool, T.B., 2008. Mapping the phase diagram of the writhe of DNA nanocircles using atomistic molecular dynamics simulations. Nucleic Acids Res. 36, 21–29.

Horowitz, D.S., Wang, J.C., 1984. Torsional rigidity of DNA and length dependence of the free energy of DNA supercoiling. J. Mol. Biol. 173, 75–91.

Irobalieva, R.N., Fogg, J.M., Catanese, D.J., Sutthibutpong, T., Chen, M., Barker, A.K., Ludtke, S.J., Harris, S.A., Schmid, M.F., Chiu, W., Zechiedrich, L., 2015. Structural diversity of supercoiled DNA. Nat. Commun. 6, 8440.

LaMarr, W.A., Sandman, K.M., Reeve, J.N., Dedon, P.C., 1997. Large scale preparation of positively supercoiled DNA using the archaeal histone HMf. Nucleic Acids Res. 25, 1660–1661.

Lankas, F., Lavery, R., Maddocks, J.H., 2006. Kinking occurs during molecular dynamics simulations of small DNA minicircles. Struct. Lond. Engl. 1993 14, 1527–1534.

Lionberger, T.A., Demurtas, D., Witz, G., Dorier, J., Lillian, T., Meyhöfer, E., Stasiak, A., 2011. Cooperative kinking at distant sites in mechanically stressed DNA. Nucleic Acids Res. 39, 9820–9832.

Liu, L.F., Wang, J.C., 1987. Supercoiling of the DNA template during transcription. Proc. Natl. Acad. Sci. U. S. A. 84, 7024–7027.

Ludtke, S.J., 2016. Single-Particle Refinement and Variability Analysis in EMAN2.1. Methods Enzymol. 579, 159–189.

Ma, J., Wang, M.D., 2016. DNA supercoiling during transcription. Biophys. Rev. 8, 75–87.

Manning, G.S., 1969a. Limiting Laws and Counterion Condensation in Polyelectrolyte Solutions I. Colligative Properties. J. Chem. Phys. 51, 924–933.

Manning, G.S., 1969b. Limiting Laws and Counterion Condensation in Polyelectrolyte Solutions II. Self-Diffusion of the Small Ions. J. Chem. Phys. 51, 934–938.

Marini, J.C., Levene, S.D., Crothers, D.M., Englund, P.T., 1982. Bent helical structure in kinetoplast DNA. Proc. Natl. Acad. Sci. U. S. A. 79, 7664–7668.

Marko, J.F., Neukirch, S., 2013. Global force-torque phase diagram for the DNA double helix: structural transitions, triple points, and collapsed plectonemes. Phys. Rev. E Stat. Nonlin. Soft Matter Phys. 88, 062722.

McClendon, A.K., Osheroff, N., 2006. The geometry of DNA supercoils modulates topoisomerase-mediated DNA cleavage and enzyme response to anticancer drugs. Biochemistry 45, 3040–3050.

McClendon, A.K., Rodriguez, A.C., Osheroff, N., 2005. Human topoisomerase IIalpha rapidly relaxes positively supercoiled DNA: implications for enzyme action ahead of replication forks. J. Biol. Chem. 280, 39337–39345.

Milne, J.L.S., Borgnia, M.J., Bartesaghi, A., Tran, E.E.H., Earl, L.A., Schauder, D.M., Lengyel, J., Pierson, J., Patwardhan, A., Subramaniam, S., 2013. Cryo-electron microscopy--a primer for the non-microscopist. FEBS J. 280, 28–45.

Peck, L.J., Wang, J.C., 1981. Sequence dependence of the helical repeat of DNA in solution. Nature 292, 375–378.

Randall, G.L., Pettitt, B.M., Buck, G.R., Zechiedrich, E.L., 2006. Electrostatics of DNA-DNA juxtapositions: consequences for type II topoisomerase function. J. Phys. Condens. Matter Inst. Phys. J. 18, S173–S185.

Randall, G.L., Zechiedrich, L., Pettitt, B.M., 2009. In the absence of writhe, DNA relieves torsional stress with localized, sequence-dependent structural failure to preserve B-form. Nucleic Acids Res. 37, 5568–5577.

Rybenkov, V.V., Vologodskii, A.V., Cozzarelli, N.R., 1997. The effect of ionic conditions on DNA helical repeat, effective diameter and free energy of supercoiling. Nucleic Acids Res. 25, 1412–1418.

Shore, D., Baldwin, R.L., 1983. Energetics of DNA twisting. I. Relation between twist and cyclization probability. J. Mol. Biol. 170, 957–981.

Shore, D., Langowski, J., Baldwin, R.L., 1981. DNA flexibility studied by covalent closure of short fragments into circles. Proc. Natl. Acad. Sci. U. S. A. 78, 4833–4837.

Singleton, C.K., Wells, R.D., 1982. The facile generation of covalently closed, circular DNAs with defined negative superhelical densities. Anal. Biochem. 122, 253–257.

Strick, T.R., Allemand, J.F., Bensimon, D., Croquette, V., 2000. Stress-induced structural transitions in DNA and proteins. Annu. Rev. Biophys. Biomol. Struct. 29, 523–543.

Strick, T.R., Allemand, J.F., Bensimon, D., Croquette, V., 1998. Behavior of supercoiled DNA. Biophys. J. 74, 2016–2028.

Sutthibutpong, T., Matek, C., Benham, C., Slade, G.G., Noy, A., Laughton, C., K Doye, J.P., Louis, A.A., Harris, S.A., 2016. Long-range correlations in the mechanics of small DNA circles under topological stress revealed by multi-scale simulation. Nucleic Acids Res. 44, 9121–9130.

Timsit, Y., Duplantier, B., Jannink, G., Sikorav, J.L., 1998. Symmetry and chirality in topoisomerase II-DNA crossover recognition. J. Mol. Biol. 284, 1289–1299.

Timsit, Y., Moras, D., 1994. DNA self-fitting: the double helix directs the geometry of its supramolecular assembly. EMBO J. 13, 2737–2746.

Timsit, Y., Shatzky-Schwartz, M., Shakked, Z., 1999. Left-handed DNA crossovers. Implications for DNA-DNA recognition and structural alterations. J. Biomol. Struct. Dyn. 16, 775–785.

Vinograd, J., Lebowitz, J., Radloff, R., Watson, R., Laipis, P., 1965. The twisted circular form of polyoma viral DNA. Proc. Natl. Acad. Sci. U. S. A. 53, 1104–1111.

Vologodskii, A., Cozzarelli, N.R., 1996. Effect of supercoiling on the juxtaposition and relative orientation of DNA sites. Biophys. J. 70, 2548–2556.

Vologodskii, A.V., Cozzarelli, N.R., 1994. Conformational and thermodynamic properties of supercoiled DNA. Annu. Rev. Biophys. Biomol. Struct. 23, 609–643.

Wang, J., Byrne, J.D., Napier, M.E., DeSimone, J.M., 2011. More effective nanomedicines through particle design. Small Weinh. Bergstr. Ger. 7, 1919–1931.

Wang, Q., Irobalieva, R.N., Chiu, W., Schmid, M.F., Fogg, J.M., Zechiedrich, L., Pettitt, B.M., 2017. Influence of DNA sequence on the structure of minicircles under torsional stress. Nucleic Acids Res. 45, 7633–7642.

Wang, Q., Myers, C.G., Pettitt, B.M., 2015. Twist-induced defects of the P-SSP7 genome revealed by modeling the cryo-EM density. J. Phys. Chem. B 119, 4937–4943.

Wang, Q., Pettitt, B.M., 2016. Sequence Affects the Cyclization of DNA Minicircles. J. Phys. Chem. Lett. 7, 1042–1046.

Watson, J.D., Crick, F.H., 1953. Molecular structure of nucleic acids; a structure for deoxyribose nucleic acid. Nature 171, 737–738.

Xu, Y.C., Bremer, H., 1997. Winding of the DNA helix by divalent metal ions. Nucleic Acids Res. 25, 4067–4071.

Zimm, B.H., Levene, S.D., 1992. Problems and prospects in the theory of gel electrophoresis of DNA. Q. Rev. Biophys. 25, 171–204.

Department of Molecular Virology and Microbiology, Verna and Marrs McLean Department of Biochemistry and Molecular Biology

Department of Pharmacology and Chemical Biology, Baylor College of Medicine, Houston, TX 77030

Contemporary Mathematics
Volume **746**, 2020
https://doi.org/10.1090/conm/746/15000

Characterizing the topology of kinetoplast DNA using random knotting

P. Liu, R. Polischuk, Y. Diao, and J. Arsuaga

ABSTRACT. Kinetoplast DNA (kDNA), the mitochondrial of DNA of trypanosomes, consists of thousands of small circular DNA molecules, called *minicircles*, that are highly condensed and that link together topologically to form a medieval chain-mail like structure. This system poses a number of interesting mathematical and physical questions, whose answers may facilitate our understanding of the structure and function of kDNA, and which may also have applications to the formation of olympic gels in material science. In our work we have proposed to use the general framework of random knotting to study the relationship between the topological properties of minicircle networks and the densities of minicircles. In this chapter we review the results obtained by our group and conclude by interpreting these results in the context of kDNA.

1. Introduction

The mitochondrial DNA of disease-causing organisms such as *Trypanosoma brucei* and *Trypanosoma cruzi* is called kinetoplast DNA (kDNA). It consists of thousands of circular DNA molecules called *minicircles* and a few larger molecules called *maxicircles*. The functional role of maxicircles and minicircles has been determined. Maxicircles contain genes associated with mitochondrial function (reviewed in [**23**]) and minicircles transcribe RNA that edits the transcription products of the maxicircles (reviewed in [**19**]). Both minicircles and maxicircles vary in size among organisms and, sometimes, within an organism. In most trypanosome species, minicircles range in size from 0.5 kb to 2.5 kb and maxicircles range from 20 kb to 40 kb [**40**]. Minicircles and maxicircles cluster spatially within a small cylindrical volume, which is called the *kinetoplast*, whose dimensions are correlated with the size of the species' minicircles (reviewed in [**27, 41**]). It has been estimated that the DNA density inside the kinetoplast is similar to that of the bacterial nucleoid [**40**].

kDNA has a unique network structure, formed by topologically linked—rather than chemically bonded—minicircles. The origin of this network is unknown. However, two hypotheses (which are not mutually exclusive) have been proposed to explain the origin of the network. The first hypothesis, of a biological nature, proposes that the network structure helps prevent minicircles from being lost through

2010 *Mathematics Subject Classification.* Primary 57M25 and 92B99.

P. Liu contributed equally to this work.

R. Polischuk contributed equally to this work.

Correspondence should be addressed to J. Arsuaga (jarsuaga@ucdavis.edu).

random segregation [**7**, **38**]. The second, of a biophysical nature, proposes that minicircle networks are driven by condensation [**13**]. Our work aims at testing the latter hypothesis, which we call the *confinement hypothesis.*

There is published data that helps our efforts. In a series of studies, the Englund Laboratory studied the topology of the the insect parasite *Crithidia fasciculata*'s kDNA network, and it found that the link type between minicircles and the mean valence of the network (*i.e.* the average number of minicircles topologically linked to any chosen minicircle) provide good descriptions of the topological complexity of the network. The Englund Lab showed that minicircles are linked by Hopf links (discussed shortly) [**34**] and that the value of the mean minicircle valence prior to replication is approximately three [**8**].

These parameters, however, are only local measures of the complexity of the network and do not provide a global view of the topological structure of the full network. Therefore, we introduced two new parameters: the critical percolation density and the mean saturation densitie [**13**]. Intuitively, during network formation, as the density of minicircles increases, the initial small minicircle clusters join together into larger clusters until a "large" cluster forms. This density is called the *critical percolation density* [**16**]. Similarly, as the density of minicircles increases beyond the percolation density, a network in which most of the minicircles ($> 90\%$) are linked into a single structure forms. This new density is called the *mean saturation density.*

In this chapter we begin by reviewing the experimental work that led to the characterization of kDNA topology in *C. fasciculata.* We then discuss our mathematical "random knotting" model and the quantitive parameters that it can estimate, as well as natural extensions of the model. After this, we compare how the linking probability, the mean valence, and the percolation and saturation densities change with these extensions of the model. We conclude by discussing how our modeling approach can help better understand the topological structure of kDNA.

2. Review of previous works

Because they are disease-causing parasites, trypanosomes attracted significant interest among biologists well before their complex mitochondrial DNA was discovered. The invention of transmission-electron microscopy in the 1930s and the development of effective ultra-microtomes in the early 1950s [**33**] provided biologists with adequate tools to study the sub-cellular structures of trypanosomes. In 1958, Meyer and colleagues reported the first thin-section images of *Trypanosoma cruzi* [**30**], the parasite responsible for Chagas disease. Although the kinetoplast had already been identified via light microscopy, it's purpose was still unknown. Meyer and colleagues showed that the kinetoplast contains electron-dense material; combined with a study by Steinert and colleagues in the same year showing the incorporation of ^{3}H thymidine [**43**], biologists concluded that the kinetoplast contains DNA.

In the 1960s, Delain showed that kDNA is comprised of circular DNA molecules, which could be topologically linked, or "catenated" [**10**, **11**].

2.1. Minicircles are linked via Hopf links. Rauch and colleagues [**34**] used restriction assays to determine the link type of the links found in kDNA. They used SstII, an enzyme known to cleave minicircles only once, to fragment the network. Network fragments, or *oligomers*, were separated using gel electrophoresis and observed under EM after RecA coating. Their results consistently showed minicircle dimers made of Hopf links. Figure 1 is an electron micrograph taken by our group showing two minicircles linked as a Hopf link. The minicircle on the left is part of a larger network fragment while the one on the right is isolated.

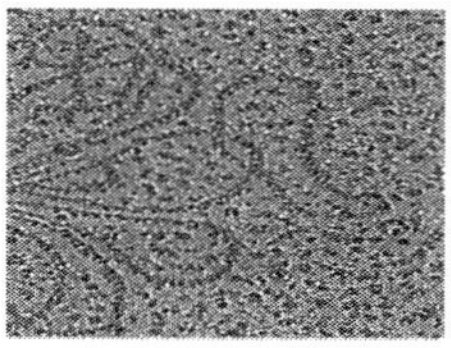

FIGURE 1. Two minicircles forming a Hopf link. The minicircle on the left is part of a larger fragment of the network.

2.2. Minicircles are relaxed. Prokaryote and mitochondrial DNA in eukaryotic cells is negatively supercoiled due to an excess of the torsional strain of the double helix. The typical superhelix density in a DNA molecule is about -0.06 [**6**]. In the study described in the previous section, Rauch and colleagues estimated the degree of supercoiling in *C. fasciculata* minicircles by comparing minicircles released from the network (via sonication) with free, radiolabeled, minicircles that had been previously relaxed by topoisomerase-I. After sonication they found that more than 90% of the unlabeled minicircles were either (1) nicked or (2) released from the network (as intact circular monomers or linear molecules). The authors separated the products by gel electrophoresis and quantified the amount of kDNA belonging to each oligomer band. The linking number between the minicircles that were released from the network and the relaxed ^3H-labelled minicircles, which serve as a control. The authors found that the average superhelical density is $+0.002$, which denotes right-handed over-twisting over its relaxed state.

2.3. Minicircles are organized into a single layer. Multiple arrangements of minicircles are consistent with the electrondensity observed in kDNA. Microscopy studies determined that the *C. fasciculata* kinetoplast is approximately 0.35 μm in thickness and 1μm in diameter[**18**]. These measurements lead to two possible models: one, proposed by Delain and Riou [**10**], in which the center of the minicircles are coplanar and minicircles are stretched like "rubber bands" [**27**], and a second model in which the centers of the minicircles are dispersed throughout the 3D structure of the kinetoplast. The second model was ruled out by Perez-Morga and Englund by visualizing purified networks using electron microscopy. Their data showed that kDNA is organized as an elliptical single-layer sheet of linked plasmids. The size of these purified ellipses was 15 μm long on the major axis and 10 μm on the minor axis [**32**]. Figure 2 shows a section of a purified network where minicircles and maxicircles are linked forming a network.

The *in vivo* size measurements and the *in vitro* imaging of expanded networks left open the possibility that networks could be two dimensional when isolated but folded into several layers inside the organism. To rule this out Ferguson and colleagues used the fact that replicated minicircles are attached to the periphery of the kDNA disk and that replicated minicircles contain nicked and gapped regions [**17, 18**]. If kDNA was folded, they argued, then nicked and gapped minicircles

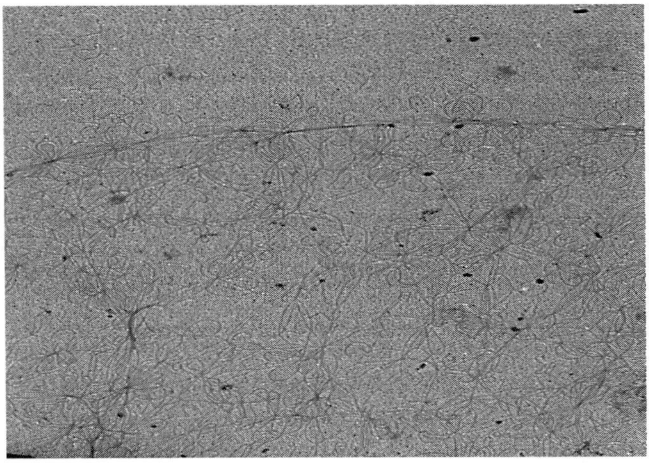

FIGURE 2. Electron micrograph of an isolated *C. fasciculata* kinetoplast (scale bar 2 μm)

would be spread throughout kDNA in purified networks. To test this hypothesis, they fluorescently labelled A-tracks in nicked and gapped regions of minicircles and found them to be localized on the periphery of the kinetoplast only [**18**]. They concluded that minicircles had to be monolayer structures like those observed in isolated networks.

2.4. Minicircle networks have mean valence three. As mentioned earlier, the topology of a topological minicircle network is determined by the type of link that connects any two minicircles and the mean valence, that is, the average number of minicircles that are linked to any given minicircle. Since the link type had been previously identified, the only missing parameter was the mean valence. To estimate it, Chen and colleagues [**8**] partially digested kDNA networks and quantified the type and concentration of oligomer populations. In order to use these measurements to build a model for kDNA the following assumptions were made:

Biological assumption 1: Maxicircles in kDNA do not play a significant role in the topology of the minicircle network.
Biological assumption 2: The kDNA network is a monolayer
Biological assumption 3: Minicircles have homogeneous valence.
Biological assumption 4: Every minicircle in the network has an equal probability of being cleaved by the restriction enzyme.

The first assumption is supported by electron micrograph observations that kinetoplasts in which maxicircles have either been removed via restriction enzymes [**32**], or are naturally not present (as it is observed in the *Trypanosoma evansi* [**21**], do not show any visible alterations in the structure of the kinetoplast. The second assumption is based on the previously discussed work by Ferguson. The third assumption is primarily a simplifying choice, but it does rely on the fact that few holes (regions with circles missing) are seen among isolated kinetoplasts and assumes that little hyperlinking is present [Biological Assumption 3][**8**].

Assuming that each minicircle is linked to v other minicircles and that each minicircle is linearized with a probability p, then v is related to the total number N of minicircles in the network and the total number M of intact monomers released from the network after linearization by the formula:

$$v = \frac{\log M - \log N - \log q}{\log p},$$

where $q = 1 - p$. Using this formula, Chen *et al.* estimated that the $v = 2.96 \pm 0.19$ based on a sample of $n = 42$ networks. To further validate their result, they identified the topology of the oligomer populations and verified that these topologies were consistent with theoretical global arrangements of minicircles in the plane. In order to classify these, they mapped networks to graphs by representing each minicircle by a vertex and each topological link by an edge. Under this mapping, classifying the global arrangement of minicircles is equivalent to classify tessellations of the plane. They observed that the oligomer populations were consistent with their four proposed valence-3 tessellations and with none of the valence-4 or valence-6 tessellations. They further reduced this possibilities by showing that the trimer data was consistent with the four valence-3 tessellations.

3. Mathematical models for testing the confinement hypothesis

In this section we describe the construction of our mathematical and computational models to test the confinement hypothesis. This hypothesis postulates that the topological complexity of the kDNA network is the product of the clustering of minicircles into a small volume. This hypothesis is consistent with the theory and experiments of other macromolecules whose topological complexity increases upon confinement [**4**]. We start by stating the basic assumptions of our model.

Biological assumption 1: The network is a monolayer.

Biological assumption 2: Maxicircles do not contribute to the overall topological structure of the minicircle network and are excluded from the model.

Mathematical assumption 1: The centers of the minicircles are positioned at the vertices of a regular square lattice.

Mathematical assumption 2: Minicircles are represented by geometric circles of radius one.

The first two assumptions are those proposed in the work of Chen and colleagues [**8**]. The mathematical assumptions are included to facilitate the development of the mathematical models described below. The first mathematical assumption imposes a distribution on the positions of minicircles. Since the true distribution of the centers of minicircles is unknown, we chose to start with a simplified model: the square lattice. This simplification is extremely helpful in developing our theoretical analysis framework and in fact, as we will show, does not significantly affect our results. The second mathematical assumption allows for two different interpretations. The first reflects the actual shape of the minicircles and it is based on two observations: kDNA minicircles are relaxed rather than highly supercoiled; and the only observed type of link is the Hopf link (Figure 1), the only possible link that can be formed with geometrical minicircles. The second interpretation assumes that minicircles represent the radius of gyration of a polymer model (wormlike chain or freely jointed chain) of the minicircle and the topological complexity of the network is estimated by the linking of its radius of gyration. The minicircles are normalized to unit circles so that the radius being unit helps simplify the model.

3.1. The "random knotting model" for kDNA: MC-kDNA. Our random knotting/linking model is Monte-Carlo based and was first published in [**12**, **13**]. We call this method Monte-Carlo kDNA (MC-kDNA). A sample configuration of the simulated kDNA network is shown in Figure 3. To generate a configuration of minicircles, a finite set of minicircles is uniquely mapped to the vertices of a finite square region of the square lattice (which is called a *grid* throughout this chapter) in a one-to-one and onto fashion. In this mapping each minicircle has an orientation defined by the normal vector to the plane containing the minicircle, which has two degrees of freedom: the *tilting angle* and the *azimuthal angle*. The orientation together with the center of the minicircle completely determine the spatial position of the minicircle in the grid (see Figure 4). The tilting angle, θ, is the angle between the normal vector and the positive z-axis, and the azimuthal angle, ϕ, is the angle between the x-axis and the projection of the normal vector onto the xy-plane. In our simplest model, minicircles are oriented randomly, with their respective normal vectors uniformly sampled from the unit sphere. As such, θ is chosen from $[0, \pi]$ (using the well-known result that $\cos\theta$ is uniformly distributed in $[-1, 1]$), and ϕ is taken uniformly from $[0, 2\pi]$.

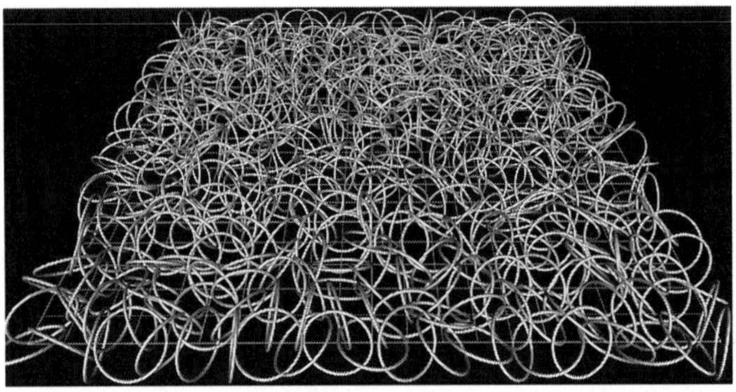

FIGURE 3. MC-kDNA: A grid generated by MC-kDNA

A key adjustable parameter in this model is the *density of minicircles* in the network, which is defined as the number of minicircles per unit area. It is important to note that since the minicircles are modeled by (normalized) unit circles, the distance a between two adjacent lattice points in the square lattice determines the density D of the minicircles by the equation $D = 1/a^2$ for the square lattice. We shall call a the *lattice constant.* We call the combination of a lattice grid with a given parameter a and minicircles (with given orientations) placed in the grid a *minicircle grid.* Two grids are the same if and only if they have the same dimension and lattice constant, and two minicircle grids based on the same grid differ only by the orientations of the minicircles in them.

3.1.1. *Linking of minicircles.* Let C_1 and C_2 be two disjoint simple closed and oriented curves in $\mathbb{R}^3$. We say that C_1 and C_2 form an unsplittable link if no topological 2-sphere separates them. Let $\mathbf{r}_1(t_1)$ and $\mathbf{r}_2(t_2)$ be parameterizations of

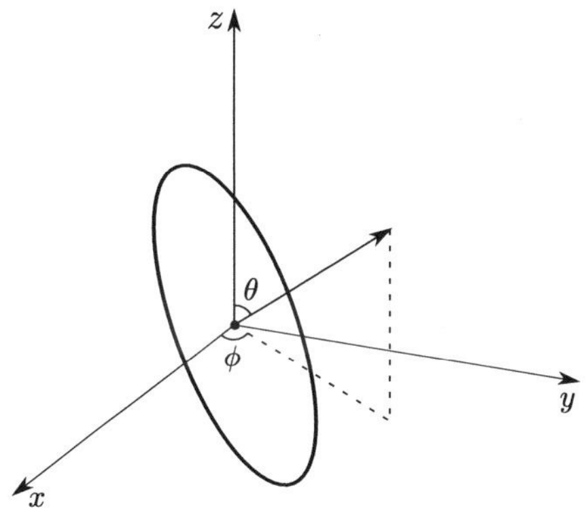

FIGURE 4. Orientation of a minicircle

C_1 and C_2 respectively with $t_1,\ t_2 \in [0, 2\pi]$. The Gauss linking number $\ell(C_1, C_2)$ of C_1 and C_2 is defined by

$$\ell(C_1, C_2) = \frac{1}{4\pi}\int_0^{2\pi}\int_0^{2\pi}\frac{(\mathbf{r}_2 - \mathbf{r}_1, \mathbf{r}_1', \mathbf{r}_2')}{|\mathbf{r}_2 - \mathbf{r}_1|^3}dt_1 dt_2,$$

where the numerator of the fraction is given by the triple product of the two vectors $\mathbf{r}_1', \mathbf{r}_2'$ tangent to the curves, and the difference of the position vectors [**35**]. It is well known that the linking number is a topological invariant. In particular, if $\ell(C_1, C_2) \neq 0$, then C_1 and C_2 are linked.

In the case that C_1 and C_2 are both circles of the same radius r, the linking between them can be determined geometrically as follows. Let P and Q be the centers of C_1 and C_2 respectively. If the normal vectors of the two planes containing C_1 and C_2 are uniformly distributed over the unit sphere, then with probability 1, these two planes and the plane that perpendicularly bisects the line segment PQ intersect at a single point x. C_1 and C_2 form a Hopf link if and only if $|Px| < r$. See Figure 5.

A significant implication of this geometric determination of linking is that if C_1 and C_2 are linked, then (with their orientations fixed) moving their centers closer to each other will not change that fact. Thus if we increase the minicircle density (to infinity) of a minicircle grid by decreasing the lattice constant a (to zero) with the orientations of the minicircles in the grid fixed, two minicircles that were not linked before will eventually become linked (with probability one), and two minicircles that were linked already will stay linked.

3.1.2. *Quantitative description of minicircle networks.* We have identified three measurements that help describe the topology of a network made of rigid geometrical minicircles: the mean minicircle valence, the critical percolation density and the mean saturation density.

i. The mean minicircle valence. The mean minicircle valence was first identified by Chen and colleagues and it is a parameter that can be measured experimentally. In our simulations, we estimate the mean minicircle valence by generating large samples of small networks (*e.g.* 7×7), computing the mean valence of the minicircle at the center and averaging over the whole sample.

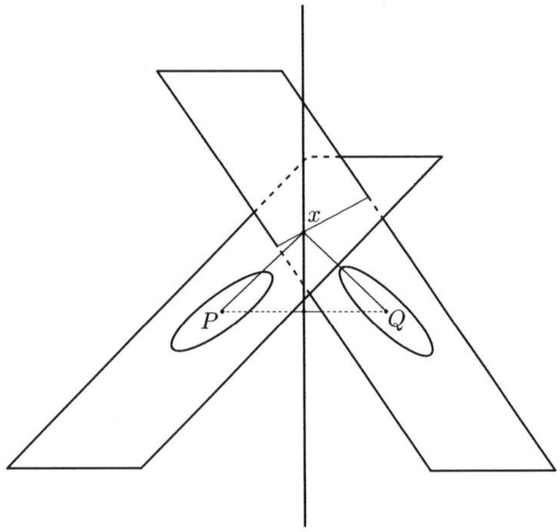

FIGURE 5. [**13**] Geometric determination of linking between two circles of the same radius.

ii. The critical percolation density. Percolation theory was motivated by the physical question of how a liquid (such as water) passes through a porous media (such as a porous slab of concrete). The questions asked revolved around identifying characteristics of the porous media and dynamics of the movement of the liquid. One example is to determine the probability that the liquid will make its way (percolate) from one side of the media to the other side. The simplest model to answer this question in the 2-dimensional case is based on a $n \times n$ grid of the square lattice. Here each line segment (called a *bond*) joining two adjacent grid points (called *sites*) models a "channel" between two sites for the liquid to go through. The channel is in either an "open" state or a "closed" state. The open state means that there is a channel from one site (represented by a lattice point) to an adjacent site (represented by an adjacent lattice point) for the liquid to flow through. Each bond is open (*i.e.* in the open state) with probability p, and the event that a given bond is open or closed is independent from the events whether other bonds are open or not. In this model, the above question becomes the following: for a given p, what is the probability that an open path exists from one side of the media (a grid of the square lattice) to the opposite side of the media? Of course, one can easily generalize this model to higher dimensional cases, and one can also use different lattices depending on the nature/properties of the media. Furthermore, different bonds can have different opening probabilities if the media is non uniform.

If an open path exists from one side of the grid (media) to its opposite side, we say that the system "percolates". In the case of a lattice grid model, if every bond has an opening probability $p > 0$, then the liquid has a positive probability of percolating, which is dependent on the size of the grid (media). The asymptotic case where the grid become unbounded is of particular interest. In this case, the corresponding question is, for the given p, does an infinite open path exist? When such an infinite open cluster exists, we say that the system *percolates.* A classical result of the percolation theory states that there exists a critical value of p (denoted by p_c) such that the percolation probability is always 0 if $p < p_c$ and the percolation probability is always 1 if $p > p_c$. For the square lattice bond model, it was shown by Kesten [24] that $p_c = 1/2$. An immediate consequence of this result is the following: if $p > 1/2$, then under the 2d bond percolation model, there exists a positive constant $\alpha > 0$, such that for a lattice square of dimension $d \times d$, an open path connecting one side of the square to its opposite side exists with a probability at least α, regardless the value of d. On the other hand, if $p < 1/2$, then as $d \to \infty$, the probability of such an open path exists goes to zero. Interested reader may refer to [42] for detailed discussions in percolation theory.

A percolation phenomenon naturally arises in our minicircle network model since minicircles are mapped to vertices in the lattice and topological links to bonds. One defines the critical percolation density to be the density at which the system percolates. Because linking is possible between minicircles whose centers are not immediate neighboring grid points on the lattice, this defines a percolation system slightly different from the classical percolation system modeled using the same lattice.

iii. The mean saturation density. kDNA networks are very dense and most minicircles seem to belong to a single linked cluster. To describe the density at which minicircles link into a single linked cluster, we introduced the concept of *saturation network* and its corresponding *mean saturation density.* We measure saturation as a percentage of the total number of minicircles in the grid since minicircles in the boundary of the grid or with specific relative orientations may drive a 100% saturation density artificially high.

Before we present our results, we will describe some extensions/generalizations of this basic model aimed at either capturing the effect of a few bio-physical factors other than the minicircle density on the topology of the minicircle networks, or removing some artificial constraints due to the use of geometric circles as minicircles.

3.2. Extensions of the basic random knotting model for kDNA. *i. Angle restrictions.* Here, the restriction on the orientation of the minicircle is used as an adjustable parameter. In [2] we proposed a restriction on the orientation of the minicircles in the form of $\theta \in [\theta_0, \pi - \theta_0]$ and $\phi \in [\phi_0, \pi - \phi_0] \cup [\pi + \phi_0, 2\pi - \phi_0]$ for some positive constants $0 \leq \theta_0, \phi_0 < \pi$. We called θ_0 the *tilting angle restriction* and ϕ_0 the *azimuthal angle restriction.* In the work discussed here we will consider only the restriction on the tilting angle but not the azimuthal angle.

ii. Volume exclusion effects. In this case we added excluded volume to our model to capture the effect due to electrostatic interactions. The excluded volume can also be used as an adjustable parameter to estimate the concentration of histone like proteins and ionic conditions in kDNA. In [13], the excluded-volume effects are achieved by placing a hard cylinder around each kDNA minicircle which are modeled as discrete rigid circles. This simple model has been shown to approximate well electrostatic interactions [37]. We observed that most of the conformations generated using standard random sampling were invalid and proposed to address this problem using an imputation approach. The idea of this imputation approach is to estimate the various conditional linking probabilities using a simulated annealing algorithm on small (10×10) minicircle grids (see below). These empirically obtained linking probabilities were then used to generate larger grids.

iii. Effects of the flexibility of minicircles. Due to the consideration that minicircles are not rigid and that their overall shape is affected by DNA flexibility induced by either the wormlike nature of the genome [45], the sequence of the kDNA molecule [29], or by the possible condensing (or DNA binding) factors [1, 5, 20], we carried out a study in which rigid minicircles [mathematical assumption 2] are replaced with flexible conformations of *freely jointed chains* [**3**]. Freely jointed chains however have their own limitations because they capture the entropic contribution to minicircle linking (both between pairs of circles and throughout the grid), but do not include an energetic term. The set of minicircle conformations was created using the generalized hedgehog algorithm [**44**] as implemented in Knotplot [**39**]. The topological state of each simulated freely jointed chain was computed using the Alexander polynomial at -1, denoted by $\Delta(-1)$. Minicircles with $\Delta(-1) \neq 0$ were rejected. Minicircles were positioned on the lattice and a random orientation selected. Once all of the $d \times d$ polygons are positioned, the topology of the minicircle grid is determined by calculating the Gauss linking number as explained in Klenin and Langowski's formulation of the Gaussian integral [**26**].

iv. Lattice effects. As discussed earlier, we chose the square lattice to perform our initial work [mathematical assumption 1]. A concern here is whether such a choice on the lattice may significantly impact the results and conclusions. We therefore also tested other planar lattices. These included the triangular, the hexagonal and the random lattices [**12, 36**]. Sample conformations are shown in Figure 6. As in the square lattice, the minicircle density, defined as the number of minicircles per unit area, is given by $(\sqrt{3}/6)r^{-2}$ in a triangular lattice and $(\sqrt{3}/9)r^{-2}$ in a hexagonal lattice [**12**]. We generated random lattices by identifying the edges of the lattice (*i.e.* generating a torus) and randomly displacing the vertices of the square lattice uniformly around the position of the minicircle.

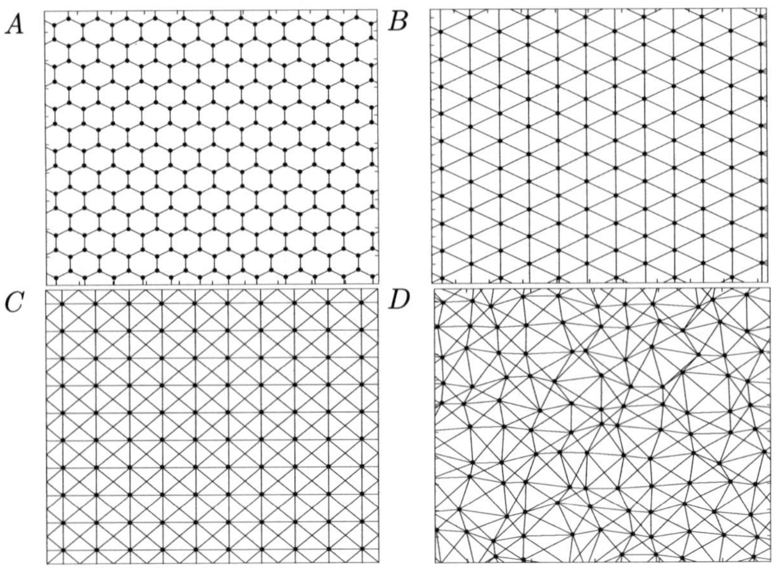

FIGURE 6. The hexagonal, triangular, square and random lattices.

4. Results: The linking probability of two minicircles

First we need to determine how the linking probability of two minicircles change with the distance between their center of masses and how the different assumptions of the model affect these probabilities. The following theorem is due to Diao and van Rensburg in [**16**].

THEOREM 4.1. *For two geometrical minicircles of unit radius whose centers are at a distance $2r$ $(r > 0)$ apart and their orientations are uniformly distributed on the unit sphere, the probability p_r for them to form an unsplittable link is given by:*

$$p_r = \begin{cases} 1 - r, & 0 < r < 1, \\ 0, & r \geq 1. \end{cases} \tag{1}$$

4.1. Changes in linking probability for restricted angles. A restriction in the tilting angle in the form of $\theta \geq \theta_0$ is applied to force the minicircles become more vertical on average (with respect to the plane of the lattice). This restriction decreases the linking probability between two minicircles. Figure 7 shows such relationship between tilting-angle restriction and linking probability between two minicircles at fixed separation becoming closer $P(r) = 1 - r$ when minicircles are very close or very far and having a more pronounced difference when the distance is between 0.2 and 0.6.

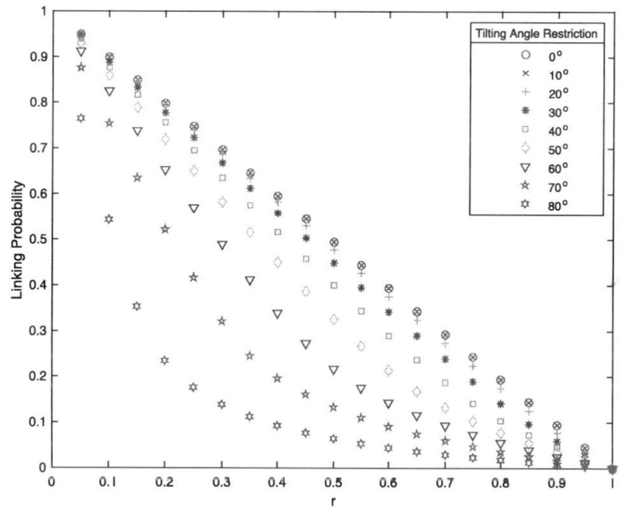

FIGURE 7. The linking probability between two minicircles whose centers are of a distance r for various values of θ_0 (in degrees as indicated in the legend). Data from [**2**]

4.2. Changes in linking probability due to volume exclusion. We used the hard cylinder model explained above to estimate the linking probability as a function of minicircle separation for various levels of volume exclusion. Interestingly we observed that although the probability is mostly monotonically decreasing as a function of r, the distance between the centers of the minicircles, the slopes of the linear portions of the curves increase as the thicknesses of the minicircles increase. In other words, as the extent of volume exclusion increases, the linking probability becomes more sensitive to minicircle separation. At very small separations, the function becomes nonlinear (and less steeply sloped) as fewer volume-excluding conformations result in a topological link (see Figure 8).

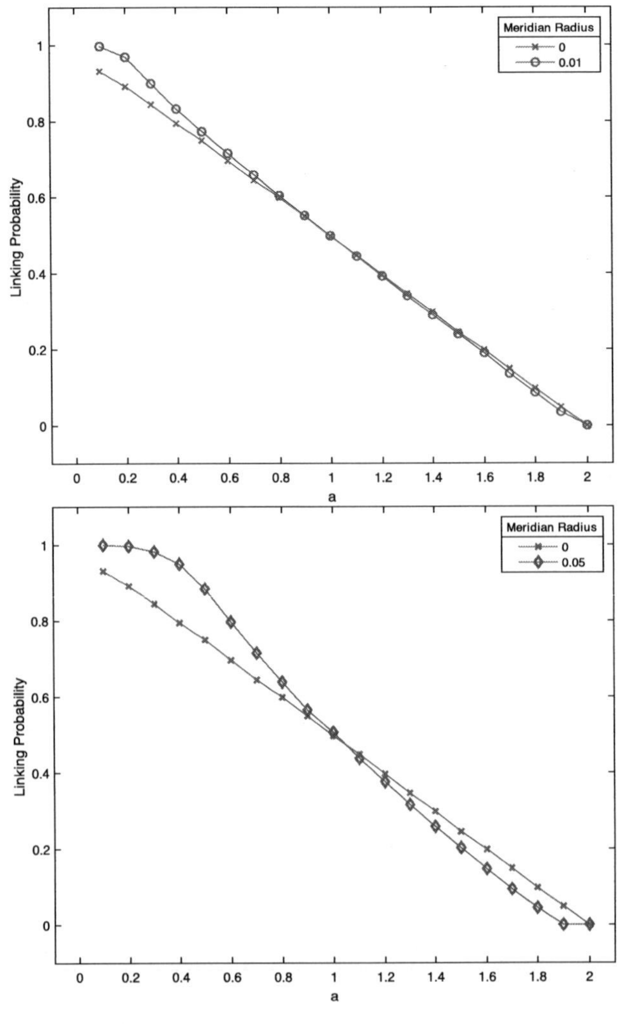

FIGURE 8. The linking probability (y-axis) is plotted against the separation (x-axis for pairs of minicircles with volume exclusion. Each panel shows one curve for thin circles (no volume exclusion) and one curve for the indicated value of cylinder thickness. The slopes of the linking-probability functions increase as the degree of volume exclusion increases. Each data point corresponds to 10,000 samples. Data from [**14**]

4.3. Changes in linking probability due to chain flexibility. Klenin and colleagues computationally studied the linking probability of two closed, freely jointed chains as a function of r, the distance between their respective centers of mass [**25**].

# of segments	Estimated linking probability
20 segments	$P(r) = 0.682\exp(-0.214r^{2.61})$
40 segments	$P(r) = 0.767\exp(-0.063r^{2.75})$
80 segments	$P(r) = 0.871\exp(-0.048r^{2.24})$

A parallel study for chain length in the range of 16 to 20 was done in [**3**]. The results are summarized in Figure 9 where an exponential decay phenomenon is also apparent.

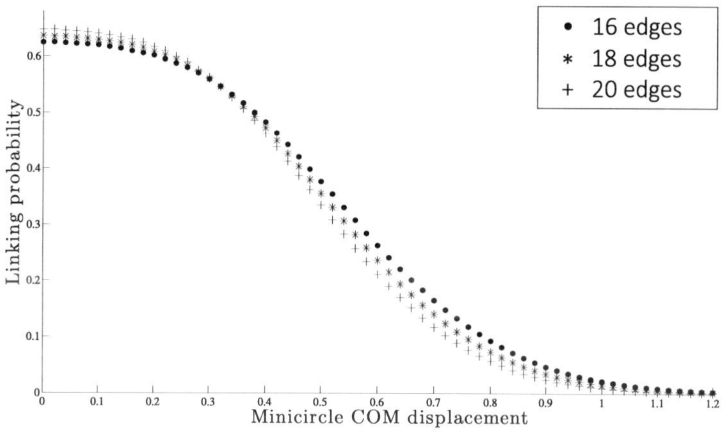

FIGURE 9. [**3**] Linking probability between two polygonal minicircles modeled by closed freely jointed chains as a function of the distance between their centers of mass (COM displacement).

It is worth noting that flexible minicircles may form topological links more complicated than a Hopf link. The question is, given the relatively short length of the minicircles (with respect to the Kuhn length), how likely it is for links other than the Hopf link to form and how often these links can occur? To understand this question, we studied the distribution of linking numbers among pairs of freely jointed chains [**3**]. Our results indicate that for the polygons we considered with $n = 16$, 18 and 20 of segments, most linked polygon pairs have absolute linking number 1. For example, in a grid dimension 100×100 with polygon size $n = 16$ edges, we found that less than 15 percent of linked polygon pairs have absolute linking number larger than 1 with densities near the mean percolation density. These percentages only increased slightly when the density increased to values near the mean saturation density, indicating that most of the linked pairs we encounter are indeed Hopf links.

5. Results: The mean valence of a minicircle network

Next, we describe how the mean valence changes as we change the different parameters of the simulation. As before we first describe the behavior of the mean valence for the initially proposed model and then present our results on how this behavior changes. Unless otherwise noted, the numerical results presented in this chapter were obtained using sample grids of dimensions ranging from 100×100 to 1000×1000 with a sample size of 1000.

5.1. The mean valence is heterogeneous. For our lattice based models where the minicircles are modeled by unit circles, the following theorem tells us that the mean valence grows linearly with the density of minicircles. This reproduces observation on kDNA replication that show that the mean valence doubles (from 3 to 6) when the number of minicircles doubles.

THEOREM 5.1. *If a minicircle network is based on a planar lattice with its minicircles modeled by unit circles, then the mean valence $E(V)$ of any given minicircle in the network is of the order of $O(D)$ where D is the minicircle density. More specifically, there exist positive constants $a < b$ such that $aD \leq E(V) \leq bD$.*

In the case of the square lattice, we have $.9D \leq E(V) \leq \frac{16}{3}D$ when D is large and computer simulations yield a linear regression of the form $E(V) = 4.1732D - 0.8716$.

5.2. Effect of the lattice structure on the mean valence. We studied the relationship between the mean valence and the underlying (regular) lattice [**2**]. The following table is a summary of the findings there. It shows that the choices of the lattice have a rather limited impact on this generally linear relationship as stated in Theorem 5.1. In all cases $R^2 > 0.9$

Lattice	Linear Predictor
Hexagonal	$y = 4.157x - 0.825$
Triangular	$y = 4.1451x - 0.8329$
Square	$y = 4.1732x - 0.8716$

5.3. Effect of minicircle orientation on the mean valence. In [**2**] we considered the tilting angle restriction θ_0 be set to $0, 30, 60$ and 87 degrees and no restriction on the azimuthal angle. The numerical results show that there is a strong linear relation between the mean valence and the minicircle density as shown in Figure 10.

Our simulations show that as the tilting-angle restriction increases, the mean valence decreases. However, for a fixed tilting-angle restriction, the relationship between mean valence and density remains linear as shown by Figure 10. The linear regression equations (all of them with $R^2 > 0.9$) are given below.

Angle restriction	Linear Predictor
$\theta_0 = 0$ (no restriction)	$y = 4.1732x - 0.8716$
$\theta_0 = 30$	$y = 3.604x - 0.847$
$\theta_0 = 60$	$y = 2.042x - 0.063$

Mathematically, when minicircle orientations are sampled from the entire unit sphere, the minicircle model predicts a mean valence that is much larger than the experimentally determined value in Chen *et al.* for physiological densities [**2**]. On the biological side, minicircles seen under electron microscopy tend to be oriented perpendicular to the plane of the kinetoplast disk [**15**].

5.4. Effect of minicircle flexibility on the mean valence. When we allowed for chain flexibility we still found a linear behavior of the mean valence, see Figure 11. However a few features are observed here that were not present in the rigid model. First, there is a non-zero probability of linking under the 0.5 minicircle density, density at which rigid minicircles do not link anymore. Our results also

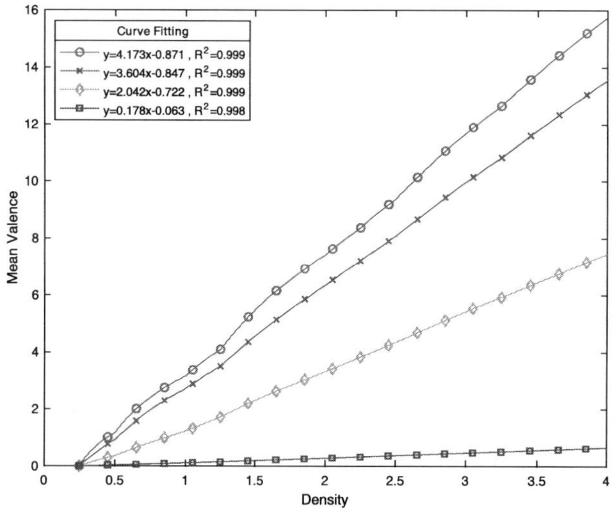

FIGURE 10. The minicircle mean valence is linear in density for each value of the tilting-angle restriction. These values are—from top to bottom—0, 30, 60, and 87 degrees. [2]

show that at high densities the minicircle valence is higher suggesting that flexible minicircles have a higher chance of linking with non-neighboring minicircles. Consistent with this observations longer chains have a higher mean valence than shorter ones, see Figure 11.

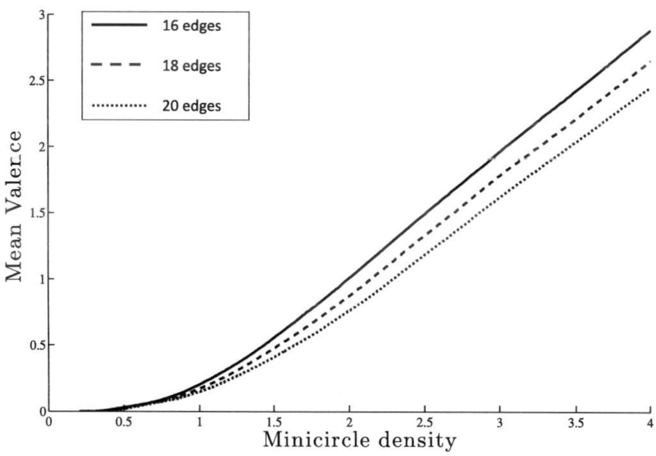

FIGURE 11. [3] Mean valence of a minicircle as the function of the minicircle density.

6. Results: Percolation and saturation phenomena in a minicircle network

6.1. General theoretical results. Percolation theory implies that at low densities minicircles form no linking clusters or small linking clusters until a critical density is reached at which point a large linked cluster (which span over the entire network) emerges. To distinguish this large cluster from the small clusters that appear before the critical percolation density, we call this large linked cluster the *percolating chain*. As discussed earlier, the critical percolation density provides an overall measure of the topological structure of the network. If a minicircle network has a minicircle density that is less than the critical percolation density, then we are quite certain that the system does not possess a percolating chain. On the other hand, if the minicircle density is above the critical percolation density, then it is likely that the system contains a percolating chain, which places the network in a globally linked state. The following theorem describes the dynamics of network formation as the density of minicircles increases, it is a generalization of a similar theorem stated in [**13**] for the case of the square lattice.

THEOREM 6.1. *If a minicircle network is based on a planar lattice with its minicircles modeled by unit circles, then there exists a critical percolation density* D_c *for the network. That is, if* $D > D_c$*, then the minicircle network percolates with a positive probability. More precisely, there exists a positive constant* $\alpha > 0$ *such that the probability that there exists an unsplittable chain of minicircles that cross the minicircle grid from one boundary to its opposite boundary is at least* α*, regardless the value of* d*. On the other hand, if* $D < D_c$*, then as* $d \to \infty$*, the probability of such an unsplittable chain of minicircles exists goes to zero.*

Interestingly, for the square lattice $D_c > 1/4$. First, since the density $D = (1/4)r^{-2}$ and the linking probability is zero once $r > 1$ then $D_c \geq 1/4$. Next, one could argue that percolation should occur at $D = 1/4$ since this is the case for the standard bond percolation model. This however is not the case because the linking of two circles is not independent of the linking of other circles neighboring them. If one considers $r = 1 - \delta$, $\delta > 0$ then the probability that the minicircles are linked will be close to 0, independently of the positions of the nearby minicircles. Hence one concludes that $D_c > 1/4$. A similar argument applies to other lattices taking into account the specific geometries of each lattice. For instance, for the hexagonal lattice $D_c > \sqrt{3}/9$ and for the triangular lattice $D_c > \sqrt{3}/6$ since no minicircles can be linked when $r > 1$. In the case of the mean saturation density, we have the following theorem, which is also a generalization of a similar theorem stated in [**13**] for the case of the square lattice.

THEOREM 6.2. *If a minicircle network is based on a grid of dimension* $d \times d$ *over a planar lattice with its minicircles modeled by unit circles, then the probability for the network to be completely saturated is bounded below by* $1 - e^{-\gamma_d D}$*, where* $\gamma_d > 0$ *is a constant depends only on* d*. It follows that at any given saturation percentage* p_0*, the mean saturation density* $D_s(p_0)$ *is at most* $\gamma_d^{-1} \ln(1 - p_0)^{-1}$*.*

6.2. Effect of the lattice structure on the percolation and saturation densities. Figure 12 shows the numerical results of the estimations of D_c and $D_s(.99)$ for the case of square lattice.

Similarly to the case of the mean valence, we studied the relationship between the critical percolation density, as well as the mean saturation density, and the

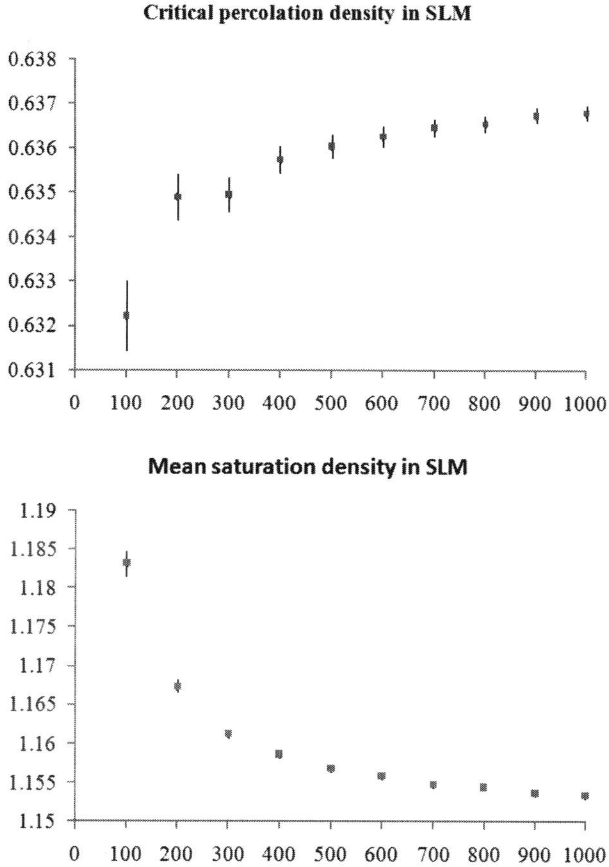

FIGURE 12. [**13**] Numerical estimations of the critical percolation density and the mean saturation density at the 99% saturation level for the SLM model. The x axis shows the dimension of the minicircle grid ($x \times x$) and the y axis the densities at which percolation and saturations are achieved.

underlying regular lattices [**2**]. The following table is a summary of the findings there. It again shows that the choices of the lattices have very limited impacts on critical percolation density and the saturation density.

Lattice	Critical Percolation Density	Mean Saturation Density
Hexagonal	0.726 ± 0.001	1.094 ± 0.001
Triangular	0.637 ± 0.001	1.157 ± 0.001
Square	0.637 ± 0.001	1.153 ± 0.001

Interestingly the critical percolation densities for the square and triangular are almost identical but the density for the hexagonal lattice is a slightly higher. This implies the critical percolation density does not change significantly when minicircles are placed in the triangular lattice and hexagonal lattice.

In [**36**], we also studied the effect of the displacement of the minicircles into the random lattice (a measurement on how far away a minicircle is allowed to deviate from a regular lattice point) on the critical percolation density. Our numerical results (given in the following table) showed that the impact on the critical percolation density is slightly more significant for larger displacements.

M_{disp}	0.5	1.0	1.5	2.0	2.5	3.0
D_c	0.7282	0.7959	0.8154	0.8239	0.8282	0.8309

6.3. Effect of minicircle orientation on the percolation and saturation densities. Next, let us examine how the critical percolation density and the mean saturation density are affected by a tilting-angle restriction. This problem was numerically studied in [**2**] based on the square lattice. Since the linking probability between any pair of minicircles decreases as the tilting angle θ_0 increases, one would naturally expect an increase in the critical percolation and mean saturation density with an increased tilting-angle restriction in the form of $\theta \geq \theta_0$. In the numerical study carried out in [**2**], the tilting angle restriction θ_0 is set from 0 to 80 degrees with an increment of 10 degrees and there is no restriction on the azimuthal angle. In the case of percolation density, an asymptotic behavior is observed at $\theta_0 = 90$. With $R^2 > 0.9$, the proposed predictor function is $y = 0.3 + \frac{29.66}{90-x}$. A similar result holds for the 99% saturation density where the predictor function is $y = 0.8 + \frac{265.7}{(90-x)^{1.5}}$ with $R^2 = 0.998$. See Figure 13. It is apparent that a restriction angle can have a much larger impact on the critical percolation density as well as the saturation density.

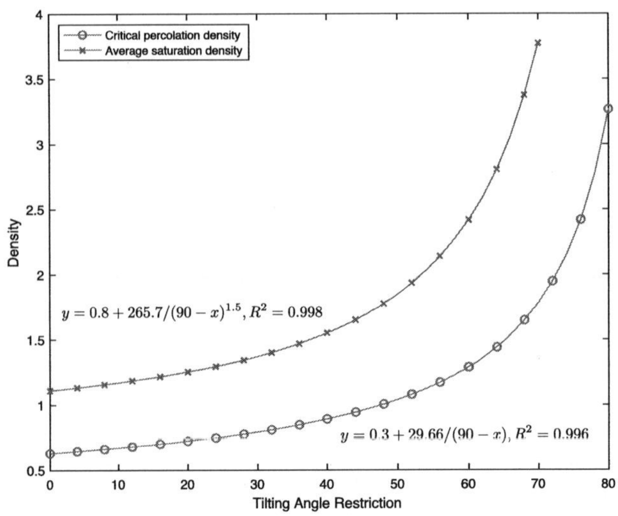

FIGURE 13. [**2**] The critical percolation density of minicircle grids increases as the tilting-angle restriction increases. The equation of a functional fit is displayed on the plot. Average saturation density behaves like the critical percolation density, increasing with increasing tilting-angle restriction.

6.4. Effect of minicircle flexibility on the percolation and saturation densities. Finally, let us examine the effect of minicircle flexibility on the critical percolation density and the mean saturation density. This problem was studied in [**3**], where minicircles are modeled by closed freely jointed chains with 16, 18 and 20 edges (segments). The numerically estimated values of the corresponding critical percolation densities and mean saturation densities are shown in the table below for chains consisting of 16, 18 and 20 edges based on grids of dimensions 100×100, 200×200 and 300×300.

D_c	100×100	200×200	300×300
16 edges	$3.40 \pm .07$	$3.44 \pm .04$	$3.45 \pm .03$
18 edges	$3.69 \pm .08$	$3.72 \pm .04$	$3.73 \pm .04$
20 edges	$3.96 \pm .08$	$4.00 \pm .05$	$4.01 \pm .04$
99% D_s	100×100	200×200	300×300
16 edges	6.40 ± 0.10	6.32 ± 0.07	$6.29 \pm .04$
18 edges	6.90 ± 0.10	6.78 ± 0.07	$6.75 \pm .05$
20 edges	7.30 ± 0.10	7.23 ± 0.07	$7.19 \pm .05$

One may wonder how the critical percolation density is estimated in the above mentioned work. It turns out that the critical percolation density is a very robust characteristic of the minicircle network and the system undergoes a rapid phase transition at this density. Figure 14 shows the percentage of minicircle networks (based on a minicircle grid with dimension 100×100, 200×200 and 300×300) with a percolating chain at each density for 16, 18 and 20 edges, respectively. The graphs show apparent phase transitions.

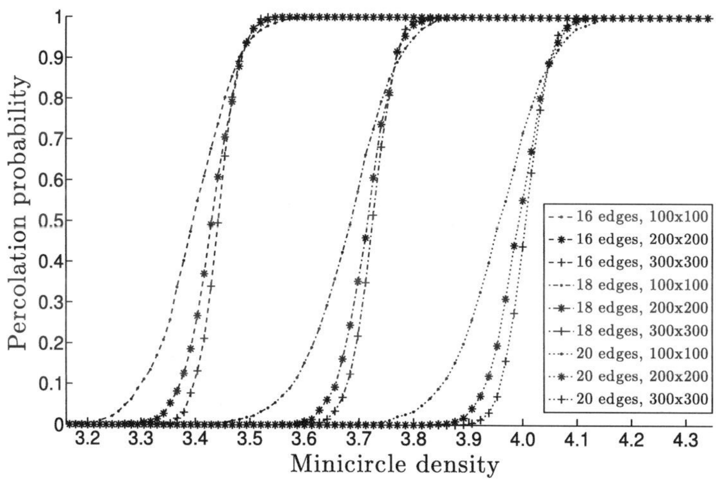

FIGURE 14. [**3**] Estimated mean percolation densities at grid with dimensions 100×100, 200×200 and 300×300 with 16, 18 and 20 edges respectively. Each data point in the graph is based on a sample of sample size 1000.

7. Applications to kDNA

We conclude this chapter by discussing some of the biological knowledge and hypotheses that these random-knotting based approaches have generated. Our random-knotting models suggest the following:

7.1. The biology of kDNA is consistent with the confinement hypothesis. Two of our observations, which are consistent with experimental observations, support this statement. First, our results show that a topological network is the most likely configuration of a set of minicircles in a confined volume. In fact, as we have shown in Theorem 4, this probability grows exponentially with the density of minicircles. Of course this equilibrium state would not be reached without the presence of agents that facilitate the crossing between strands. It is well known however that enzymes such type II topoisomerases are present in kDNA (reviewed in [**23**]) and may facilitate this process. Second, the mean valence grows linearly with the number of minicircles (Theorem 2). This was first reported by Chen et al. [**9**], when they observed that the mean valence of replicated networks is twice that of pre-replicated networks.

7.2. The network of minicircles was evolutionary assembled through a percolation process. Theorem 3 states that there is a critical density of minicircles at which the network percolates, transitioning from a collection of small clusters to a singly large cluster (or network). Interestingly, it has been observed that the mitochondrial DNA of the free-living organism *Bodo saltans* (an ancestor of *T. brucei*) consists of minicircles that are found in small linked clusters, rather than in a single structure [**28**]. These observation also support the formation of kDNA by confinement and in that case a transition through a percolation phenomenon. A natural question to ask is whether kDNA in trypanosomes is at the critical percolation density or not. In [**31**], Micheletto and colleagues suggested tested this hypothesis computationally and suggested that this was the case. Our model however does not support this hypothesis since the density at which mean valence three is reached and the percolation density as very different and instead suggest that this event happen at a different stage during evolution. In an recent study we mathematically explored whether this critical percolation density could be estimated experimentally using restriction reactions. This however may not be the case and additional information is needed [**22**].

7.3. Minicircles in the network are vertical with respect to the grid plane. The original minicircle model predicts that the mean valence of a minicircle is two orders of magnitude larger than its experimentally observed value. In [**15**] we tested what factors (of those presented in the previous section) could reduce the topological complexity while increasing the minicircle density. With a minimal set of assumptions, we observed that the only parameter that can significantly reduce the mean valence, while preserving the minicircle density, is the tilting of minicircles. Our model suggests that an angle of approximately 88 degrees provides the smallest mean valence value that is consistent with the saturation density values. See Figure 15.

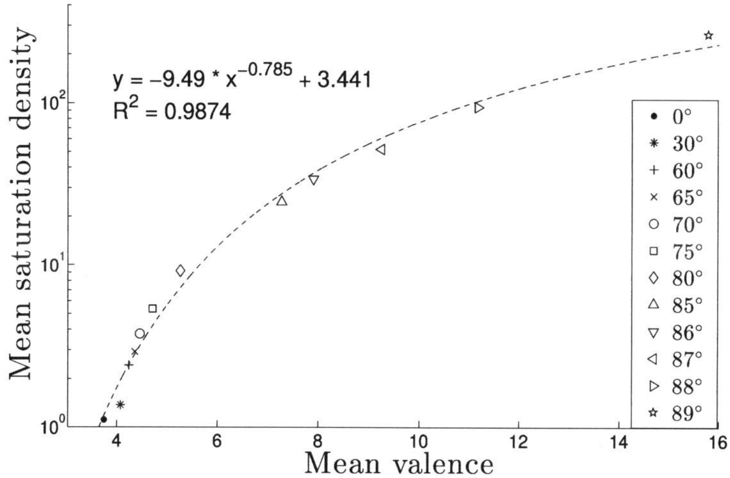

FIGURE 15. [**15**] Estimated relationship between mean valence, saturation density and tilting angle

8. Discussion

Genomic three-dimensional organization has attracted significant attention in recent years due to its connection to functional aspects of the cell. Modern experimental techniques such as chromosome-conformation capture assays and new advances in electron tomography have provided insights into chromatin architecture at unprecedented resolutions. However, in spite of these advances, an understanding of the topological properties of any genome remains elusive.

Topological considerations are important because they enable us to explore the downstream influence of a genome's 3-D structure on a variety of biological processes. Theoretical studies have shown that DNA molecules in confined volumes are entangled in knots and links. However, *in-vitro* assays show that these knots and links inhibit basic DNA-processing functions such as replication and transcription. This apparent paradox suggests that mechanisms have evolved in cells to perform DNA-processing functions in topologically complex environments.

The kDNA genome is one of the few to have been topologically characterized to date. We know that the elegant topological structure of kDNA is essential for completion of the parasite's life cycle, and we also know that this structure allows for independent replication of minicircles and maxicircles, as well as a coordinated regulation of transcription. However, information potentially important for drug development, such as the origin, function and maintenance of the network, is still lacking. The studies and results presented in this chapter aim, in part, at developing models that can be used for modeling potential drug targets.

In this chapter we discussed our efforts to study several key topological characteristics of minicircle networks, namely the mean valence, the critical percolation density and the mean saturation density and explored how various bio-physical parameters can affect these characteristics by studying various minicircle models that take into account of these parameters.

A key finding among all of these studies is that minicircle density seems to be the most important driving force in minicircle network formation, with minicircle-orientation restrictions being an important contributor, as well. Other factors, such as the distribution of the minicircles in the k-disk, the flexibility of the minicircles, and the volume exclusion effect, playing a secondary role. Our modeling approach and our results can be refined if more information is discovered about the specific shapes and orientations of the minicircles in the k-disk (even only in the average sense), and can be potential research topics in the future.

Acknowledgments

Figures 2 and 3 were produced by L. Ibrahim in the Arsuaga laboratory. The *C. fasciulata* cells used to produce these images were grown by M. Klinbeil (University of Massachusetts). P. Liu was supported by the grant of the Federal Government of Canada's Canada 150 Research Chair program to Dr. Caroline Colijn. R. Polischuk was partially supported by NSF grant DMS-1817156. J. Arsuaga was partially supported by NFS grant DMS-1817156 and by grant to Dr. O. Jenda NSF-DMS-1343651 (US-Africa Masamu Advanced Study Institute (MASI) and Workshop Series in Mathematical Sciences).

References

[1] Abu-Elneel K, Kapeller I, and Sholomai J 1999, *J. Biol. Chem* **274** 13419–26.
[2] Arsuaga J, Diao Y and Hinson K 2012, *Journal of Statistical Physics*, **146**(2), 434–45.
[3] Arsuaga J, Diao Y, Klingbeil M and Rodriguez V 2014, *Molecular Based Mathematical Biology* **2**(1) 98–106.
[4] Arsuaga J, Tan RK, Vazquez M, Sumners DW and Harvey S 2002, *Biophys. Chem.* **101** 475–84.
[5] Avrahami D, Tzfati Y and Shlomai J 1995, *Proc. Natl. Acad. Sci. USA* **92**10511–5.
[6] Bates AD, Maxwell A. DNA Topology (Oxford Biosciences). DNA. 2005;1500:23946.
[7] P. Borst, Why kinetoplast DNA networks? *Trends in Genetics* 7, 139-41 (1991).
[8] Chen J, Rauch CA, White JH, Englund PT and Cozzarelli NR 1995, *Cell* **80**(1) 61–9.
[9] Chen J, Englund PT and Cozzarelli NR 1995, *EMBO J* **14**(24) 6339–47.
[10] Delain E and Riou G 1969, *C. R. Acad. Sci. Ser. D* **268** 1225–7.
[11] Delain E and Pauletti C 1967, *J. Mol. Biol* **28**(2) 377–82.
[12] Diao Y, Arsuaga J and Hinson K 2012, *J. Phys. A: Math. Theor* **45** 035004.
[13] Diao Y, Hinson K, Kaplan R, Vazquez M and Arsuaga J 2011, *J Math Biology* **64**(6) 1087–108.
[14] Diao Y, Hinson K, Sun Y and Arsuaga J 2015, *J. Phys. A: Math. Theor.* **48** 435202.
[15] Diao Y, Rodriguez V, Klingbeil M, and Arsuaga J 2015, *PLoS One* **10**(6) e0130998.
[16] Diao Y and Rensburg Janse Van 1998, *Topology and Geometry in Polymer Science* (eds Whittington S G et al), IMA Volumes in Mathematics and its Applications **103** 79–88.
[17] Englund PT 1978, *Cell* **14**(1) 157–68.
[18] Ferguson MF *et al.* 1992, *Cell* **70** 621–9.
[19] Goringer HU. 2012 *Annual review of microbiology.*; **66**:65–82.
[20] Hines JC and Ray DS 1998, *Mol. Biochem Parasitol* **94** 41–52.
[21] Hoeijmakers J and Weijers P 1980, *Plasmid* **4** 97–116.
[22] Ibrahim L, Liu P, Diao Y, Klingbeil M and Arsuaga J 2018 *J. Phys A* **52** 034001.
[23] Jensen, R.E, Simpson, L, Englund P.T. 2008 *Trends Parasitol.* **24**(10) 428–431.
[24] Kesten H 1982, *Percolation theory for mathematicians*, Birkhauser.
[25] Klenin KV *et al.* 1988, *J. Biomol. Struct. Dyn.* **5** 1173–85.
[26] Klenin KV and Langowski J 2000, *Biopolymers* **54**(5) 307–17.
[27] Liu B *et al.* 2005, *Trends Parasitol* **21**(8) 363–9.
[28] Lukeš J *et al.* 2002, *Eukaryotic Cell* **1**(4) 495–502.
[29] Marini JC, Levene SD, Crothers DM, and Englund PT 1983, *Cold Spring harb. Symp. Quant. Biol.* **47**(1) 279–83.

[30] Meyer H, Oliveira M and Andrade M 1958, *Parasitology* **48** 1–8.
[31] Michieletto D, Marenduzzo D, Orlandini E. 2015 *Physical biology* **12** (3):036001.
[32] Perez-Morga D and Englund PT 1993, *J. Cell. Biol.***123** 1069–79.
[33] Porter K and Blum J 1953, *Anat. Rec.* **117** 685–92.
[34] Rauch CA, Perez-Morga D, Cozzarelli NR and Englund PT 1993, *EMBO J.* **12** 403–11.
[35] Ricca R and Nipoti B 2011, *J. Knot. Theor. Rami.* **20** 1325–43.
[36] Rodriguez V, Diao Y and Arsuaga J 2013, *J Phys: Conference Series* **454** 01207.
[37] Rybenkov VV, Cozzarelli NR and Vologodskii AV 1993, *Proc Natl Acad Sci USA* **90** 5307–11.
[38] Savill JS and Higgs PG 1999, *Proceedings of the Royal Society B* **226** 1419.
[39] Scharein R, *KnotPlot* (software), `https://knotplot.com/`.
[40] Shapiro TA and Englund PT 1995, *Annu. Rev. Microbiol.* **49**, 117–43.
[41] Shlomai J 2004, *Curr. Mol. Med.* **4**(6) 623–47.
[42] Stauffer D and Aharony A 1994, *Introduction to Percolation Theory*, CRC Press, New York.
[43] Steinert G, Firket H and Steinert M 1958, *Exp. Cell. Res.* **15** 632–5.
[44] Varela R, Hinson K, Arsuaga J and Diao Y 2009, *J. Phys A: Math Gen* **42**(9) 095204.
[45] Vologodskii A 2006, *Computational Studies of RNA and DNA*, Springer, 579–604.

DEPARTMENT OF MATHEMATICS AND STATISTICS, UNIVERSITY OF NORTH CAROLINA AT CHARLOTTE, CHARLOTTE, NC 28223

Current address: Department of Mathematics, Simon Fraser University, Burnaby, BC V5A 4Y8, Canada

Email address: `pengyu_liu@sfu.ca`

DEPARTMENT OF PHYSICS, UNIVERSITY OF CALIFORNIA AT DAVIS, DAVIS, CA 95616

Email address: `rdpolischuk@ucdavis.edu`

DEPARTMENT OF MATHEMATICS AND STATISTICS, UNIVERSITY OF NORTH CAROLINA AT CHARLOTTE, CHARLOTTE, NC 28223

Email address: `ydiao@uncc.edu`

DEPARTMENT OF MATHEMATICS, DEPARTMENT OF MOLECULAR AND CELLULAR BIOLOGY, UNIVERSITY OF CALIFORNIA AT DAVIS, DAVIS, CA 95616

Email address: `jarsuaga@ucdavis.edu`

Contemporary Mathematics
Volume **746**, 2020
https://doi.org/10.1090/conm/746/15001

Did sequence dependent geometry influence the evolution of the genetic code?

Alex Kasman and Brenton LeMesurier

Abstract. The genetic code is the function from the set of codons to the set of amino acids by which a DNA sequence encodes proteins. Since the codons also influence the shape of the DNA molecule itself, the same sequence that encodes a protein also has a separate geometric interpretation. A question then arises: How well-duplexed are these two "codes"? In other words, in choosing a genetic sequence to encode a particular protein, how much freedom does one still have to vary the geometry (or vice versa). A recent paper by the first author addressed this question using two different methods. After reviewing those results, this paper addresses the same question with a third method: the use of Monte Carlo and Gaussian sampling methods to approximate a multi-integral representing the mutual information of a variety of possible genetic codes. Once again, it is found that the genetic code used in nuclear DNA has a slightly lower than average duplexing efficiency as compared with other hypothetical genetic codes. A concluding section discusses the significance of these surprising results.

The first author's talk at the AMS Special Session on the Topology of Biopolymers explored the mathematical relationship between two different roles that a DNA sequence serves in living cells: encoding proteins to be produced and influencing the shape of the DNA molecule itself. Those results were subsequently published as a journal article [**5**]. After briefly summarizing the main results of that published paper, this article takes them a step further using a more sophisticated approach to the numerical computation of the mutual information. By combining Gaussian and Monte Carlo sampling methods with a new geometric inversion formula for computing the geometries, this new approach provides a more reliable result which strengthens and reconfirms the previously announced conclusions.

1. Measuring the efficiency of duplexed codes

1.1. A motivating example. Consider the following unlikely situation: You will soon need to send a text message conveying a two letter word to your friend Georgina *and* you also have to send a two letter word by text message to your friend Fred. However, because of your restrictive data plan, you must achieve this by sending a single two character message to both of them at the same time.

2010 *Mathematics Subject Classification.* Primary 92B05, 94A17; Seconday 65D30.

c	$f_1(c)$	$f_2(c)$	$g(c)$
0	H	H	N
1	O	O	H
2	O	H	N
3	I	O	H
4	I	I	O
5	H	I	O

TABLE 1. The functions f_1, f_2 and g used in this introduction to illustrate duplexing and mutual information.

You can hope to achieve this by teaching Fred one of the two functions f_i and teaching Georgina the function g shown in Table 1. Each of those functions turns one of the integers from 0 to 5 into a letter and can therefore be used as a simple "code". For example, since Georgina knows the function g you can send her the numerical message "24" and she would interpret it as "$g(2)g(4) = \texttt{NO}$". Alternatively, she would interpret the message "53" as the exclamation "OH". Similarly, using either of the two functions f_1 or f_2, Fred could recognize the signal "04" as the greeting "HI".

The really interesting thing is that you could send the *same* two digit message to both Georgina and Fred and they would interpret it differently. That is the defining characteristic of *duplexed codes*, that the same signal has two different interpretations.

Let us first suppose that Fred has memorized f_1 and Georgina knows the code g. If you wanted to send Georgina a message that will be interpreted as "NO", you have four different choices of signal which would convey that message to her and each one would mean something different to Fred. For instance, you could send "04" which Fred will interpret as "HI" or you could send "25" which has the same interpretation for Georgina but which Fred will interpret as "OH". In this scenario, you have the freedom to send different messages to Fred while still sending the desired message to Georgina at the same time.

In contrast, things would be different if Fred had learned f_2 as his code instead. Even though you would still have a choice of four signals to send Georgina that would be interpreted as "NO", you would have not be able to separately control the message that was sent to Fred because all four of the signals that mean "NO" under the code g would be interpreted as "HI" using code f_2. There would be no way to send Fred the message "OH" or any other message besides "HI" if Georgina's message is to be interpreted as "NO". Even though there is nothing wrong with the code f_2 on its own, there is something unfortunate about its relationship to g which creates an obstruction to sending the message "NO" to Georgina while simultaneously sending the message "OH" to Fred.

Loosely speaking, we say that two codes are *well-duplexed* if such obstructions to encoding two messages simultaneously are rare. Conversely they are *poorly-duplexed* if the choice of a message for one recipient severely restricts the messages that can be sent to the other recipient with the same signal. A more rigorous and quantifiable method of determining whether two codes are well-duplexed or poorly duplexed is by using the concept of *mutual information* that is part of the branch of mathematics knowns as information theory.

1.2. Duplexed codes and mutual information. Let us say that f and g are *duplexed codes* whenever $f : \mathcal{C} \to \mathcal{X}$ and $g : \mathcal{C} \to \mathcal{Y}$ are two functions with the same domain. The terminology makes sense when one imagines sending a single "signal" $c \in \mathcal{C}$ to two recipients each of whom knows one of those two codes. The goal of this section is to introduce a number associated to any duplexed codes which measures how much freedom you have to send different messages to one recipient even after the message for the other recipient is fixed.

For a randomly selected element $c \in \mathcal{C}$, let $P_f(x)$ denote the probability that $f(c) = x \in \mathcal{X}$, $P_g(y)$ be the probability that $g(c) = y$, and $P_{f,g}(x \text{ and } y)$ be the probability that both $f(c) = x$ and $g(c) = y$.

For example, using the functions defined in Table 1 with domain $\mathcal{C} = \{0, 1, 2, 3, 4, 5\}$, we see $f_1(c) = \texttt{N}$ is true for two of the six possible values of c and so $P_{f_1}(\texttt{N}) = 2/6 = 1/3$. Moreover, $P_{f_1,g}(\texttt{H}, \texttt{N}) = 1/6$ since the only way that $f_1(c) = \texttt{H}$ and $g(c) = \texttt{N}$ could both be true is if $c = 0$. However, $P_{f_2,g}(\texttt{H}, \texttt{N}) = 1/3$ since both $c = 0$ and $c = 2$ satisfy $f_1(c) = \texttt{H}$ and $g(c) = \texttt{N}$.

The mutual information (measured in bits) of the duplexed codes $f : \mathcal{C} \to \mathcal{X}$ and $g : \mathcal{C} \to \mathcal{Y}$ is defined to be[1]

$$M(f,g) = \sum_{y\in\mathcal{Y}} \sum_{x\in\mathcal{X}} P_{f,g}(x \text{ and } y) \log_2 \left(\frac{P_{f,g}(x \text{ and } y)}{P_f(x)P_g(y)} \right). \tag{1.1}$$

It is easy to see that $0 \le M(f,g)$ is true for any two codes f and g. The minimum possible value of 0 occurs when $P_{f,g}(x \text{ and } y) = P_f(x)P_g(y)$ for all choices of x and y. A familiar fact from probability theory is that the joint probability is equal to the product of the two probabilities precisely when the events are independent. Indeed, the same idea applies here, although we now interpret it in terms of the independence of the two codes. If the mutual information of two codes is zero then this tells us that the codes are very well-duplexed in that the selection of a message to one recipient does not restrict the message that can be sent to the other.

Since a mutual information of 0 represents the best possible duplexing of codes, larger mutual information means that the codes are *not* as well-duplexed. For example, we can compute that

$$M(f_1, g) \approx 0.584963 \qquad \text{and} \qquad M(f_2, g) \approx 1.58496.$$

for the codes f_1, f_2 and g from Table 1 in the previous section. The combination of functions f_2 and g is a bad choice for duplexing since if we were using those as codes for message to send Fred and Georgina then we could not separately choose a message for each recipient. In contrast, f_1 and g work better as a combination because even after we have chosen the message for one of the intended recipients we still have a choice of message that can be sent to the other. This is reflected here in the fact that $M(f_1, g) < M(f_2, g)$; the mutual information when using f_1 is closer to zero and therefore closer to being optimal for duplexing.

1.3. Comparisons with expected values. Let $F : S \to \mathbb{R}$ be a real-valued function on the finite set $S = \{\sigma_1, \dots, \sigma_n\}$. Then define the expected value

---

[1] When $P_{f,g}(x \text{ and } y) = 0$ it is understood that $P_{f,g}(x \text{ and } y) \log_2 \left(\frac{P_{f,g}(x \text{ and } y)}{P_f(x)P_g(y)} \right) = 0$.

$E_S(F(\sigma))$ by the familiar formula

$$E_S(F(\sigma)) = \frac{1}{n} \sum_{\sigma \in S} F(\sigma).$$

You will probably notice that this is nothing other than the *mean* of the values that F takes. The terminology "expected value" taken from probability theory is a notion analogous to the average in the context of random variables. The way to interpret it here is to imagine an experiment in which you randomly select an element σ from S and make a measurement of it to find the value $F(\sigma)$. Then $E_S(F(\sigma))$ is the expected value in the sense that it would be the average of the measurements after a large number of experiments. In particular, if for a particular $\hat{\sigma} \in S$ one has

$$F(\hat{\sigma}) < E_S(F(\sigma))$$

then one can say that the value of $F(\hat{\sigma})$ is lower than the value one would *expect* for a randomly selected element of S.

For example, using the functions f_1, f_2 and g from the motivating example above, we can consider the mutual information $M(f_\sigma, g)$ as a real-valued function on the index set $S = \{1, 2\}$. Then

$$0.584963 \approx M(f_1, g) < E_S(M(f_\sigma, g)) \approx \frac{0.584963 + 1.58496}{2} \approx 1.0849615,$$

tells us that the duplexing of the code f_1 with g is better than average for codes selected from $\{f_1, f_2\}$. Although we already knew that in this case simply by comparing the individual mutual information values, this notation will prove useful below where we will be doing something similar but with a very large index set.

2. A natural example of duplexed codes associated to DNA

2.1. The genetic code. Let $\mathcal{B} = \{\mathtt{A}, \mathtt{C}, \mathtt{G}, \mathtt{T}\}$ be the set of DNA bases. Because DNA sequences of length 2 and 3 will play special roles in this paper, let us introduce the following terminology and notation: The set of *dimers* (length two sequences) is $\mathcal{D} = \{b_1 b_2 \ : \ b_i \in \mathcal{B}\}$. and the set of *codons* (length three sequences) is $\mathcal{C} = \{b_1 b_2 b_3 \ : \ b_i \in \mathcal{B}\}$.

A genetic code is simply a function f_I from the set $\mathcal{C}$ of codons to the set $\mathcal{X}$ of amino acids (and the word "stop"):

$$\mathcal{X} = \{\text{I,L,V,F,M,C,A,G,P,T,S,Y,W,Q,N,H,E,D,K,R,Stop}\}.$$

The genetic code used by the nuclear DNA in humans is shown in Table 2, and this is the same genetic code used by nearly all known living organisms [**8, 9**]. We will refer to the particular genetic code given in Table 2 as "the natural genetic code" so as to distinguish it from other hypothetical codes that are not found in biology but will be used for comparison later in the paper.

However, it is important to realize that there are other genetic codes that are used by biological organisms (notably, mitochondria use a different genetic code) and that scientists have also introduced artificial genetic codes which nevertheless seem to function well enough to support life [**2, 6, 7, 10, 12–14**]. So, there is no *physical law* requiring this to be the genetic code. In theory, the genetic code could have been different and it is reasonable to ask the question "Why do nearly all living organisms use this particular genetic code?"

Codon (c)	Amino Acid ($f_I(c)$)
ATT , ATC , ATA	I
CTT , CTC , CTA , CTG , TTA , TTG	L
GTT , GTC , GTA , GTG	V
TTT , TTC	F
ATG	M
TGT , TGC	C
GCT , GCC , GCA , GCG	A
GGT , GGC , GGA , GGG	G
CCT , CCC , CCA , CCG	P
ACT , ACC , ACA , ACG	T
TCT , TCC , TCA , TCG , AGT , AGC	S
TAT , TAC	Y
TGG	W
CAA , CAG	Q
AAT , AAC	N
CAT , CAC	H
GAA , GAG	E
GAT , GAC	D
AAA , AAG	K
CGT , CGC , CGA , CGG , AGA , AGG	R
TAA , TAG , TGA	Stop

TABLE 2. This table defines the genetic code $f_I : \mathcal{C} \to \mathcal{X}$. *Each of the codons from* $\mathcal{C}$ in the first column is a pre-image of the corresponding amino acid (or "Stop") in the second.

There is evidence to support the hypothesis that the natural genetic code is the result of a combination of coincidences and evolutionary pressures (see [**15**] and references therein). For example, two codons for the same amino acid differ only in the third base much more frequently than would be predicted by chance if the genetic code was to be constructed entirely randomly. This has evolutionary advantages in that it decreases the likelihood that a mutation or mis-pairing of mRNA and tRNA will produce a different protein [**1, 3**]. It is therefore presumed that this feature is not a coincidence but an example of the effect of natural selection on the formation of the genetic code.

2.2. Sequence dependent DNA geometry. When shown in illustrations, DNA often looks like a perfectly straight double-helix, a twisted ladder with "rungs" that are the base pairs carrying the genetic sequence. However, real DNA is not straight; it is bent and twisted into compact shapes that fit into living cells.

It is perhaps not surprising that the way that a DNA molecule bends is affected by the sequence of bases which make it up. After all, A, C, G, and T in $\mathcal{B}$ are not just abstract mathematical symbols. They represent actual chemical structures that form the base pairs in a DNA molecule. Hence, the electrical repulsion and attraction between successive "rungs" in the DNA ladder will vary with that sequence.

Olson et all [**11**] experimentally determined the geometry of each of the 16 dimers $d \in \mathcal{D}$ by repeatedly measuring the configurations of DNA strands that were two base pairs long. They computed the average and standard deviation of each of the six Hassan-Calladine dimer step parameters (see [**4, 5**]). Their results are shown in Table 3

Assuming that the geometric configuration of each dimer in a longer sequence has the same expected values and standard deviations as the isolated dimers in those experiments, it is possible to make a similar table for the geometric configurations

d	$\bar{\Delta}_1(d)$ $(\hat{\Delta}_1(d))$	$\bar{\Delta}_2(d)$ $(\hat{\Delta}_2(d))$	$\bar{\Delta}_3(d)$ $(\hat{\Delta}_3(d))$	$\bar{\theta}_1(d)$ $(\hat{\theta}_1(d))$	$\bar{\theta}_2(d)$ $(\hat{\theta}_2(d))$	$\bar{\theta}_3(d)$ $(\hat{\theta}_3(d))$
AA	−0.03 (0.57)	−0.08 (0.45)	3.27 (0.22)	−1.4 (3.3)	0.07 (5.4)	35.1 (3.9)
AC	0.13 (0.59)	−0.58 (0.41)	3.36 (0.23)	−0.1 (3.1)	0.7 (3.9)	31.5 (4.2)
AG	0.09 (0.69)	−0.25 (0.41)	3.34 (0.23)	−1.7 (3.3)	4.5 (3.4)	31.9 (4.5)
AT	0 (0.57)	−0.59 (0.31)	3.31 (0.21)	0 (2.5)	1.1 (4.9)	29.3 (4.5)
CA	0.09 (0.55)	0.53 (0.89)	3.33 (0.26)	0.5 (3.7)	4.7 (5.1)	37.3 (6.5)
CC	−0.05 (0.76)	−0.22 (0.64)	3.42 (0.24)	0.1 (3.7)	3.6 (4.5)	32.9 (5.2)
CG	0 (0.87)	0.41 (0.56)	3.39 (0.27)	0 (4.2)	5.4 (5.2)	36.1 (5.5)
CT	0.28 (0.46)	0.09 (0.7)	3.37 (0.26)	1.5 (3.8)	1.9 (5.3)	36.3 (4.4)
GA	−0.28 (0.46)	0.09 (0.7)	3.37 (0.26)	−1.5 (3.8)	1.9 (5.3)	36.3 (4.4)
GC	0 (0.61)	−0.38 (0.56)	3.4 (0.24)	0 (3.9)	0.3 (4.6)	33.6 (4.7)
GG	0.05 (0.76)	−0.22 (0.64)	3.42 (0.24)	−0.1 (3.7)	3.6 (4.5)	32.9 (5.2)
GT	−0.09 (0.55)	0.53 (0.89)	3.33 (0.26)	−0.5 (3.7)	4.7 (5.1)	37.3 (6.5)
TA	0 (0.52)	0.05 (0.71)	3.42 (0.24)	0 (2.7)	3.3 (6.6)	37.8 (5.5)
TC	−0.09 (0.69)	−0.25 (0.41)	3.34 (0.23)	1.7 (3.3)	4.5 (3.4)	31.9 (4.5)
TG	−0.13 (0.59)	−0.58 (0.41)	3.36 (0.23)	0.1 (3.1)	0.7 (3.9)	31.5 (4.2)
TT	0.03 (0.57)	−0.08 (0.45)	3.27 (0.22)	1.4 (3.3)	0.07 (5.4)	35.1 (3.9)

TABLE 3. This table shows the mean $(\bar{\Delta}_i)$ and standard deviation $(\hat{\Delta}_i)$ of each of the Hassan-Caladine step parameters for each dimer as determined experimentally in [**11**].

associated to each of the 64 codons in $\mathcal{C}$. The function

$$\bar{g} : \mathcal{C} \to \mathbb{R}^3 \tag{2.1}$$

shown in Table 4 which associates to each codon $c \in \mathcal{C}$ a 3-tuple of numbers $\bar{g}(c)$ which gives the location of the center of the top of the codon in Angstroms if its base is located at the origin and if each of the dimers takes exactly the expected geometry according to Olson et al. (Note: In [**5**], this role is played by a function $\bar{\Gamma} : \mathcal{C} \to \mathbb{R}^6$ whose image has six components because it has angular information as well, but for simplicity in this note we are considering only the first three components which encode the location of the center of the third rung and not the way it is tilted.)

Figure 1 shows just the projection of $\bar{g}(c)$ onto its first two coordinates for each of the 64 codons $c \in \mathcal{C}$. You can imagine that a codon (a DNA sequence of length 3) is coming straight out of the xy-plane at you. Each point in this figure represents a codon and they all start out at the origin, but because the expected dimer step parameters depend on the particular bases involved, by the time they get up to their third rung they are in slightly different positions. In particular, the points indicate the locations of the center of the third rung (with units given in Angstroms) if each of the dimer step parameters takes its expected values in agreement with the experiments of Olson et al.

As you can see, the different codons do have slightly different expected geometries. It is important to realize that these small differences can combine in dramatic ways when considering longer sequences made up of many successive codons. For instance, Figure 2 shows the expected geometry for two different DNA sequences. Clearly, the sequence

$S_1 =$ AAAAACGGGCAAAAACGGGCAAAAACGGGCAAAAACGGGCAAAAACGGGCAAAAACGGGC

bends significantly more than sequence

$S_2 =$ AAGAATGGGCAGAAGCGTGCGAAGACTGGAAAGAATGGCCAGAAGCGTGCAAAAACGGGT.

So, geometrically they are quite different. But, consider how each of these two sequences is translated into a protein according to the natural genetic code. The

Codon (c)	$\bar{g}(c)$	Codon (c)	$\bar{g}(c)$
AAA	$(-0.0200699, -0.0157676, 6.54172)$	GAA	$(-0.162531, 0.126559, 6.6418)$
AAC	$(0.507057, -0.214617, 6.63862)$	GAC	$(0.371275, -0.0603654, 6.72489)$
AAG	$(0.259583, 0.0783038, 6.60773)$	GAG	$(0.11697, 0.227073, 6.69827)$
AAT	$(0.435648, -0.32513, 6.59098)$	GAT	$(0.300772, -0.172859, 6.68065)$
ACA	$(0.088445, 0.0062801, 6.68549)$	GCA	$(-0.133882, 0.138511, 6.72633)$
ACC	$(0.544148, -0.604157, 6.77731)$	GCC	$(0.343321, -0.455178, 6.81874)$
ACG	$(0.131333, -0.110103, 6.74604)$	GCG	$(-0.0871259, 0.0235416, 6.78672)$
ACT	$(0.521928, -0.212063, 6.72169)$	GCT	$(0.307084, -0.0639479, 6.76442)$
AGA	$(0.248608, -0.0483119, 6.71895)$	GGA	$(0.169153, -0.187748, 6.79938)$
AGC	$(0.78757, -0.206103, 6.71786)$	GGC	$(0.710666, -0.336536, 6.7972)$
AGG	$(0.76381, 0.0120173, 6.72551)$	GGG	$(0.683221, -0.11919, 6.81159)$
AGT	$(0.101016, 0.416757, 6.66124)$	GGT	$(0.0142188, 0.276429, 6.75535)$
ATA	$(0.271807, -0.439449, 6.72675)$	GTA	$(0.138917, 0.758201, 6.73073)$
ATC	$(0.483576, -0.728264, 6.64251)$	GTC	$(0.383801, 0.499296, 6.63932)$
ATG	$(0.577547, -1.03579, 6.6656)$	GTG	$(0.520989, 0.208472, 6.66249)$
ATT	$(0.350045, -0.605603, 6.57626)$	GTT	$(0.230872, 0.600458, 6.57895)$
CAA	$(0.329419, 0.573579, 6.58491)$	TAA	$(0.272116, 0.0943844, 6.68384)$
CAC	$(0.869997, 0.394174, 6.63371)$	TAC	$(0.812838, -0.080167, 6.74664)$
CAG	$(0.610102, 0.677742, 6.62709)$	TAG	$(0.55067, 0.201059, 6.73324)$
CAT	$(0.798832, 0.281315, 6.59152)$	TAT	$(0.743802, -0.193659, 6.70265)$
CCA	$(0.0575601, 0.353981, 6.74685)$	TCA	$(0.146711, 0.202584, 6.67568)$
CCC	$(0.532134, -0.244268, 6.82172)$	TCC	$(0.611685, -0.405631, 6.72668)$
CCG	$(0.106247, 0.239328, 6.8063)$	TCG	$(0.194597, 0.0857918, 6.73151)$
CCT	$(0.497544, 0.14655, 6.7635)$	TCT	$(0.583038, -0.0130668, 6.67772)$
CGA	$(0.0878838, 0.434499, 6.76845)$	TGA	$(-0.153175, -0.669182, 6.73584)$
CGC	$(0.636519, 0.316015, 6.75001)$	TGC	$(0.384081, -0.831342, 6.75681)$
CGG	$(0.597463, 0.531588, 6.76423)$	TGG	$(0.361217, -0.613645, 6.77368)$
CGT	$(-0.0939686, 0.889059, 6.72712)$	TGT	$(-0.294599, -0.200635, 6.69264)$
CTA	$(0.455558, 0.20792, 6.78231)$	TTA	$(0.113284, -0.0740582, 6.68867)$
CTC	$(0.698918, -0.0510585, 6.6871)$	TTC	$(0.353916, -0.33747, 6.59885)$
CTG	$(0.829989, -0.345276, 6.70098)$	TTG	$(0.478231, -0.634527, 6.61433)$
CTT	$(0.550076, 0.055828, 6.62638)$	TTT	$(0.209316, -0.226962, 6.53448)$

TABLE 4. The expected location of the center for the third rung of each codon relative to the position of the first rung in Angstroms. (See Figure 1 for a visual representation of this data and [**5**] for mathematical details.)

first codon in S_1 (AAA) and the first codon in S_2 (AAG) both encode the amino acid K. Similarly, the second codon in each encode the amino acid N.

In fact, the corresponding codons in each sequence always are mapped by the natural genetic code to the same amino acid. So, S_1 and S_2 encode exactly the same protein according to the natural genetic code, according to the function $\bar{g}$, one of them exhibits a much greater curvature than the other.

2.3. The geometric pressure hypothesis. Note that the last example of two DNA sequences with very different expected geometries is in some ways similar to the opening example of a simple duplexed code. Just as Fred and Georgina can have different interpretations of the same signal of numbers, a sequence of codons can be interpreted either as encoding a protein or as influencing the shape of the DNA molecule.

The shape of a DNA molecule is relevant to its biological function. It must bend in the right places and be straight in the right places in order for the enzymes and RNA responsible for transcription to be able to occur. This gives biological importance to the question of how well-duplexed the genetic and geometric codes discussed in the previous two sections are. For instance, if they are very poorly-duplexed, then it could often be the case that a DNA molecule cannot encode the protein that a creature needs unless it bends in a bad way. Conversely, it would be to a creature's advantage for the codes to be well-duplexed because then it would always be able to simultaneously encode whatever protein and geometry are optimal.

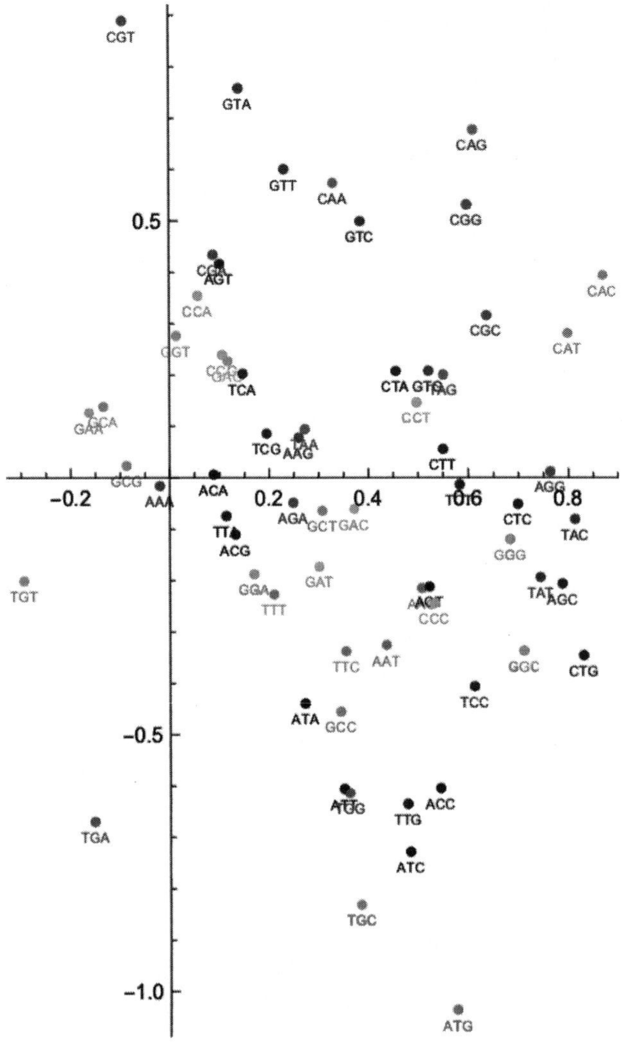

FIGURE 1. This graphic shows the x and y-coordinates of the images of the function $\bar{g}(c)$ (Table 4) for each codon $c \in \mathcal{C}$. They show the ways that the DNA molecule carrying that sequence would likely bend. If one was looking down at all 64 codons, each with its bottom rung fixed at the origin and with each dimer exhibiting its expected geometry (see Table 3) as it comes up out of the page towards you, the projections of the centers of the top rungs would be located at the locations indicated (with axes measured in Angstroms).

The geometric code was not something that evolution could act upon since it is determined by the laws of physics and chemistry. However, as we have seen, the genetic code could have been different and likely was influenced by natural selection. In [**5**], it was hypothesized that one of the factors that influenced that selection was pressure to ensure that the genetic and geometric codes were well-duplexed.

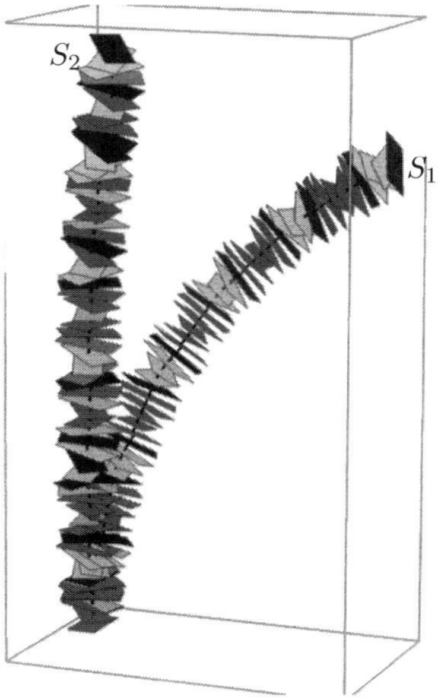

FIGURE 2. The sequences of the two DNA molecules shown encode the same protein, but due to the differences in expected codon geometry, one of them is noticeably more bent than the other.

3. How well duplexed is the real genetic code as compares with alternatives?

To test the "Geometric Pressure Hypothesis" (GPH), two different measures of duplexing efficiency were developed in [**5**]. Then, the duplexing efficiency of the geometric code with the natural genetic code was compared with its average duplexing efficiency with a large set of reasonable alternative genetic codes.

3.1. Alternative genetic codes. Let $\mathrm{Sym}(\mathcal{C})$ be the group of permutations on the set $\mathcal{C}$ of codons. Then for any $\sigma \in \mathrm{Sym}(\mathcal{C})$, $f_\sigma = f_I \circ \sigma$ is the function from $\mathcal{C}$ to $\mathcal{X}$ which first replaces the codon $c \in \mathcal{C}$ with its image $\sigma(c)$ and then applies the natural genetic code function f_I to that. (In other words, the function f_σ would be represented by a table very much like Table 2 for the natural genetic code above, but the codons would be rearranged according to the permutation σ.)

Notice that no matter which permutation σ is selected, the alternative genetic code f_σ has in common with the natural genetic code f_I not only that it is a map from $\mathcal{C}$ to $\mathcal{X}$ but also that for each amino acid $a \in \mathcal{X}$ the preimages are of the same size:

$$|f_I^{-1}(a)| = |f_\sigma^{-1}(a)|.$$

However, not all of those alternative genetic codes are realistic. For most choices of permutation σ, the alternative genetic code f_σ will not have the property that two codons are more likely to encode the same amino acid (or chemically similar amino acids) when their first two bases are equal, which will have already noted is a property of the real genetic code which has evolutionary advantages. Since we only want to consider alternative codes that also have this property, the permutations considered in [**5**] were further restricted: we considered not arbitrary permutations but only ones with the property that the first two bases in the codons $\sigma(c_1)$ and $\sigma(c_2)$ are equal if and only if the first two bases of c_1 and c_2 are equal. Let S be the set of such permutations. Symbolically, we can define the restricted set $S \subset \mathrm{Sym}(\mathcal{C})$ of permutations using the map $d(b_1b_2b_3) = b_1b_2$ which projects a codon onto its initial dimer as follows:

$$S = \{\sigma \in \mathrm{Sym}(\mathcal{C}) \ : \ \forall c, c' \in \mathcal{C} \ d(\sigma(c)) = d(\sigma(c')) \Leftrightarrow d(c) = d(c')\}.$$

To test the GPH in [**5**], the duplexing efficiency of the natural genetic code with the geometric code was compared with the expected value for the duplexing efficiency of the alternatives indexed by the set S. Since there are $(4^2)!(4!)^{16}$ permutations in S, this is a very large set of permutations to consider.

3.2. Total network length. The natural genetic code is shown in Table 2 and the expected geometries of the codons is shown in Figure 1. One way to combine this information is to draw an edge on the figure between any two codons that encode the same amino acid, turning it into a graph with vertices and edges. Thus, for instance, an edge would be drawn between the vertices labeled `TAT` and `TAC` because they both encode the amino acid `Y`, while the vertex labeled `ATG` would not be connected to any other vertices.

Each connected component of the graph corresponds to an amino acid. If there are only short edges or no edges in the component associated to an amino acid that you wish to encode in a sequence, then that means you have almost no geometric choice in the DNA molecule's expected geometry at that point. On the other hand, if there are long edges in the connected component, then you would have a choice of different codons that encode that same amino acid but which would cause a very different geometric configuration of the DNA molecule.

One could do the same for an alternative genetic code f_σ. That graph would have the same number of edges as the graph for the natural genetic code, but they would not have the same lengths.

With all of this in mind, we define the "total network length" of the genetic code f_σ for any $\sigma \in S$ to be the sum of the lengths of the edges in the graph[2]:

$$T_\sigma = \sum_{a \in \mathcal{X}} \left(\sum_{c,c' \in f_\sigma^{-1}(a)} |\bar{g}(c) - \bar{g}(c')| \right).$$

In the case of the natural genetic code, T_I was found to be 45.4238 Å. Because a larger total network length indicates more geometric freedom within the genetic code, if the GPH was true, one might expect the total network length for the natural genetic code to be *large* as compared with the total network length for the alternative codes which are not found in nature.

Disappointingly, that is not what was found in [**5**]. A 95% confidence interval was constructed for the expected value of the total network length over all permutations in S. It was found that the average total network length $E_S(T_\sigma)$ is probably between 45.7655 and 45.9639 Å. If so, then

$$45.4238 = T_I < 45.7655 < E_S(T_\sigma) < 45.9639$$

[2]The second sum is over all distinct, unordered pairs in the pre-image $f_\sigma^{-1}(a)$ and the length denotes the ordinary metric on $\mathbb{R}^3$ (i.e. $|(a,b,c)| = \sqrt{a^2+b^2+c^2}$).

Contrary to the predictions of the GPH, the total network length T_I for the natural genetic code is apparently a bit *smaller* than average rather than being especially large.

3.3. Mutual information with a discretized geometric code. The previous paper [**5**] also uses the concept of mutual information to quantify the mutual information of DNA's geometric and genetic codes. Using mutual information as a measure of duplexing efficiency has two big advantages over the use of total network length as described in the previous section:

- Firstly, it is a well-known measure of duplexing efficiency which is widely studied and used, whereas total network length is an *ad hoc* approach developed only for this particular project.
- Total network length was based only on the expected values in Table 3 and therefore ignored the standard deviations that represented the flexibility of the dimers. Of course, once that flexibility is taken into account, the "geometric code" is no longer a function since there is more than one possible configuration for each codon. Because the definition of mutual information in 1.1 involves probabilities, it is well suited to address this situation.

In [**5**], the geometry of a codon was represented by a point in $\mathbb{R}^6$ where the first three numbers (like the image of $\bar{g}$ above) indicate the location of the center of the last "rung" of the codon relative to the first and the other three were angles indicating how it was tilted and twisted relative to the first. Then, $\mathbb{R}^6$ was divided into $4096(=4^6)$ subsets called 'bins'. If each bin is indexed by an element of $\mathcal{Y}=\{1,2,\ldots.,4095,4096\}$ then the geometric information is encoded into a map $g:\mathcal{C}\to\mathcal{Y}$.

Unlike any of the maps discussed earlier, g is not a *function* since given codon can be in many different possible geometric configurations due to its flexibility. Although it is more likely to be in certain configurations than others, and so $g(c)$ is a random variable for any given codon c. In order to compute the mutual information of the genetic codes f_σ with this geometric map g, we need to be able to compute the associated probabilities. In [**5**] that was done by running a computer program which looped through a large number of different configurations and recorded the number of the 'bin' in which they ended up. In other words, the probabilities were computed *empirically*, using the assumption that the dimer step parameters for are *normally* distributed with the mean and standard deviation shown in Table 3.

Using this information it is now possible to compute (or, perhaps it would be better to say "approximate") the mutual information of any of the genetic codes f_σ with this geometric code g. When this was done in [**5**], was found that the mutual information of the *natural* genetic code f_I with the geometric code g is about $M(f_I,g)\approx 0.154144$ bits. However, when the same computation was repeated for randomly selected alternative genetic codes f_σ for $\sigma\in S$ a 95% confidence interval found that the average mutual information is probably between 0.14286 and 0.148569. It it is, then

$$0.14286\le E_S(M(f_\sigma,g))\le 0.148569<M(f_I,g)\approx 0.154144.$$

Since a *smaller* mutual information (closer to the ideal value of 0) represents better duplexed codes, this means that the duplexing of the real genetic code f_I is *worse* than average. This, again, is the opposite of what would have been predicted by the GPH.

4. New results: Mutual information via a Monte-Carlo style discretization

Since the geometric parameters take values in a continuous space, the functions $P_{f,g}$ and P_g which appear in the formula for the mutual information are actually probability distribution functions whose values only become probabilities when integrated over regions of that space. In the previous paper this computation was discretized in a *rigid* way by "binning" the data into fixed and pre-determined subsets of equal size.

That approach is plausible, but not uniquely so. A more standard approach in numerical analysis is to consider a discretization based on choosing a suitable *random* sample Y_N of N geometries that are chosen taking into account the Gaussian distributions in Table 3, and replace the usual discrete mutual information by

$$M_N(f,g)=\frac{1}{W}\sum_{y\in Y_N}\sum_{a}P_{f,g}(a,y)\log_2\left(\frac{P_{f,g}(a,y)}{P_f(a)P_g(y)}\right)\text{ with }W=\sum_{y\in Y_N}\sum_{a}P_{f,g}(a,y),$$

where y is a codon geometry, a is an amino acid, $f=f_\sigma$ is one of the genetic codes mapping codons to amino acids. The normalization by factor $1/W$ is the volume element for approximate integration over the probability density $\sum_a P_{f,g}(a,y)$. It normalizes the sum appropriately, in the sense of $M_N(f,g)$ approaching a common value $M(f,g)$ as N increases.

The randomness of the sample Y_N is one of the main differences between the prior results and this new approach. Another difference is the randomness used in the approximation of the values of the probability distributions themselves. Unlike the previous approach in which the probabilities were estimated by using rigidly chosen deviations from the expected values, this time a Monte Carlo approach will be utilized. In particular, here we construct random choices for $N=64n$ geometry samples as the union of a set of n sample points for each codon, with the samples for each codon constructed from n sets of random values for the twelve Hassan-Calladine parameters for the two dimers; the randomness based on assuming that these parameters are all independent and that each is normally distributed with mean and standard deviation as in Table 3.

The final, and perhaps most interesting, difference between the previous approach and the one taken in this section is an inversion of the geometric data which directly computes the probability that a given codon will take on a given geometric conformation. The basic quantity needed is the probability distribution $P_d(\sigma)$ for the values of the dimer step σ for dimer d with Hassan-Calladine parameters $\Delta_1,\Delta_2,\Delta_3,\theta_1,\theta_2,\theta_3$. As above, this is based on the assumption that these parameters are independent and each is normally distributed, so:

(4.1)

$$P_d(\sigma)=\prod_{i=1}^{3}\frac{1}{\hat{\Delta}_i(d)\sqrt{2\pi}}\exp\left(-\frac{(\Delta_i-\bar{\Delta}_i(d))^2}{2\hat{\Delta}_i(d)^2}\right)\prod_{j=1}^{3}\frac{1}{\theta_j(d)\sqrt{2\pi}}\exp\left(-\frac{(\theta_j-\bar{\theta}_j(d))^2}{2\hat{\theta}_j(d)^2}\right).$$

Consider codon C consisting of dimers d_1 and d_2, and for a choice of their dimer steps σ_1 and σ_2, denote the resulting codon geometry y as the "step product" $\sigma_1 * \sigma_2$. Then $P_C(y)$, the probability density of the codon C having geometry y, is given by an integral over all paths to y:

$$P_C(y) = \int_{\sigma_1} \left[\sum_{\sigma_2 | \sigma_1 * \sigma_2 = y} P_{d_2}(\sigma_2) \right] P_{d_1}(\sigma_1)\, d\sigma_1$$

For each pair of values for y and σ_1, we must solve for all possible values σ_2; fortunately this can be done explicitly, and generically there are only two such paths; this is detailed in the next subsection.

For each path the quantity $P_{d_2}(\sigma_2)$ will be evaluated using Eq. (4.1). The outer integral over a six dimensional space is instead dealt with by a Monte-Carlo method: noting that it is the integral w.r.t. a probability measure, we approximate by choosing a sample of m random sets of values for the Hassan-Calladine parameters, in turn determining a set Σ_m of values for σ_1, and averaging:

$$P_C(y) \approx \frac{1}{m} \sum_{\sigma_1 \in \Sigma_m} \sum_{\sigma_2 | \sigma_1 * \sigma_2 = y} P_{d_2}(\sigma_2)$$

Then we assemble the pieces:

$$P_g(y) = \frac{1}{64} \sum_C P_C(y)$$

$$P_{f,g}(y) = \frac{1}{64} \sum_{C | f(C) = a} P_C(y)$$

and the easy one, the fraction of codons that give a specified amino acid:

$$P_f(a) = \frac{1}{64} \big| \{ C | f(C) = a \} \big|$$

4.1. Reconstructing the second dimer step. What remains is to solve $\sigma_1 * \sigma_2 = y$ for the second dimer step σ_2; that is, find the corresponding Hassan-Calladine angles θ_i and then the lengths Δ_i.

Except for one case noted below (of negligible probability), the angles θ_i, $i = 1$–3 are determined up to negation of the pair θ_1, θ_2. This comes from the formula

$$T = M R_3(\theta_3/2 - \phi) R_2(\eta) R_3(\theta_3/2 + \phi) \tag{4.2}$$

as in Section 2.2 of [**5**] (See also Eq's (9) of [**4**].) The matrices M and T are the frames respectively for the end of the first dimer and the end of the codon, R_2 and R_3 are the familiar matrices for rotations about y and z axes

$$R_2(\eta) = \begin{bmatrix} \cos\eta & 0 & -\sin\eta \\ 0 & 1 & 0 \\ -\sin\eta & 0 & \cos\eta \end{bmatrix} \qquad R_3(\theta) = \begin{bmatrix} \cos\theta & -\sin\theta & 0 \\ \sin\theta & \cos\theta & 0 \\ 0 & 0 & 1 \end{bmatrix},$$

and $\eta = \operatorname{sign}(\theta_2)\sqrt{\theta_1^2 + \theta_2^2}$, $\sin\phi = \theta_1/\eta$ with $-\pi \le \phi \le \pi$,

From this,

$$\theta_2 = \operatorname{sign}(\theta_2)\sqrt{\eta^2 - \theta_1^2} = \eta \cos\phi.$$

Let $R = M^{-1}T = R_3(\theta_3/2 - \phi)R_2(\eta)R_3(\theta_3/2 + \phi)$ be the combined rotation.

$$R_{33} = (R_3(\theta_3/2-\phi)^T e_3)^T R_2(\eta)(R_3(\theta_3/2-\phi)e_3) = e_3^T R_2(\eta)e_3 = (R_2(\eta))_{33} = \cos\eta,$$

so

$$\eta = \pm \arccos R_{33} \in [0, \pi]$$

Case 1 (Generic): $-1 < R_{33} < 1$, so $0 < |\eta| < \pi$.
Defining $\alpha = \theta_3/2 + \phi$ and $\beta = \theta_3/2 - \phi$, α is determined in $(-\pi, \pi]$ by

$$\cos\alpha = -R_{31}/\sin\eta, \quad \sin\alpha = R_{32}/\sin\eta$$

and likewise β by

$$\cos\beta = R_{13}/\sin\eta, \quad \sin\beta = R_{23}/\sin\eta$$

$\sin\eta \neq 0$, so no problems here.

Then $\theta_3 = \alpha + \beta$, $\phi = (\alpha - \beta)/2$, $\theta_1 = \eta\sin\phi$, $\theta_2 = \eta\cos\phi$.

The two choices for η likewise negate θ_1 and θ_2, but only shift θ_3 by an irrelevant increment of 2π.

Case 2. $R_{33} = 1$, so $\eta = 0$.
$\theta_1 = \theta_2 = 0$, and $R = R_3(\theta_3)$ so θ_3 is determined easily.

Case 3: $R_{33} = -1$, so $\eta = \pm\pi$.
This is the problem case, as R now depends only on ϕ, not θ_3, so the latter is not constrained at all. However, the value $\eta = \pm\pi$ is extremely unlikely: $\eta^2 = \theta_1^2 + \theta_2^2$, and as seen in Table 3, the values for the later two angles are far too small.

Reconstructing the remaining Hassan-Calladine parameters Δ_i is now straightforward; they are related to the known positions of the ends of each dimer and the Hassan-Calladine by a system of linear equations, as seen in the formulas

$$T = (\mathbf{v}_1^\top \ \mathbf{v}_2^\top \ \mathbf{v}_3^\top) = MR_3\left(\frac{\theta_3}{2} - \phi\right) R_2\left(\frac{\eta}{2}\right) R_3(\phi).$$

with the (known) positions $\mathbf{p}_2$ and $\mathbf{p}_3$ of the ends of the second and third bases related by

$$\mathbf{p}_3 = \mathbf{p}_2 + \Delta_1\mathbf{v}_1 + \Delta_2\mathbf{v}_2 + \Delta_3\mathbf{v}_3$$

from [**5**]; see also Eq's (10,11) of [**4**].

4.2. Numerical results. The most accurate calculation so far for the true genetic code is with $N = 4096$ samples and $m = 2048$ samples for each evaluation of $P_C(y)$. This gives $M_N(f_I, g) \approx 0.1834$, with standard error of the mean estimated at 0.003.

Comparisons to alternative genetic codes have been done with 16 randomly generated codes, each with $N = 2048$ and $m = 1024$; the random codes then give a mean $E_S(M_N(f_\sigma, g))$ value of 0.1776 with 95% confidence interval $[0.1708, 0.1844]$. With those same sample size parameters for the true genetic code the result was 0.1877. Much as seen in Section 3.3, this is slightly out of the 95% confidence interval, in the opposite direction to that suggested by the Geometric Pressure Hypothesis:

$$0.1708 \leq E_S(M_N(f_\sigma, g)) \leq 0.1844 < M_N(f_I, g) \approx 0.1877.$$

5. Conclusions

A given genetic sequence can be interpreted as encoding a protein and also as influencing the geometry of the DNA molecule that carries it. This is therefore a situation like the one in the Introduction where Georgina and Fred are each interpreting the same signal differently. It is therefore of interest to understand how well-duplexed these two "codes" are.

Combining the results of the previous paper [**5**] with the new results in Section 4 this has now been done in three different ways. Disappointingly, each time the answer has turned out to be "about what you'd expect if the real genetic code was just selected randomly, but maybe a little worse". In other words, contrary to the predictions of the Geometric Pressure Hypothesis (GPH), the natural genetic code does not appear to be especially well-duplexed. In other words, even if one replaced the natural genetic code with a random alternative, it is likely that there there would be more freedom in the geometry of the DNA molecule while encoding any given protein.

It is interesting to speculate about what that tells us about the evolution of the genetic code. On the one hand, it could simply imply that there is not much evolutionary advantage in having the ability to alter the shape of the DNA molecule without changing the protein it encodes. However, that is not the only possible explanation. Another intriguing possibility, which was raised during the discussion after the talk in the session for which this volume serves as the proceedings, is that the code became fixed before the molecules became large enough for it to matter. In particular, it does seem likely that the geometry of the molecule is not very important when the chromosome is very short. So, if the genetic code that we are familiar with was shaped during an early period in evolution when the genome of living creatures were all very small, the GPH might not have applied. And, since the genetic code is no longer very malleable (as demonstrated by its near ubiquity), it might no longer have been able to change once the molecules grew large enough for their geometry and topology to matter.

In any case, whatever the explanation may be, the new computations have only re-confirmed the answer found previously to the question of the title. Since the natural genetic code appears only slightly less well-duplexed with the geometric code than an average alternative, it does not appear that the evolution of the genetic code was shaped by any pressure to optimize it.

References

[1] Alexander RW, Schimmel P, (2001) "Wobble Hypothesis" in *Encyclopedia of Genetics* (S Brenner and JH Miller, eds) Elsevier.

[2] Barrell BG, Bankier AT, Drouin J. A different genetic code in human mitochondria. Nature 1979;282:189-194.

[3] Berg JM, Tymoczko JL, Stryer L, (2002) Biochemistry. 5th Edition. WH Freeman. Section 5.5.1.

[4] Hassan MA and Calladine CR "The Assessment of the Geometry of Dinucleotide Steps in Double-Helical DNA; a New Local Calculation Scheme" J. Mol. Biol. (1995) 251 648-664

[5] Alex Kasman, *The duplexing of the genetic code and sequence-dependent DNA geometry*, Bull. Math. Biol. **80** (2018), no. 10, 2734–2760, DOI 10.1007/s11538-018-0486-3. MR3856996

[6] Kawaguchi Y, Honda H, Taniguchi-Morimura J, Iwasaki S. The codon CUG is read as serine in an asporogenic yeast Candida cylindracea. Nature 1989;341:164-166.

[7] Kiga D, Sakamoto K, Kodama K, Kigawa T, Matsuda T, Yabuki T, Shirouzu M, Harada Y, Nakayama H, Takio K, et al. An engineered Escherichia coli tyrosyl-tRNA synthetase for site-specific incorporation of an unnatural amino acid into proteins in eukaryotic translation and its application in a wheat germ cell-free system. Proc. Natl Acad. Sci. USA 2002;99:9715-9720.

[8] Koonin EV and Novozhilov AS, "Origin and Evolution of the Universal Genetic Code" *Annu. Rev. Genet.* 2017. 51:45–62

[9] Lajoie MJ, Söll D and Church GM, "Overcoming challenges in engineering the genetic code" *J Mol Biol.* 2016 Feb 27; 428(5 Pt B): 1004–1021.

[10] Liu CC, Schultz PG. Adding new chemistries to the genetic code. Annu. Rev. Biochem. 2010;79:413-444.

[11] Olson WK, Gorin AA, Lu XJ, Hock LM, and Zhurkin VB "DNA sequence-dependent deformability deduced from protein-DNA crystal complexes" Proc. Natl . Acad. Sci. USA Vol. 95, pp. 11163–11168, September 1998

[12] Srinivasan G, James CM, Krzycki JA. Pyrrolysine encoded by UAG in *Archaea: Science.* 2002 May 24;296 (5572) 1459-62.

[13] Wang L, Brock A, Herberich B, Schultz PG. Expanding the genetic code of Escherichia coli. Science 2001;292:498-500.

[14] Yamao F, Muto A, Kawauchi Y, Iwami M, Iwagami S, Azumi Y, Osawa S. UGA is read as tryptophan in Mycoplasma capricolum. Proc. Natl Acad. Sci. USA 1985;82:2306-2309.

[15] Zhang Z and Yu J, "On the Organizational Dynamics of the Genetic Code", *Genomics, Proteomics & Bioinformatics* Vol. 9, 1–2, April 2011, pp. 21–29

Department of Mathematics, College of Charleston, Charleston, SC 29424
Email address: `kasmana@cofc.edu`

Department of Mathematics, College of Charleston, Charleston, SC 29424
Email address: `lemesurier@cofc.edu`

Contemporary Mathematics
Volume **746**, 2020
https://doi.org/10.1090/conm/746/15002

Topological sum rules in the knotting probabilities of DNA

Tetsuo Deguchi and Erica Uehara

ABSTRACT. We revisit a topological sum rule that the sum of the coefficients (or the amplitudes) of the knotting probabilities of self-avoiding polygons (SAP) over all prime knots is equal to 1. Here we define the coefficient as a fitting parameter of a formula expressing the knotting probability as a function of the number of segments, where the estimates of the fitting parameters are shown to be very close to those of the asymptotic expansion of the knotting probability. We numerically show these results for a model of semi-flexible ring polymers such as circular DNA consisting of cylindrical segments with radius $r_{\rm ex}$ of unit length by simulation with several values of $r_{\rm ex}$. From the sum rule we argue that SAPs with the trefoil knot and its descendants are dominant among all SAPs if the excluded volume is large. We also suggest that the knot exponent of a prime knot is smaller than 1 if the excluded volume is very small and it gradually increases to 1 as the excluded volume increases.

1. Introduction

Theoretical studies predict that long polymers have a tendency to become knotted even in thermal equilibrium [**1–5**]. If we generate random conformations of ring polymers in solution, the fraction of conformations which are topologically equivalent to the trivial knot decreases almost exponentially with respect to the number of segments [**6–11**]. However, knotted ring polymers have been observed mainly in experiments of semiflexible polymers [**12**]. For example, knotted ring polymers randomly circularized were observed in the experiments of circular DNA [**13–15**] (see also [**16**]). Here we regard DNA molecules as semiflexible polymers, since they are negatively charged and hard to bend due to electric repulsions among their segments [**17, 18**]. Experimental results support this model. Trivially knotted circular DNA molecules were employed to measure the two-point correlation functions of ring polymers in two dimensions [**19, 20**] and compared with results from Monte-Carlo simulations where each circular DNA molecule was modeled as a closed sequence of thin hard cylinders [**21**]. We call the model the cylindrical self-avoiding polygon (SAP) [**22, 23**]. It corresponds to a model of semiflexible polymers, as explained later [**24**]. Here we suggest that semiflexible polymers are much more easily or frequently knotted than flexible polymers [**12**]. Several aspects

---

2010 *Mathematics Subject Classification.* Primary 82D60; Secondary 82B41, 57M25.
Key words and phrases. Polymers, random walks, knots.
The authors were supported in part by JSPS KAKENHI Grant Number JP17H06463.

of the knotting probabilities such as bimodality have been studied in simulations of semiflexible polymers [**24**–**26**].

We define the knotting probability of a knot K for N segments by the probability for a conformation of random polygons (RPs) or self-avoiding polygons (SAPs) with N segments being topologically equivalent to the knot K. We denote it by $P_K(N)$. It plays an important role in topological entanglement effects in ring polymers: It is one of the most fundamental quantities characterizing them, i.e. its logarithm leads to the entropy of the knotted ring polymer; furthermore, it is not only defined mathematically rigorously but also relevant to experiments [**27**–**30**]. The knotting probability of a given knot K as a function of segment number N is well approximated by a simple formula for off-lattice models, in particular, for those with excluded volume [**22, 23, 31**]

$$P_K(N) = C_K \widetilde{x}^{m(K)} e^{-\widetilde{x}} \tag{1.1}$$

where variable $\widetilde{x}$ is given by

$$\widetilde{x} = \frac{N - \Delta N(K)}{N_K} \, . \tag{1.2}$$

We call the parameter C_K the coefficient of the knotting probability of the knot K or the knot coefficient. We also call it the amplitude of the knotting probability. We call the parameter $m(K)$ the exponent of the knot K or the knot exponent. It has been shown in simulation that formula (1.1) gives good fitted curves to all the data of the knotting probabilities for 145 different knots with ten values of cylindrical radius $r_{\rm ex}$, i.e. 1450 good curves, in the cylindrical SAPs [**22, 23**], which consist of hard cylindrical segments with radius $r_{\rm ex}$. It has been observed that the estimates of N_K are independent of knot types for the 145 knots [**22**]. We therefore denote it by N_0 and call it the characteristic length of the knotting probability. We have

$$N_K \approx N_0 \quad \text{for all } K \, . \tag{1.3}$$

The parameter $\Delta N(K)$ in eq. (1.2) describes finite-size effects related to the minimal number of segments necessary to form a conformation with the knot type K. Eq. (1.1) is associated with an asymptotic expansion of the partition function of knotted polygons with respect to the segment number N, as will be reviewed in §3.

Let us consider composite knots [**32**]. If a given knot K consists of n prime knots, we denote the number n by $|K|$. It was argued in a previous study [**23**] that the sum of knot coefficients C_K over knots that have $|K| = n$ satisfy simple relations for the cylindrical SAPs

$$\sum_{K:|K|=n} C_K = 1/n!. \tag{1.4}$$

We call relations (1.4) *the sum rules* or *topological sum rules* for knot coefficients. They explain several aspects of the knotting probability. In the present paper we revisit the sum rules numerically and argue relevant aspects. We present the values of the sum of coefficients over several prime knots (i.e. $n = 1$) and those of the composite knots with $n = 2$ and $n = 3$. Previously we have shown only the case for $n = 1$ [**23**]. We argue that as a consequence of the sum rule SAPs with the trefoil knot and its descendants are dominant in thick SAPs [**23**].

Interestingly, the best estimates of the parameters C_K, $m(K)$ and N_K obtained by applying formula (1.1) to the knotting probabilities of the cylindrical SAP are rather close to those of the asymptotic expansion of the logarithm of the knotting probabilities with respect to segment number N up to $O(1/N)$, which are obtained by applying the following

$$P_K(N) = \widehat{C}_K \left(N/\widehat{N}_K\right)^{\widehat{m}(K)} \exp\left(-N/\widehat{N}_K\right) \exp\left(b_K/N\right) . \tag{1.5}$$

We suggest that the parameter $\Delta N(K)$ and the coefficient b_K in eq. (1.5) play a similar role when we search good fitted curves by applying formulas (1.1) and (1.5) to the data of the knotting probability plotted againt the number of segments N. Here we say that a fitted curve is good if the χ^2 value per degree of freedom (χ^2/DF) is smaller than two. Without $\Delta N(K)$ the logarithm of eq. (1.1) corresponds to the asymptotic expansion up to $O(1)$ terms (see eq. (3.3)).

Eq. (1.1) is useful since it leads to good fitted curves to all data of the knotting probabilities, while eq. (1.5) leads to good ones only for limited cases, as shown in TABLES A.1 to A.11 of the best estimates of $\widehat{C}_K$, $\widehat{m}(K)$, $\widehat{N}_K$ and b_K in Appendix A. In TABLES A.1 to A. 11 the best estimates of the parameters in eq. (1.5) are listed only for successful fitted curves among the ten values of radius $r_{\rm ex}$. Here we say that fitted curves are successful if the estimates of parameters are finite and not divergent although the χ^2 values may be larger than two.

Furthermore, we suggest that the knot exponent $m(K)$ of a prime knot K is smaller than 1 if the excluded volume is very small, while it increases to 1 as the thickness of the polymer chain increases. We show it numerically by plotting the previous simulation results [**23**] for the cylindrical SAP.

Let us explain some physical or biological background of the cylindrical SAP model [**18**]. It produces random conformations of a closed sequence of cylindrical segments with radius $r_{\rm ex}$ of unit length, where the excluded-volume interaction is introduced by preventing overlaps among non-neighboring pairs of segments while neighboring pairs of segments are allowed to overlap each other [**13, 14, 22**]. The radius of the cylindrical segments corresponds to the effective thickness of DNA chains, which is associated with the screening length of the negatively charged stiff polymer chains surrounded by counter ions in solution [**33, 34**]. It is thus controlled in experiments by changing the concentration of counter ions in solution, as demonstrated in the experiments of circular DNA [**13, 14**].

The cylindrical SAP is only a theoretical model and does not explicitly describe any details of double-stranded DNA molecules in reality. However, it is one of the simplest coarse-grained models in statistical mechanics which describe the knotting probability of circular DNA with counter ions in solution well, and hence should be useful not only for theoretical studies but also for predicting experimental results.

We now explain that the cylindrical SAP corresponds to a model of semiflexible ring polymers through a mapping [**24**]. Consider a sequence of hard spherical beads with bending rigidity where neighboring beads are in contact, as a fundamental model of semiflexible polymers with excluded volume [**26**]. If the bending rigidity is very large, most of the conformations in the model of hard spherical beads can be approximately described as sequences of thin hard cylinders, as suggested in [**24**]. The mapping from the semiflexible ring polymers to the cylindrical SAP is useful for predicting properties of the knotting probability of the hard spherical beads model [**24**]. Here we remark that the hard spherical beads model reduces to the Kratky-Porod chain if the spherical beads have no excluded volume.

The contents of the paper are as follows. In section 2 we explain the numerical methods for generating the cylindrical SAP and detecting the knot type of a given conformation of SAP. In section 3 we review the argument for deriving the sum rules from formula (1.1) [**23**]. We then show that the sum rules for $n \geq 2$ are derived by assuming that of $n = 1$ and the factorization of knot coefficients C_K. Here, we mean by the factorization of C_K a numerical conjecture that states for a given knot K if it consists of n prime knots: $K = K_1 \# \cdots \# K_n$, the knot coefficient C_K is well approximated by the product of knot coeffcients C_{K_j} of the prime knots K_j for $j = 1, 2, \ldots, n$. We shall specify more details in section 3. In section 4 we present the data supporting the sum rules. In section 5 we illustrate that the ratios of coefficients for pairs of prime knots depend exponentially on the cylindrical radius $r_{\rm ex}$. We argue that SAPs with the trefoil knot and its descendants are dominant and also that the probability for SAP having a knot that contains a given prime knot approaches 1 as segment number N goes to infinity, both as a consequence of the sum rules. In section 6 we show that the best estimates of the fitting parameters of eq. (1.1) are very close to those of the asymptotic expansion (1.5). In section 7 we show numerically that the best estimates of the knot exponent $m(K)$ evaluated for the cylindrical SAP model [**23**] increase slightly with respect to the cylindrical radius and approach the number of constituent prime knots, which are integer values. In section 8, through the study of the cylindrical SAPs connecting ideal ring chains with real ring chains, we suggest that the behavior of the knotting probability for thick SAPs is much simpler than that of equilateral RPs.

2. Numerical methods

Let us explain the method for evaluating the knotting probability of a given knot K for the cylindrical SAP: We generate an ensemble of cylindrical SAP by the Monte-Carlo (MC) method, detect the knot type of the SAP by calculating some knot invariants, and then evaluate the knotting probability of the knot K for the cylindrical SAP.

2.1. Algorithm for generating cylindrical SAP. We construct an ensemble of SAP consisting of N cylindrical segments with radius $r_{\rm ex}$ as follows [**22**]. First, we construct an initial polygon by an equilateral regular N-gon, where the vertices have numbers from 1 to N, consecutively. We then apply two MC procedures: folding and inverting subchains. In the folding process, after we choose randomly two vertices out of the N vertices such as p_1 and p_2, we rotate a sub-chain between the vertices p_1 and p_2 around the straight line connecting them by an angle chosen randomly from 0 to 2π. We check whether the rotated subchain has any overlap with the other part of the polygon or not. If every pair of non-neighboring segments (or polygonal edges) in the polygon is separated by a distance larger than $2r_{\rm ex}$, we consider that it has no overlap. If it has no overlap, we employ the rotated configuration as the cylindrical SAP in the next MC step. If it has an overlap, we employ the previous configuration of SAP before rotation in the next MC step. In the inverting process, after we choose randomly two vertices such as p_1 and p_2, we choose a subchain between them and invert the subchain so that the order of bond vectors is reversed. Let us assume that $p_1 < p_2$ and denote the difference by $n = p_2 - p_1$. If the position vectors of vertices in the original subchain are given by $\vec{r}_{p_1+j} = \vec{r}_{p_1} + \vec{b}_1 + \vec{b}_2 + \cdots + \vec{b}_j$ for $j = 1, 2, \ldots, n$, then those of the inverted subchain are given by $\vec{r}_{p_1+j} = \vec{r}_{p_1} + \vec{b}_n + \vec{b}_{n-1} + \cdots + \vec{b}_{n+1-j}$ for $j = 1, 2, \ldots, n$. Similarly as in the case of the folding process, we check whether the inverted subchain has any overlap with the other part of the polygon or not. If the distance between any pair of non-neighboring segments (or polygonal edges) in the polygon is larger than $2r_{\rm ex}$, we find that it has no overlap. If it has no overlap, we employ the inverted configuration as the cylindrical SAP in the next MC step. If it has an overlap, we employ the previous configuration of SAP before inverting it in the next MC step.

We combine one folding move and one inverting move together. We repeat the combined MC procedure many times such as $2N$ times. We then accept the $2N$th employed conformation as an independent conformation of the cylindrical SAP of N segments and add it to the ensemble of random conformations of the cylindrical SAP. Here we remark that simulation results of [**22**] and [**23**] were obtained by repeating $2N$ times the combined MC process of one folding move and one inverting move, although it was not clearly explained there.

We now introduce the correlation between two sets of conformations of SAP at different time steps of the MC process, t_1 and t_2, respectively. Let us express the position vectors of the jth vertices in the cylindrical SAP at time step t of the MC process by $\vec{r}_j(t)$ for $j = 1, 2, \ldots, N$. We define the correlation $C(t)$ between a set of conformations of cylindrical SAP at the initial time (i.e. $t = 0$) and those at time step t by

$$C(t) = \frac{\sum_{j=1}^{N} \langle (\vec{r}_j(0) - \vec{r}_{cm}(0)) \cdot (\vec{r}_j(t) - \vec{r}_{cm}(t)) \rangle}{\sqrt{\sum_{j=1}^{N} \langle (\vec{r}_j(0) - \vec{r}_{cm}(0))^2 \rangle} \sqrt{\sum_{j=1}^{N} \langle (\vec{r}_j(t) - \vec{r}_{cm}(t))^2 \rangle}}, \tag{2.1}$$

where the position vector of the center of mass at time step t $\vec{r}_{cm}(t)$ is given by

$$\vec{r}_{cm}(t) = \sum_{j=1}^{N} \vec{r}_j(t)/N \,. \tag{2.2}$$

The estimates of the correlation $C(t)$ are plotted against time step t of the MC process in FIGURE 1 with the semi-logarithmic scale for the cylindrical SAP of $N = 100$ segments. Here the number of random conformations for the cylindrical SAP is given by 1000. We observe that the correlation $C(t)$ decays almost exponentially with respect to time step t and statistically fluctuates after it becomes as small as the order of $O(1/\sqrt{10^3})$. Here we remark that statistical error is approximately estimated as $1/\sqrt{10^3} \approx 0.03$. We also observe that it decreases to the order of $O(1/\sqrt{10^3})$ when $t = 25$ and 30 for the cylindrical SAP with $r_{\rm ex} = 0.02$ and 0.1, respectively. The time steps $t = 25$ and 30 are much smaller than the interval for sampling $2N = 200$ MC steps. We therefore interpret that the sampled conformations of SAP are statistically independent.

In the case of $r_{\rm ex} = 0$, the SAPs generated by the above algorithm are given by equilateral random polygons. The algorithm for generating cylindrical SAPs with $r_{\rm ex} = 0$ is also called the polygonal folding method (PFM) [**35**]. The ergodicity of PFM is shown in [**35, 36**] (see also [**37**]). For recent references on the ergodicity of SAPs, we refer to [**30**].

2.2. Method for evaluating the knotting probability. We detect the knot type of a given SAP by evaluating mainly the values of the two knot invariants: The absolute value of the Alexander polynomial $|\Delta_K(t)|$ evaluated at $t = -1$ and the Vassiliev invariant of the second order $v_2(K)$ for a knot K. If a given SAP has

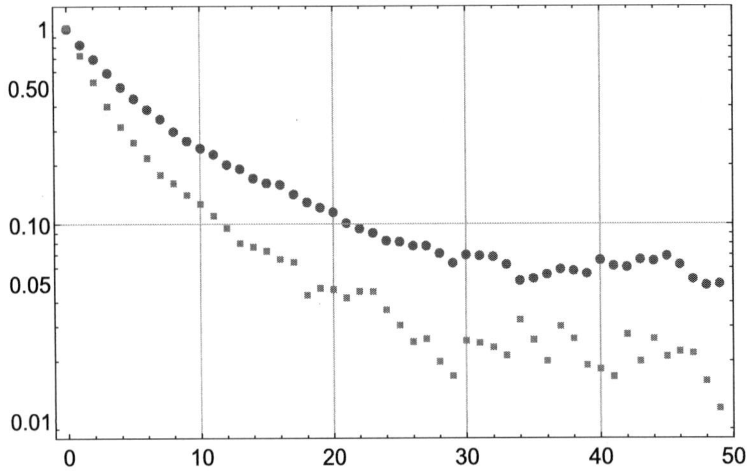

FIGURE 1. Correlation $C(t)$ between the conformations of the cylindrical SAP at time step 0 and those of time step t plotted against time step t for $N = 100$ segments with $r_{\text{ex}} = 0.02$ and $r_{\text{ex}} = 0.1$ denoted by filled squares and filled circles, respectively, in the semilogarithmic scale. Each datapoint is averaged over 1000 samples.

the same values of two knot invariants as a knot K, we assume that the topology of the polygon is given by the knot K. For some cases we also evaluate the Vassiliev invariant of the third order, as we shall see later.

Some pairs of knots have the same values of the two knot invariants in common. For example, both knot 7_4 and knot $3_1\#5_1$ have the same values $|\Delta_K(-1)| = 15$ and $v_2(K) = 4$. Therefore, we cannot distinguish between knot 7_4 and knot $3_1\sharp 5_1$ only by calculating the two knot invariants. In order to distinguish them we evaluate the Vassiliev invariants of the third order for such polygons.

The Vassiliev invariants of any order can be calculated by the method of the quasi-classical expansion of the R-matrix of the quantum group [**27**, **38**]. However, we employ the algorithm due to Polyak and Viro to calculate the Vassiliev invariants of the second order and the third order [**39**]. In fact, by the latter method the Vassiliev invariants are calculated only through the Gauss codes [**32**] (or the Dowker codes).

2.3. Number of SAPs generated in the simulation. In the present simulation we employ the same ensembles of the cylindrical SAP constructed in [**23**]. For each value of radius r_{ex} we generated 2×10^5 polygons for $N \leq 4,000$, 10^5 polygons for N satisfying $4,000 \leq N \leq 6,000$, 5×10^4 polygons for N satisfying $6,000 \leq N \leq 8,000$, and 4×10^4 polygons for N satisfying $8,000 \leq N \leq 10,000$.

In the present simulation the number of segments N is given from 100 to $3,000$ for the cylindrical SAP of zero thickness ($r_{\text{ex}} = 0$), i.e. equilateral random polygons; from 100 to 3000 for the cylindrical SAP with $r_{\text{ex}} = 0.005$ and 0.01; from 100 to 4000 with $r_{\text{ex}} = 0.02$; from 100 to $5,000$ with $r_{\text{ex}} = 0.03$; from 100 to $7,000$ with $r_{\text{ex}} = 0.04$; from 100 to $8,000$ with $r_{\text{ex}} = 0.05$; from 100 to 10^4 with $r_{\text{ex}} = 0.06$, 0.08 and 0.1.

3. Review on the formula of the knotting probability and the sum rules

3.1. Approximate expression of the knotting probability as a function of segment number. Let us briefly review the derivation of fitting formula (1.1) to the knotting probability of a knot for an off-lattice model of RPs or SAPs. We first consider the asymptotic expression of the knotting probability with respect to the number of segments N. For simplicity we consider SAPs on a lattice. We denote the number of all SAPs with N segments by $Z_{All}(N)$ and that of such SAP having a given knot type K by $Z_K(N)$. The knotting probability of the knot K for such a SAP with N segments is given by

$$P_K(N) = Z_K(N)/Z_{All}(N). \tag{3.1}$$

For the number of SAP of N segments we expect the asymptotic expansion for large N as [**40**]

$$\log Z_{All}(N) = \kappa_{All} N + m_{All} \log N + \log Z_{All}^{(0)} + O(1/N). \tag{3.2}$$

Here we have denoted the constant term in the large-N expansion of $\log Z_{All}(N)$ by $Z_{All}^{(0)}$. We thus expect that the number of SAP of N segments with the fixed knot type K is asymptotically given by

$$\log Z_K(N) = \kappa_K N + m_K \log N + \log Z_K^{(0)} + O(1/N). \tag{3.3}$$

Here we have again denoted the constant term in the large-N expansion of $\log Z_K(N)$ by $Z_K^{(0)}$. By taking the exponential of eqs. (3.2) and (3.3) and by introducing N_K and $m(K)$ with expressions: $1/N_K = \kappa_{All} - \kappa_K$ and $m(K) = m_K - m_{All}$, respectively, we have

$$P_K(N) = C_K \left(N/N_K\right)^{m(K)} \exp\left(-N/N_K\right). \tag{3.4}$$

Furthermore, in order to take into account finite-size corrections we replace N/N_K by $(N - \Delta N(K))/N_K$ we have Eq. (1.1) [**31**].

3.2. Empirical factorization properties of the knotting probability. It was shown previously [**22, 23, 27, 31**] that the estimates of exponents $m(K)$ and coefficients C_K for several off-lattice models satisfy the factorization properties:
(i) For knot exponents $m(K)$, to a given composite knot K with $|K| = 2$, i.e. $K = K_1 \# K_2$, the following approximation has been suggested [**27**]

$$m(K_1 \# K_2) = m(K_1) + m(K_2). \tag{3.5}$$

In general, for a composite knot K consisting of n prime knots it has been suggested that the exponent $m(K)$ is approximately given by the sum over those of the constituent prime knots K_j

$$m(K_1 \# K_2 \# \cdots \# K_n) = \sum_{j=1}^{n} m(K_j). \tag{3.6}$$

We shall suggest in §7 that the factorization of knot exponents should hold only approximately, for instance, in the cylindrical SAP model particularly when the excluded volume is small.
(ii) For the estimates of coefficients C_K, in particular, evaluated with eq. (1.1) for the cylindrical SAPs, we have a numerical conjecture [**23**] that if K_1 and K_2 are different prime knots, we have

$$C_{K_1 \# K_2} = C_{K_1} C_{K_2}, \tag{3.7}$$

while if they are the same we have

$$C_{K_1\#K_1} = (C_{K_1})^2/2! \,. \tag{3.8}$$

For a composite knot K consisting of n prime knots where there are n_j copies of prime knot K_j and the sum of n_j is equal to n: $\sum_j n_j = n$, we have a numerical conjecture [**23**] that the knot coefficient C_K is given by the product over those of the constituent prime knots K_j divided by n_j!

$$C_K = \prod_j (C_{K_j})^{n_j}/n_j! \,. \tag{3.9}$$

By precisely investigating the estimates of $m(K)$ obtained for the cylindrical SAP [**23**], we show numerically that the factorization properties of knot exponents $m(K)$ in eqs. (3.5) and (3.6) hold only approximately. We shall show it in §7.

It is interesting to compare the properties of off-lattice models with those of lattice models. For lattice knots, the factorization property of coefficients C_K for large N is studied numerically by making use of the asymptotic expansion of the knotting probability [**41**]. It thus seems that the behavior of lattice SAPs is much simpler than that of off-lattice SAPs for the factorization of the knotting probability. We suggest that the knot exponent of a prime knot is almost equal to 1 for lattice SAPs since we expect that knotted regions are localized when the excluded volume is large and lattice SAPs indeed have large excluded volumes, and also that the factorization of knot exponents holds better for the lattice models, for which we expect that knot localization should appear more clearly than for the off-lattice models of thin SAPs.

3.3. Topological sum rules derived from the factorization of knotting probabilities and the exponent being equal to 1 for every prime knot. We now review how we obtain the sum rules (1.4) when the approximate expression (1.1) is numerically very close to the asymptotic expansion, the knotting probability of a composite knot K with $|K| \geq 2$ is factorized into those of constituent prime knots such as (3.5), (3.7) and (3.8), and the exponent $m(K)$ in the approximate expression (1.1) of the knotting probability is equal to 1 for any prime knot K [**23**].

By neglecting the finite-size corrections $\Delta N(K)$ in the approximate formula (1.1) together with the empirical property (1.3) we can approximate $\widetilde{x}$ by x defined by $x \equiv N/N_0$ as

$$\widetilde{x} = (N - \Delta N(K))/N_K \to x \equiv N/N_0 \,. \tag{3.10}$$

We therefore have the three-parameter formula (3.4) with $N_K = N_0$ for all K.

It follows from the assumption that $m(K) = 1$ for any prime knot K and the factorization property (3.5) that we have

$$m(K_1\#K_2\#\cdots\#K_n) = n \,. \tag{3.11}$$

For a knot K consisting of n prime knots (i.e. $|K| = n$) we have with $x = N/N_0$

$$P_K(N) = C_K x^n \exp(-x) \,. \tag{3.12}$$

Since the sum of the knotting probabilities over all knots is equal to 1, we have

$$\begin{aligned} 1 &= \sum_{n=0}^{\infty} \sum_{K:|K|=n} P_K(N) \\ &= \sum_{n=0}^{\infty} \sum_{K:|K|=n} C_K x^n \exp(-x) \, . \end{aligned} \tag{3.13}$$

The symbol $\sum_{K:|K|=n}$ denotes the sum over all such composite knots that consist of n prime knots. We therefore have

$$e^x = \sum_{n=0}^{\infty} x^n \sum_{K:|K|=n} C_K \, . \tag{3.14}$$

We thus obtain an infinite number of relations (1.4) for knot coefficients C_K with $|K| = n$: The sum of knot coefficients over such knots that consist of n prime knots is given by $1/n!$. We have called them the *sum rules* or *topological sum rules.*

We remark that in the above derivation it plays a significant role that the exponent $m(K)$ for any prime knot is equal to 1.

3.4. Sum rules derived from that of prime knots when the factorization of C_K holds. We now show that if the factorization properties of coefficients C_K hold, the sum rules for composite knots K with $|K| \geq 2$ are derived from the sum rule for prime knots. We should emphasize that in the derivation it is not necessary to assume that the exponent $m(K)$ is equal to 1 for every prime knot K.

For an illustration, let us consider the case of $n = 2$. By expressing a given composite knot K with $|K| = 2$ as $K = K_1 \# K_2$ we have

$$\begin{aligned} \sum_{|K|=2} C_K &= \sum_{(K_1,K_2):K_1 \neq K_2;prime} C_{K_1 \# K_2} + \sum_{K_1=K_2:prime} C_{K_1 \# K_2} \\ &= \frac{1}{2} \sum_{K_1:prime} \sum_{K_2 \neq K_1:prime} C_{K_1 \# K_2} + \sum_{K_1=K_2:prime} C_{K_1 \# K_1} \\ &= \frac{1}{2} \sum_{K_1:prime} \sum_{K_2 \neq K_1:prime} C_{K_1} C_{K_2} + \frac{1}{2} \sum_{K_1=K_2:prime} (C_{K_1})^2 \, . \end{aligned} \tag{3.15}$$

In the first line of eq. (3.15) the symbol (K_1, K_2) denotes a pair of prime knots K_1 and K_2 and the sum is over all pairs of two different prime knots. In the last line we have made use of the factorization properties of knot coefficients (3.7) and (3.8). We therefore obtain the sum rule for composite knots with $|K| = 2$

$$\sum_{K:|K|=2} C_K = \left(\sum_{K_1:prime} C_{K_1} \right)^2 /2! . \tag{3.16}$$

We can show the sum rules (1.4) for such knots with $|K| = n$ for any integer n, similarly. The result has not been discussed in [**23**] and should be new.

4. Numerical support for the sum rules

4.1. Data of the sum rules for knots K with $|K| = 1$, 2 and 3. We now present the numerical estimates of the sum of knot coefficients over prime knots and those of composite knots consisting of two prime knots and three prime knots, respectively. Here we recall that the knot coefficients C_K are evaluated in the Monte-Carlo simulation performed for the cylindrical SAP model with ten different values of cylindrical radius [**23**].

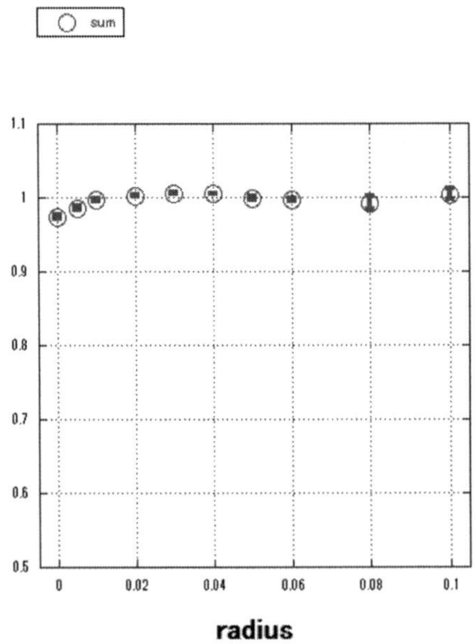

FIGURE 2. Sum rule for prime knots: the sum of coefficients over prime knots plotted against the radius of cylindrical segments r_{ex}.

r_{ex}	$\sum_K C_K$	# prime knots
0.0	0.973753 ± 0.005195	14
0.005	0.985518 ± 0.005164	14
0.01	0.996749 ± 0.003934	14
0.02	1.002151 ± 0.002922	14
0.03	1.005664 ± 0.002850	14
0.04	1.005100 ± 0.002769	14
0.05	0.999070 ± 0.002245	7
0.06	0.996975 ± 0.002398	7
0.08	0.992127 ± 0.004922	4
0.1	1.003953 ± 0.016912	4

TABLE 1. Sum rule for prime knots in the cylindrical SAP with radius r_{ex}, i.e. eq. (1.4) for $n = 1$: The sum $\sum_{K:\text{prime}} C_K$ over several prime knots K. If the number of prime knots is 14, the sum $\sum_K C_K$ is taken over the 14 prime knots up to seven minimal crossings such as 3_1, 4_1, 5_1, 5_2, 6_1, 6_2, 6_3, 7_1, 7_2, 7_3, 7_4, 7_5, 7_6, and 7_7; if it is 7, we take the sum over prime knots up to six minimal crossings; if it is 4, up to five minimal crossings.

In FIGURE 2 the estimates for the sum of coefficients C_K over several prime knots up to seven crossings are plotted against cylindrical radius $r_{\rm ex}$. In TABLE 1 they are listed for the ten values of cylindrical radius $r_{\rm ex}$. The number of prime knots which contribute to the sum is up to 4, 7, and 14, depending on the value of radius $r_{\rm ex}$. We remark that the coefficients of such knots that have large minimal crossing numbers are numerically much smaller particularly for large values of cylindrical radius $r_{\rm ex}$. We thus take the sum of knot coefficients up to the knots with only five minimal crossings for $r_{\rm ex} = 0.08$ and 0.10, as shown in TABLE 1.

We observe in FIGURE 2 that the estimates of the sum $\sum_K C_K$ over prime knots K are all almost equal to 1 with respect to errors, at least except for the cases of small radius such as $r = 0$ and $r = 0.005$. In particular, the sum rule of prime knots holds well for larger values of cylindrical radius such as $r_{\rm ex} = 0.08$ and 0.10. For $r = 0$ and $r = 0.005$ the sum does not reach 1.0 within the error bars. However, we expect that if we take into account all the small contributions from other knot types with larger minimal numbers of crossings, the sum will become much closer to the value of 1. We remark that the errors are estimated by taking the sum of the errors in the coefficients C_K for prime knots taken into account.

$r_{\rm ex}$	$\sum_K C_K$	# knots $\lvert K\rvert = 2$
0	0.445290 ± 0.011392	15
0.005	0.482257 ± 0.011936	15
0.01	0.498757 ± 0.009530	13
0.02	0.520886 ± 0.006346	13
0.03	0.509611 ± 0.007167	7
0.04	0.514030 ± 0.006679	7
0.05	0.512432 ± 0.007047	4
0.06	0.511393 ± 0.009850	4

TABLE 2. Sum rule for composite knots with $|K| = 2$ in the cylindrical SAP with radius $r_{\rm ex}$. The sum $\sum_{K:|K|=2} C_K$ over 15 composite knots K with $|K| = 2$ such as $3_1\#3_1$, $3_1\#4_1$, $3_1\#5_1$, $3_1\#5_2$, $3_1\#6_1$, $3_1\#6_2$, $3_1\#6_3$, $3_1\#7_1$, $3_1\#7_2$, $3_1\#7_3$, $3_1\#7_5$, $3_1\#7_6$, $4_1\#4_1$, $4_1\#5_1$, and $4_1\#5_2$. In the case of 13 knots we have all knots of 15 knots except for $4_1\#5_1$ and $4_1\#5_2$.

It is quite interesting that the sum rule for prime knots holds for different values of cylindrical radius such as from $r_{\rm ex} = 0.0$ to $r_{\rm ex} = 0.10$. Here we recall the empirical fact that exponent $m(K)$ for any prime knot is equal to 1 plays a significant role in the derivation of the sum rules, as shown in subsection 2.3. However, the best estimates of exponent $m(K)$ for any given prime knot is smaller than 1.0, in particular, for small values of cylindrical radius $r_{\rm ex}$, as shown in section 6.

4.2. Sum rules for composite knots K with $|K| \geq 2$. In TABLE 2 we observe that the estimates of the sum of the knot coefficients over several composite knots consisting of two prime knots, i.e., for $n = 2$, are equal to 1/2 with respect to errors except for the two cases where the values of radius are very small such as $r_{\rm ex} = 0.0$ and 0.005. We suggest that for $r_{\rm ex} = 0.0$ and 0.005 we can recover the value of 1/2 if we take into account the knot coefficients with larger minimal crossing numbers. Here we remark that in TABLE 2 the estimates of the sum of the knot coefficient over several composite knots consisting of two prime knots up to seven minimal crossings are listed for eight different values of cylindrical radius r. [1]

$r_{\rm ex}$	$\sum_K C_K$	# of knots with $\lvert K\rvert = 3$
0	0.103923 ± 0.008382	8
0.005	0.128356 ± 0.008898	8
0.01	0.14482 ± 0.008897	8
0.02	0.165077 ± 0.009478	8
0.03	0.149652 ± 0.009302	4
0.04	0.161485 ± 0.009726	4

TABLE 3. Sum rule for composite knots with $|K| = 3$ in the cylindrical SAP with radius $r_{\rm ex}$. The sum $\sum_{K:|K|=3} C_K$ over 8 composite knots K with $|K| = 3$ such as $3_1\#3_1\#3_1$, $3_1\#3_1\#4_1$, $3_1\#3_1\#5_1$, $3_1\#3_1\#5_2$, $3_1\#3_1\#6_1$, $3_1\#3_1\#6_2$, $3_1\#3_1\#6_3$, and $3_1\#4_1\#4_1$.

We observe in TABLE 3 that the estimates of the sum of knot coefficients over several composite knots consisting of three prime knots, i.e. $n = 3$, are equal to $1/3! = 0.1666$ with respect to errors except for the three cases of small radius such as $r_{\rm ex} = 0.0$, 0.005 and 0.01. We suggest that if we evaluate the knot coefficients of larger minimal crossing numbers we can recover the value of 1/6 also for the three cases. Here we recall that in TABLE 3 the estimates of the sum over the knot coefficients of composite knots consisting of three prime knots up to seven minimal crossings are listed for six different values of cylindrical radius r. [2]

It should be nontrivial that the sum rules for $n = 2$ and 3 hold for almost all values of cylindrical radius as shown numerically in TABLES 2 and 3. We recall that we have shown in subsection 3.4 that all the sum rules for $n \geq 2$ are derived if the sum rule for prime knots and the factorization of knot coefficients C_K hold. However, it is also nontrivial whether the factorization of knot coefficients C_K holds for the cylindrical SAP model. Here we recall that the best estimates of coefficient C_K are evaluated by four-parameter fitting formula (1.1) and not by three-parameter formula (3.4) [**23**].

5. Trefoil knot dominance in thick ring polymer chains

5.1. Ratio of knot coefficients as a function of cylindrical radius. The ratios of knot coefficients C_{K_1}/C_{K_2} for pairs of prime knots K_1 and K_2 are plotted in FIGURE 3 against the radius of cylindrical segments $r_{\rm ex}$ in the semi-logarithmic scale for the cylindrical SAP model. For any pair of prime knots the plot can be approximated by almost a straight line. It shows that the ratios of knot coefficients C_K depend exponentially on the radius of cylindrical segments $r_{\rm ex}$. In fact, the knot

[1] In the case of 7 crossings, we have $3_1\#3_1$, $3_1\#4_1$, $3_1\#5_1$, $3_1\#5_2$, $3_1\#6_1$, $3_1\#6_2$, and $3_1\#6_3$. For 4 knots we have $3_1\#3_1$, $3_1\#4_1$, $3_1\#5_1$, and $3_1\#5_2$.

[2] In the case of 4 composite knots we have $3_1\#3_1\#3_1$, $3_1\#3_1\#4_1$, $3_1\#3_1\#5_1$, and $3_1\#3_1\#5_2$.

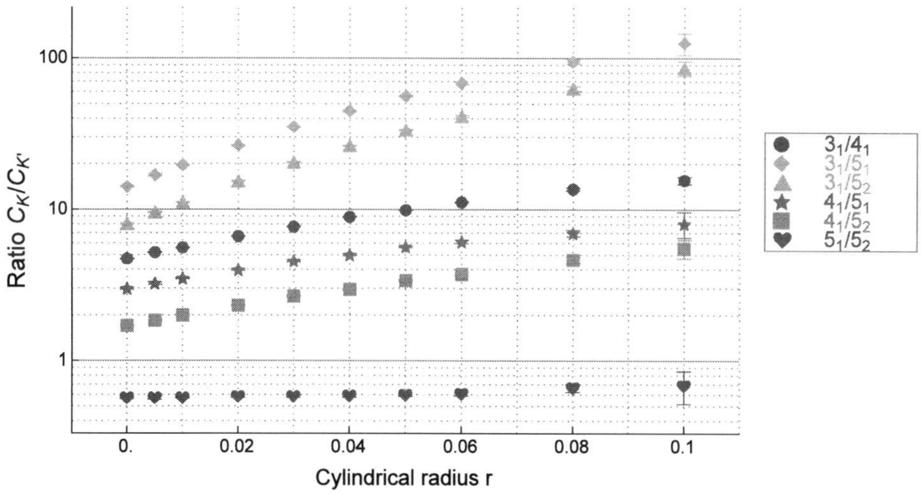

FIGURE 3. Ratios of coefficients C_{K_1}/C_{K_2} versus cylindrical radius $r_{\rm ex}$ for pairs of prime knots K_1 and K_2 in the cylindrical SAP. The estimates of knot coefficients C_K are evaluated in [**23**].

coefficients except for that of the trefoil knot are well approximated by exponentially decaying functions of cylindrical radius $r_{\rm ex}$ [**23**].

In FIGURE 3 for any pair of prime knots with the same minimal number of crossings the ratio C_{K_1}/C_{K_2} does not change with respect to cylindrical radius, as shown for 5_1 and 5_2 knots. For pairs of prime knots whose minimal numbers of crossings are different by one such as 3_1 and 4_1, 4_1 and 5_1, and 4_1 and 5_2, the plots increase with respect to the radius. For pairs of prime knots whose minimal numbers of crossings are different by two such as 3_1 and 5_1, 3_1 and 5_2, the plots increase with respect to the radius with larger gradients than those of the differences being equal to one.

The ratios of knot coefficients C_{K_1}/C_{K_2} for some pairs of prime knots K_1 and K_2 are given in [**42**] by

$$\begin{array}{lll} C_{3_1}/C_{4_1} = 28(\pm 1) & C_{3_1}/C_{5_1} = 400(\pm 20) & C_{3_1}/C_{5_2} = 280(\pm 20) \\ & C_{4_1}/C_{5_1} = 15(\pm 1) & C_{4_1}/C_{5_2} = 9(\pm 1) \\ & & C_{5_1}/C_{5_2} = 0.67(\pm 3)\,. \end{array} \tag{5.1}$$

We observe that the estimated ratios are consistent with those of FIGURE 3 around at $r_{\rm ex} = 0.1$ or values slightly larger than 0.1.

It is conjectured in [**23**] that around at $r_{\rm ex} = 1/8$ the ratio C_{K_1}/C_{K_2} approaches its universal value evaluated for lattice SAPs in [**42**].

5.2. Thickness dependence of the knot coefficients C_K. It has been shown through numerical simulation [**23**] that for any prime knot other than the trefoil knot 3_1 the thickness dependence of the coefficient of the knotting probability C_K is given by an exponentially decaying function of cylindrical radius $r_{\rm ex}$

$$C_K(r_{\rm ex}) = a_1(K)\exp(-b_1(K)r_{\rm ex})\,. \tag{5.2}$$

For the trefoil knot 3_1, however, we express the coefficient of knotting probability, C_{3_1}, as a function of cylindrical radius $r_{\rm ex}$ by the following function:

$$C_{3_1}(r_{\rm ex}) = a_0(3_1)(1 - a_1(3_1)\exp(-b_1(3_1)r_{\rm ex}))\,. \tag{5.3}$$

Here we recall that the parameters $a_0(3_1)$, $a_1(3_1)$ and $b_1(3_1)$ are evaluated as $a_0(3_1) = 0.919 \pm 0.003$, $a_1(3_1) = 0.327 \pm 0.002$ and $b_1(3_1) = 33.1 \pm 0.8$ with χ^2 per degree of freedom (DF) is given by 1.8 [**23**].

Knot type K	$a_1(K)$	$b_1(K)$	χ^2/DF
4_1	0.1357 ± 0.0013	8.82 ± 0.27	8.70
5_1	0.04387 ± 0.00042	20.81 ± 0.34	3.56
5_2	0.07741 ± 0.00046	21.91 ± 0.22	3.02
6_1	0.02234 ± 0.00029	34.30 ± 0.56	3.96
6_2	0.02389 ± 0.00050	35.97 ± 0.82	7.82
6_3	0.01580 ± 0.00037	40.8 ± 1.1	5.56
7_1	0.003022 ± 0.000038	47.02 ± 0.79	0.42
7_2	0.00665 ± 0.00010	47.52 ± 0.82	0.76
7_3	0.00538 ± 0.00011	48.0 ± 1.2	1.17
7_4	0.002866 ± 0.000071	50.2 ± 1.5	0.90
7_5	0.00778 ± 0.00014	51.22 ± 0.99	1.83
7_6	0.009159 ± 0.000071	52.02 ± 0.44	0.41
7_7	0.00624 ± 0.00021	55.6 ± 1.7	1.80

TABLE 4. Best estimates of parameters in eq. (5.2) which expresses the coefficients C_K of prime knots other than the trefoil knot as a function of cylindrical radius $r_{\rm ex}$. Reproduced from E. Uehara and T. Deguchi, J. Chem. Phys. **147**, 094901 (2017).

5.3. Trefoil knot coefficient increases with respect to the thickness. We now argue that the fraction of the trefoil knot and its descendants increases as the thickness of the polymer chain increases. Here we assume the sum rule for prime knots.

Suppose that the sum rule for prime knots holds for the cylindrical SAP model with any value of radius $r_{\rm ex}$. Then the knot coefficient of the trefoil knot is given by

$$C_{3_1} = 1 - C_{4_1} - C_{5_1} - C_{5_2} - \cdots\,. \tag{5.4}$$

Here we recall that the knotting probabilities other than the trefoil knot are well approximated by exponentially decreasing functions of cylindrical radius $r_{\rm ex}$. Since the largest term other than that of the trefoil knot is given by that of the figure-eight knot 4_1, we may approximate the knot coefficient of the trefoil knot as follows.

$$C_{3_1} = 1 - a_1(4_1)\exp(-b_1(4_1)r_{\rm ex}) - \cdots\,. \tag{5.5}$$

It thus suggests a rough but useful approximation to the knot coefficient C_{3_1}.

The approximation explains the possible reasons why the coefficient of the trefoil knot is close to 1 and increases with respect to the cylindrical radius. However, we should remark that the knot coefficient of the trefoil knot approaches a constant $a_0(3_1) = 0.919$ as cylindrical radius $r_{\rm ex}$ goes to infinity. The estimate of

$a_0(3_1) = 0.919$ is close to but slightly smaller than 1. Therefore, the approximation (5.5) is only rough and not completely valid.

5.4. Dominance of the trefoil knot and its descendants. We now evaluate the sum of the knotting probabilities over the trefoil knot and all such composite knots consisting of only trefoil knots. For simplicity, we assume that the knot exponent $m(3_1)$ of the trefoil knot is equal to 1. In terms of $x = N/N_0$ we have

$$\begin{aligned} P_{3_1\text{family}}(x) &= P_{3_1}(x) + P_{3_1\#3_1}(x) + P_{3_1\#3_1\#3_1}(x) + \cdots \\ &= C_{3_1} x e^{-x} + \frac{1}{2!}(C_{3_1})^2 x^2 e^{-x} + \frac{1}{3!}(C_{3_1})^3 x^3 e^{-x} + \cdots \\ &= e^{-x}\left(C_{3_1} x + \frac{1}{2!}(C_{3_1} x)^2 + \frac{1}{3!}(C_{3_1} x)^3 + \cdots \right) \\ &= e^{-x}\left(\exp\left(C_{3_1} x\right) - 1\right) . \end{aligned} \tag{5.6}$$

The graph of $P_{3_1\text{family}}$ versus x is plotted against the reduced number of segments $x = N/N_0$ in FIGURE 4. The peak height $P_{\max}$ of the graph of $P_{3_1\text{family}}$ is given by

$$P_{\max} = (1 - C_{3_1}(r))^{(1-C_{3_1}(r))/C_{3_1}(r)}\, C_{3_1}(r) \tag{5.7}$$

at the peak position $x_{\max}$

$$x_{\max} = -\log\left(1 - C_{3_1}(r)\right)/C_{3_1}(r) . \tag{5.8}$$

Both of them are expressed in terms of the knot coefficient $C_{3_1}(r)$.

If the coefficient of the trefoil knot approaches 1 as the cylindrical radius increases, then the peak value P_{max} approaches 1 while the peak position $x_{\max}$ increases infinitely. We can show it by taking the logarithm of the right-hand side of eq. (5.7) and sending coefficient C_{3_1} to 1.

For the thick cylindrical SAPs with large cylindrical radius such as $r_{\text{ex}} = 0.125$ the graph of the fraction of SAPs consisting of only trefoil knots versus variable $x = N/N_0$ has its maximum $P_{\max} = 0.738$ at $x_{\max} = 2.745$, as shown in FIGURE 4. Thus, in the case of thick semi-flexible polymers such as radius $r_{\text{ex}} = 0.125$ almost

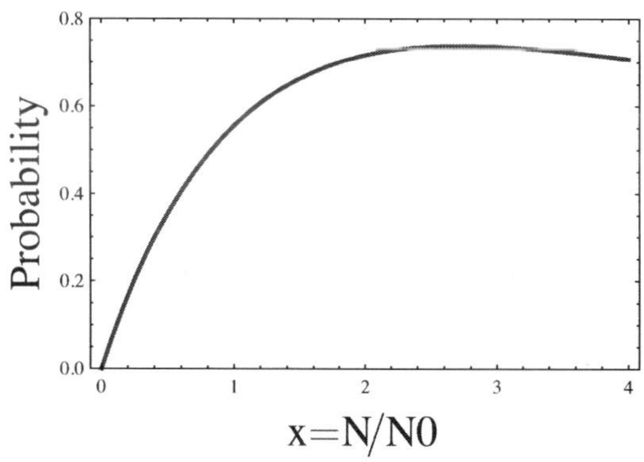

FIGURE 4. $P_{3_1\text{family}}$ versus $x = N/N_0$: the sum of the knotting probabilities over all composite knots consisting of only trefoil knots plotted against x

75 percentages of SAP are given by the trefoil knot and its descendants consisting only of trefoil knots. if the number of segments N are given by $N = 2.75N_0$.

5.5. Probability of a composite knot containing a given prime knot. We now argue that if the number of segments is large, the fraction of SAP consisting of at least one trefoil knot and other prime knots are dominant. We evaluate the fraction of such SAP that are descendants of the trefoil knot with possible other prime knots are given by

$$\begin{aligned} & P_{\text{including } 3_1} \\ & = P_{3_1}(x) + P_{3_1\#3_1}(x) + P_{3_1\#4_1}(x) + P_{3_1\#5_1}(x) + P_{3_1\#3_1\#3_1}(x) + \cdots \\ & = e^{-x}\left(C_{3_1}x + \frac{1}{2!}(C_{3_1}x)^2 + (C_{3_1}x)(C_{4_1}x) + \frac{1}{3!}(C_{3_1}x)^3 + \cdots \right) \\ & = e^{-x}\left\{ e^{x(C_{3_1}+C_{4_1}+\cdots)} - e^{x(C_{4_1}+\cdots)} \right\} \end{aligned} \tag{5.9}$$

For simplicity, we further assume that the sum rule for prime knots holds. We then have

$$\begin{aligned} P_{\text{including } 3_1} & = e^{-x}\left\{ e^{x(C_{3_1}+C_{4_1}+C_{5_1}+\cdots)} - e^{x(C_{4_1}+C_{5_1}+\cdots)} \right\} \\ & = e^{x(-1+C_{3_1}+C_{4_1}+C_{5_1}+\cdots)} - e^{x(-1+C_{4_1}+C_{5_1}+\cdots)} \\ & = 1 - e^{-C_{3_1}x}. \end{aligned} \tag{5.10}$$

We therefore conclude that the fraction of SAP with composite knots that consist of at least a trefoil knot becomes dominant and approaches 1.0 as the number of segments increases.

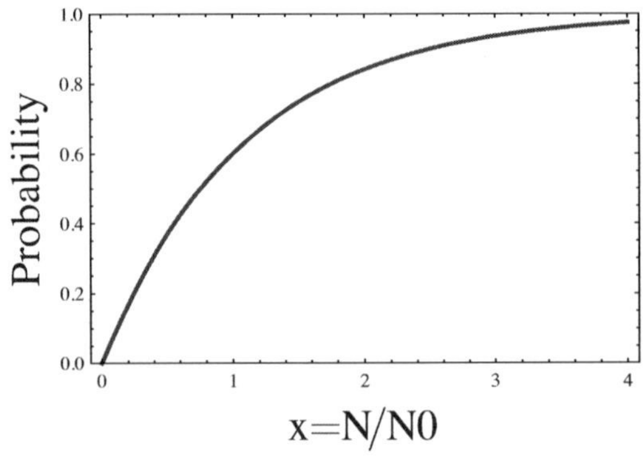

FIGURE 5. $P_{\text{including } 3_1}$ versus $x = N/N_0$: the sum of the knotting probabilities over all composite knots containing at least one trefoil knot

Similarly, for any prime knot, for instance, say, 4_1, we can show that the fraction of SAP with composite knots that consist of at least a figure-eight knot is given by

$$P_{\text{including } 4_1} = 1 - e^{-C_{4_1}x}. \tag{5.11}$$

As shown in FIGURE 3 the ratio C_{3_1}/C_{4_1} is much larger than 10 for $r_{\text{ex}} = 0.1$, and hence the fraction of SAP with composite knots that consist of at least a figure-eight knot becomes as large as 0.5 only if the value of x is very large such as $x > 10$.

REMARK 5.1. If the sum rule of prime knots holds, the exponent is equal to 1 for every prime knot and the factorization of knot coefficients holds, the probability for a cylindrical SAP with N segments to have such a composite knot K that contains a given prime knot K_1 as a constituent knot is approximately given by

$$P_{\text{including } K_1} = 1 - \exp\left(-C_{K_1} N/N_0\right) . \tag{5.12}$$

6. Fitting parameters being close to those of the asymptotic expansion

6.1. Estimates of $m(K)$ and $N(K)$ by eqs. (1.1) and (1.5) being close. We now show numerically that the best estimates of the fitting parameters $m(K)$, C_K and N_K given by (1.1) are very close to those of the parameters $\widehat{m}(K)$, $\widehat{C}_K$ and $\widehat{N}_K$ of eq. (1.5) of the asymptotic expansion up to $O(1/N)$ with respect to the number of segments. The best estimates of the parameters $\widehat{m}(K)$, $\widehat{C}_K$ and $\widehat{N}_K$ of eq. (1.5) are listed in TABLES A. 1 to A. 11 of Appendix A. They are evaluated by applying the asymptotic formula of eq. (1.5) to the data points of the knotting probabilities versus cylindrical radius r_{ex} for the cylindrical SAP model.

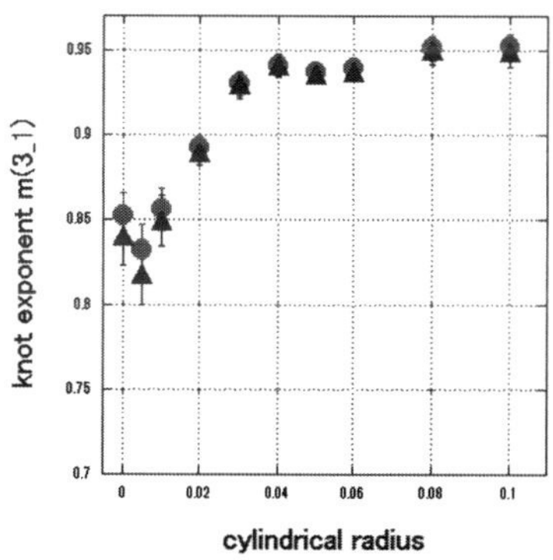

FIGURE 6. Estimates of exponent $m(3_1)$ versus cylindrical radius r_{ex} evaluated by fitting formula (1.1) and asymptotic formula (1.5) depicted by filled circles and filled upper triangles, respectively.

In FIGURE 6, we plot against cylindrical radius r_{ex} the best estimates of the knot exponent $m(3_1)$ evaluated by the fitting formula (1.1) and those of the asymptotic expansion of the logarithm of the knotting probability evaluated by eq. (1.5). The former and latter data points are depicted by filled circles and upper triangles, respectively. They are almost the same values with respect to error bars for each value of the cylindrical radius.

For example, let us compare two estimates of the knot exponent for the trefoil knot, which are plotted in FIGURE 6. At $r_{\text{ex}} = 0$ we have $m(3_1) = 0.852 \pm 0.014$ in Table II of [**23**] and $\widehat{m}(3_1) = 0.840 \pm 0.017$ shown in TABLE A. 1 of Appendix A. The difference 0.012 is much smaller than the estimate of $m(3_1)$ and similar to the errors 0.014. Furthermore, at $r_{\text{ex}} = 0.04$ we have $m(3_1) = 0.9414 \pm 0.0064$ in Table II of [**23**] and $\widehat{m}(3_1) = 0.9411 \pm 0.0067$ shown in TABLE A. 1 of Appendix A. The difference 0.0003 is much smaller than the estimate of $m(3_1)$ and the errors 0.0064.

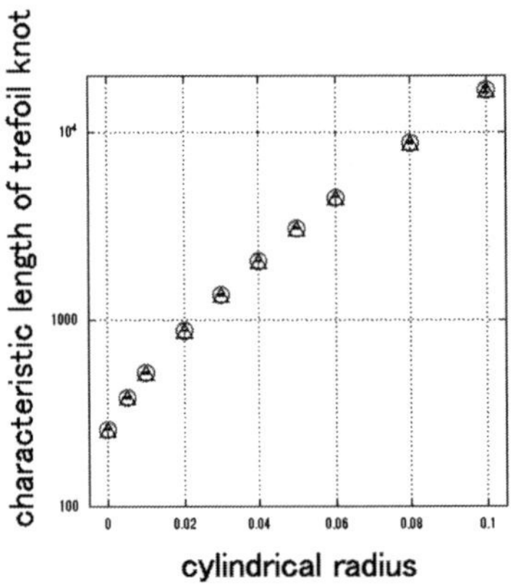

FIGURE 7. Estimates of characteristic length N_{3_1} of trefoil knot (3_1) versus cylindrical radius r_{ex} evaluated with fitting formula (1.1) and asymptotic expansion (1.5) depicted by open circles and upper triangles, respectively, in the semi-logarithmic plot.

For characteristic lengths N_K of the knotting probabilities the estimates of the four-parameter fitting formula (1.1) and those of the asymptotic expansion up to $O(1/N)$ are completely the same with respect to error bars. In FIGURE 7 the estimates of the characteristic length N_K for the trefoil knot (i.e. 3_1) evaluated by the fitting formula (1.1) and that of the asymptotic expansion up to $O(1/N)$ (1.5) are plotted against cylindrical radius r_{ex} for the cylindrical SAP model. We observe that they are almost identical with respect to errors.

6.2. Estimates of coefficients C_K evaluated by eqs. (1.1) and (1.5) are numerically close to each other. The best estimates of knot coefficients C_K evaluated with eq. (1.1) and those of the asymptotic expansion (1.5) are plotted against the cylindrical radius for knots $K = 3_1$ and 4_1 in FIGURES 8 and 9, respectively.

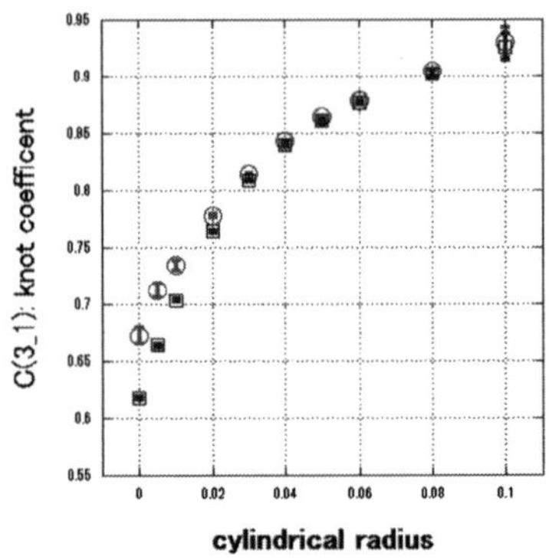

FIGURE 8. Coefficient C_{3_1} of trefoil knot (3_1) versus cylindrical radius r_{ex} evaluated by fitting formula (1.1) and asymptotic formula (1.5) depicted by circles and upper-triangles, respectively.

In both FIGURES 8 and 9, the estimates of the fitted curves and those of asymptotic expansion are quite similar to each other. However, for small values of cylindrical radius $r_{\rm ex}$, the estimates of the fitted curves slightly smaller than those of the asymptotic expansions.

For example, let us consider two estimates of the knot coefficient for the trefoil knot, which are plotted in FIGURE 8. At $r_{\rm ex} = 0$ we have $C_{3_1} = 0.6183 \pm 0.0012$ in Table II of [**23**] and $\widehat{C}_{3_1} = 0.6727 \pm 0.0071$ shown in TABLE A. 1 of Appendix A. The difference 0.0544 is smaller than the value of C_{3_1} but larger than the errors 0.0071. However, at $r_{\rm ex} = 0.04$ we have $C_{3_1} = 0.8404 \pm 0.0014$ in Table II of [**23**] and $\widehat{C}_{3_1} = 0.8432 \pm 0.0012$ shown in TABLE A. 1 of Appendix A. The difference 0.0028 is much smaller than the value of C_{3_1} and similar to the errors 0.0014.

In FIGURE 8 the coefficient C_K of the trefoil knot (3_1) increases with respect to cylindrical radius $r_{\rm ex}$. The increasing behavior is characteristic to the trefoil knot. In fact, in FIGURE 9 the coefficient C_K of the figure-eight knot (4_1) decreases with respect to cylindrical radius $r_{\rm ex}$. The decreasing behavior is common to all prime knots other than the trefoil knot.

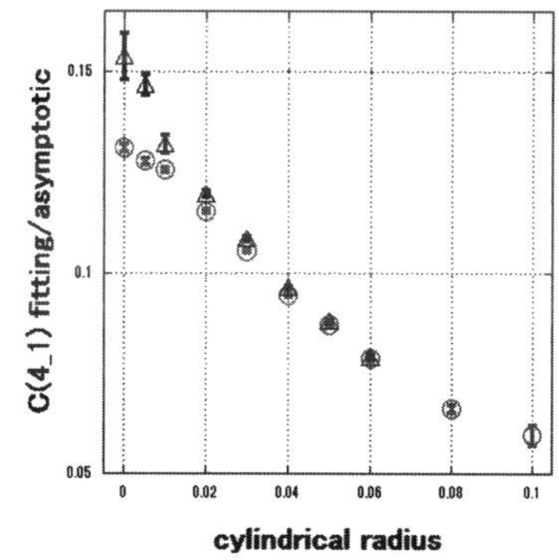

FIGURE 9. Coefficient C_{4_1} of figure-eight knot (4_1) versus cylindrical radius $r_{\rm ex}$ evaluated with fitting formula (1.1) and asymptotic formula (1.5) depicted by circles and upper-triangles, respectively.

7. Gradual monotonic flow of the knot exponent toward that of the large excluded volume

7.1. Knot exponent $m(K)$ as a function of the cylindrical radius. Let us now study how the knot exponent $m(K)$ should depend on the excluded-volume parameter, i.e. the radius of hard cylindrical segments, $r_{\rm ex}$, for the cylindrical SAP. After many trials and errors, we suggest the following expression for describing the knot exponent $m(K)$ as a function of radius $r_{\rm ex}$.

$$m(K) = |K| - \alpha_K \exp(-\beta_K \sqrt{r_{\rm ex}}), \tag{7.1}$$

where α_K and β_K are fitting parameters. We recall that the symbol $|K|$ denotes the number of constituent prime knots of knot K.

The empirical formula (7.1) suggests that for a given knot K the knot exponent $m(K)$ approaches the integer value $|K|$ as the cylindrical radius $r_{\rm ex}$ becomes very

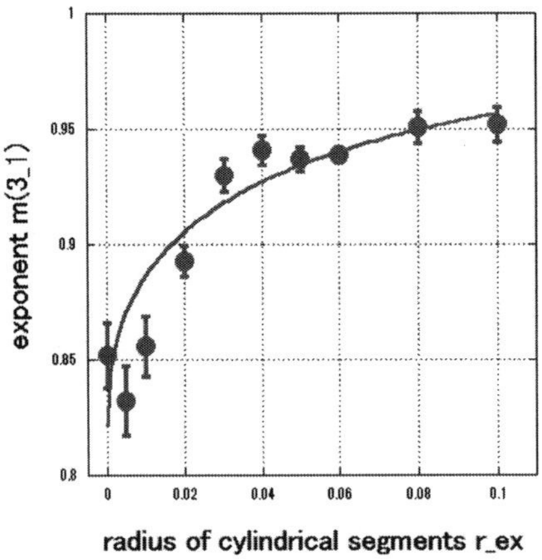

FIGURE 10. Best estimates of knot exponent $m(3_1)$ for the trefoil knot versus cylindrical radius r_{ex}. The fitted curve is given by $\alpha_K = 0.178 \pm 0.011$, and $\beta_K = 4.46 \pm 0.32$ with $\chi^2/\text{DF} = 3.5$.

large. For RPs the exponent $m(K)$ becomes smaller than the integer $|K|$: $m(K) \to |K| - \alpha_K$ as $r_{\text{ex}} \to 0$ where we have $\alpha_K > 0$ and $\alpha_K \approx 0.2$.

Formula (7.1) gives some plausible fitted curves with not very large χ^2 values, for several nontrivial knots, as shown shortly later. It can be a surprise, since there is no theoretical support for deriving eq. (7.1). In particular, we cannot explain the reason why it depends on the square root of the cylindrical radius.

We now plot the estimates of the knot exponent $m(K)$ against cylindrical radius r_{ex}, which are evaluated in the simulation of the cylindrical SAP for some knots. In FIGURE 10 the best estimates of knot exponent $m(3_1)$ for the trefoil knot 3_1 are plotted against cylindrical radius r_{ex}. They are evaluated by applying formula (1.1) to the knotting probabilities of the cylindrical SAP. In FIGURE 11 the best estimates of knot exponent $m(3_1\#3_1)$ for a composite knot $3_1\#3_1$ are plotted against cylindrical radius r_{ex}. They are evaluated by applying formula (1.1) to the knotting probabilities of the cylindrical SAP.

We observe in FIGURES 10 and 11 that formula (7.1) gives good fitted curves for the trefoil knot and the composite knot of the two trefoil knots. We also observe that the estimates of the parameters α_K and β_K are given by approximately the same values, respectively: $\alpha_K \approx 0.2$ and $\beta_K \approx 4$.

If the radius-dependence of the knot exponent is given by eq. (7.1), it is suggested that the factorization of the knot exponent such as shown in eqs. (3.5) and (3.6) should be valid only approximately. If we express the exponents $m(K_1)$ and $m(K_2)$ for two given knots K_1 and K_2 as

$$\begin{aligned} m(K_1) &= |K_1| - \Delta m(K_1), \\ m(K_2) &= |K_2| - \Delta m(K_2), \end{aligned} \tag{7.2}$$

then we have

$$m(K_1) + m(K_2) = |K_1| + |K_2| - \Delta m(K_1) - \Delta m(K_2). \tag{7.3}$$

Clearly we have $|K_1| + |K_2| = |K_1\#K_2|$, while it is not likely to have $\Delta m(K_1) + \Delta m(K_2) = \Delta m(K_1\#K_2)$.

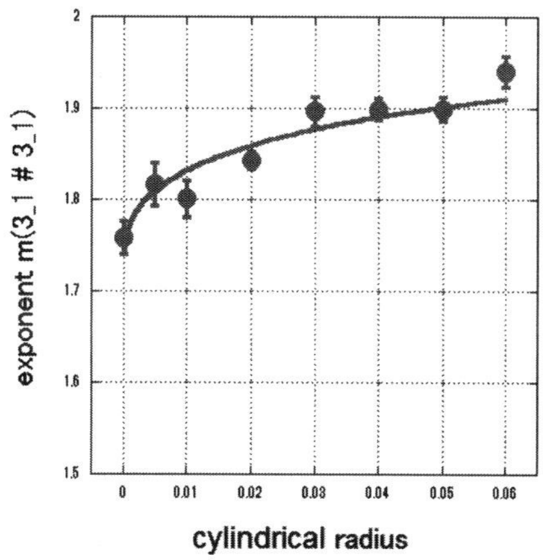

FIGURE 11. Best estimates of knot exponent $m(3_1\#3_1)$ for a composite knot $3_1\#3_1$ versus cylindrical radius $r_{\rm ex}$. The fitted curve is given by $\alpha_K = 0.257 \pm 0.016$ and $\beta_K = 4.31 \pm 0.45$ with χ^2/DF $= 1.9$.

7.2. Derivation of the sum rule for prime knots from the factorization. If the exponent $m(K)$ is equal to 1 for prime knots and to n for such a composite knot that consists of n prime knots, arguments for deriving the sum rules are simple, as shown in §3.3. However, according to the previous simulation results, the best estimates of the knot exponents $m(K)$ are smaller than the number of constituent prime knots, $|K|$.

In order to explain the numerical fact that the sum rules hold even when the estimates of the knot exponents $m(K)$ are smaller than the integer values of $|K|$: $m(K) < |K|$, let us assume a simple formula for describing how the knot exponent depends on the cylindrical radius $r_{\rm ex}$. For simplicity we assume that the shift of the knot exponent from the integer value $|K|$ is given by the same value Δm, i.e. we have

$$m(K) - |K| - \Delta m \tag{7.4}$$

where Δm is given by

$$\Delta m = \alpha \exp(-\beta \sqrt{r_{\rm ex}}) \tag{7.5}$$

for any knot K. Then, we can derive the sum rule for prime knots by making use of the factorization properties of knot coefficients C_K. The sum of the knotting probabilities over all knots should be given by 1, and we have

$$\begin{aligned}\sum_K P_K(x) &= P_{0_1}(x) + P_{3_1}(x) + P_{4_1}(x) + P_{3_1\#3_1}(x) + P_{3_1\#4_1}(x) + P_{4_1\#4_1}(x) + \cdots \\ &= C_{0_1} e^{-x} x^{m(0_1)} + C_{3_1} x^{m(3_1)} e^{-x} + C_{4_1} x^{m(4_1)} e^{-x} + \frac{1}{2!} \left(C_{3_1}\right)^2 x^{m(3_1\#3_1)} e^{-x} \\ &\quad + C_{3_1} C_{4_1} x^{m(3_1\#4_1)} e^{-x} + \frac{1}{2!} \left(C_{4_1}\right)^2 x^{m(4_1\#4_1)} e^{-x} + \cdots\end{aligned}$$

Here we assume that $C_{0_1} = 1$, for simplicity. Noting $m(0_1) = -\Delta m$ we have

$$\begin{aligned}\sum_K P_K(x) &= x^{-\Delta m} e^{-x} \Big\{ \Big(1 + C_{3_1} x + \frac{1}{2!}\left(C_{3_1} x\right)^2 + \cdots \Big) \\ &\quad \times \Big(1 + C_{4_1} x + \frac{1}{2!}\left(C_{4_1} x\right)^2 \Big) + \cdots \Big\} \\ &= x^{-\Delta m} e^{-x} \exp\left(C_{3_1} x + C_{4_1} x + \cdots\right) \\ &= \exp\left\{ \left(-1 + \sum_{K_1:\text{prime}} C_{K_1} \right) x - \frac{\Delta m}{x} \log x \right\}. \end{aligned} \tag{7.6}$$

Therefore, the sum of knot coefficients over all prime knots should be evaluated logarithmically with respect to x as follows.

$$\sum_{K_1:\text{prime}} C_{K_1} = 1 + \frac{\Delta m}{x} \log x. \tag{7.7}$$

If we evaluate the sum at some large value of x, we obtain the sum rule (1.4) for the prime knots (i.e. $n = 1$).

8. Discussion

We have studied several nontrivial properties of the parameters of the approximate formula (1.1) expressing the knotting probability as a function of the number of segments N for the cylindrical SAP model. The cylindrical SAP model makes connections between ideal ring chains such as equilateral RPs and real ring chains such as thick SAPs by changing the radius of cylindrical segments r_{ex} from 0 to 0.1 or larger values.

We suggest that the behavior of the knotting probability for thick SAPs is much simpler than that of equilateral RPs in many aspects. For instance, we expect that the knot exponent $m(K)$ of a given knot K approaches the number of constituent prime knots of the knot K in the limit of sending the excluded volume to infinity. Here we recall the suggested approximation of eq. (7.1). We also expect that the sum rules for knot coefficients should hold completely when the radius r_{ex} is very large.

Finally we suggest that the study on the nontrivial behavior of the knotting probability as a function of the excluded volume should be useful to understand the knotting probability of circular DNA screened with counter ions in solution.

Appendix A. Estimates of the parameters in asymptotic expansion (1.5)

The best estimates of the parameters $\widehat{C}_K$, $\widehat{m}(K)$, $\widehat{N}_K$ and b_K evaluated with eq. (1.5) of the asymptotic expansion of the knotting probability up to $O(1/N)$ applied to the data of the knotting probability of the cylindrical SAP model. Hereafter we denote $\widehat{C}_K$, $\widehat{m}(K)$ and $\widehat{N}_K$ by C_K, $m(K)$ and N_K, respectively, for simplicity.

r_{ex}	C_K	$m(K)$	N_K	b_K	χ^2/DF
0	0.6727 ± 0.0071	0.840 ± 0.017	257.7 ± 1.3	-19.6 ± 2.6	0.56
0.005	0.7126 ± 0.0063	0.818 ± 0.018	381.7 ± 2.6	-23.6 ± 3.2	1.11
0.01	0.7346 ± 0.0043	0.849 ± 0.015	518.7 ± 3.4	-19.7 ± 3.0	1.21
0.02	0.7781 ± 0.0017	0.8899 ± 0.0074	868.6 ± 3.5	-14.0 ± 1.8	0.68
0.03	0.8145 ± 0.0014	0.9292 ± 0.0078	1348.9 ± 7.1	-9.1 ± 2.3	1.12
0.04	0.8432 ± 0.0012	0.9411 ± 0.0067	$2047. \pm 11.$	-6.5 ± 2.3	1.15
0.05	0.8649 ± 0.0011	0.9358 ± 0.0056	$3053. \pm 17.$	$-14. \pm 2.2$	0.93
0.06	0.87936 ± 0.00099	0.9372 ± 0.0038	$4459. \pm 22.$	-14.6 ± 1.7	0.42
0.08	0.9041 ± 0.0031	0.9495 ± 0.0076	$8800. \pm 150.$	-19.7 ± 3.5	0.97
0.1	0.93 ± 0.013	0.9487 ± 0.0088	$16830. \pm 610.$	-27.1 ± 4.5	1.04

TABLE A.1. Trefoil knot 3_1

r_{ex}	C_K	$m(K)$	N_K	b_K	χ^2/DF
0	0.1538 ± 0.0058	0.813 ± 0.059	258.5 ± 4.6	-38.0 ± 9.4	1.51
0.005	0.1468 ± 0.0026	0.749 ± 0.039	387.3 ± 5.6	-47.1 ± 6.9	0.96
0.01	0.1321 ± 0.0022	0.919 ± 0.039	494.0 ± 7.9	-23.2 ± 7.7	1.36
0.02	0.11975 ± 0.00078	0.874 ± 0.022	$865. \pm 10.$	-30.7 ± 5.6	0.89
0.03	0.10854 ± 0.00042	0.885 ± 0.018	$1366. \pm 16.$	-38.3 ± 5.4	0.72
0.04	0.0963 ± 0.0004	0.879 ± 0.019	$2108. \pm 33.$	-39.8 ± 6.7	0.99
0.05	0.08798 ± 0.00029	0.928 ± 0.014	$3023. \pm 43.$	-31.8 ± 5.8	0.57
0.06	0.07918 ± 0.00054	0.927 ± 0.023	$4440. \pm 130.$	$-30. \pm 10.$	1.36

TABLE A.2. Figure-eight knot 4_1

r_{ex}	C_K	$m(K)$	N_K	b_K	χ^2/DF
0	0.0552 ± 0.0031	0.814 ± 0.088	260.0 ± 6.8	$-59. \pm 14.$	1.09
0.005	0.0478 ± 0.0019	0.705 ± 0.091	$396. \pm 14.$	$-68. \pm 16.$	1.62
0.01	0.03829 ± 0.0008	0.932 ± 0.047	496.7 ± 9.7	-30.3 ± 9.5	0.58
0.02	0.03045 ± 0.00054	0.911 ± 0.056	$845. \pm 25.$	$-39. \pm 14.$	1.39
0.03	0.02415 ± 0.00023	0.849 ± 0.043	$1373. \pm 40.$	$-62. \pm 13.$	0.89
0.04	0.01932 ± 0.00023	0.904 ± 0.055	$2034. \pm 89.$	$-56. \pm 20.$	1.58

TABLE A.3. Knot 5_1

r_{ex}	C_K	$m(K)$	N_K	b_K	χ^2/DF
0	0.0999 ± 0.0035	0.803 ± 0.054	255.0 ± 4.1	$-65. \pm 8.6$	0.69
0.005	0.0802 ± 0.0023	0.83 ± 0.057	374.1 ± 7.6	$-51. \pm 10.$	1.09
0.01	0.0678 ± 0.0013	0.937 ± 0.041	$485. \pm 8.$	-34.3 ± 8.2	0.74
0.02	0.05279 ± 0.00054	0.876 ± 0.034	$853. \pm 16.$	-46.2 ± 8.8	0.86
0.03	0.04134 ± 0.00038	0.881 ± 0.041	$1364. \pm 38.$	$-59. \pm 13.$	1.38
0.04	0.03274 ± 0.00019	0.932 ± 0.027	$1992. \pm 42.$	$-46.5 \pm 10.$	0.64

TABLE A.4. Knot 5_2

r_{ex}	C_K	$m(K)$	N_K	b_K	χ^2/DF
0	0.0345 ± 0.0019	0.607 ± 0.096	281.9 ± 8.4	$-113. \pm 16.$	0.79
0.005	0.0246 ± 0.0014	0.75 ± 0.12	$393. \pm 16.$	$-99. \pm 22.$	1.23
0.01	0.01892 ± 0.00042	0.829 ± 0.054	$515. \pm 12.$	$-84. \pm 11.$	0.32
0.02	0.01207 ± 0.00029	0.802 ± 0.086	$906. \pm 43.$	$-87. \pm 23.$	1.15
0.03	0.00828 ± 0.00015	0.976 ± 0.069	$1256. \pm 53.$	$-65. \pm 22.$	0.71
0.04	0.00572 ± 0.00013	0.800 ± 0.095	$2160. \pm 170.$	$-124. \pm 38.$	1.26

TABLE A.5. Knot 6_1

r_{ex}	C_K	$m(K)$	N_K	b_K	χ^2/DF
0	0.0303 ± 0.0026	0.90 ± 0.13	252.5 ± 9.1	$-58. \pm 20.$	1.22
0.005	0.025 ± 0.0013	0.85 ± 0.1	$369. \pm 13.$	$-77. \pm 19.$	0.95
0.01	0.01981 ± 0.00095	0.85 ± 0.11	$506. \pm 23.$	$-82. \pm 23.$	1.43
0.02	0.01296 ± 0.00028	0.767 ± 0.078	$886. \pm 38.$	$-106. \pm 21.$	1.01
0.03	0.00849 ± 0.0002	0.79 ± 0.11	$1420. \pm 100.$	$-113. \pm 34.$	1.71
0.04	0.00561 ± 0.00012	0.996 ± 0.094	$1820. \pm 120.$	$-70. \pm 37.$	1.15

TABLE A.6. Knot 6_2

r_{ex}	C_K	$m(K)$	N_K	b_K	χ^2/DF
0	0.0255 ± 0.0019	0.58 ± 0.13	$274. \pm 11.$	$-123. \pm 22.$	0.9
0.005	0.0154 ± 0.0011	1.02 ± 0.12	$340. \pm 13.$	$-56. \pm 21.$	0.79
0.01	0.01252 ± 0.00073	0.86 ± 0.13	$501. \pm 28.$	$-82. \pm 28.$	1.3
0.02	0.0076 ± 0.00025	0.76 ± 0.12	$901. \pm 61.$	$-112. \pm 33.$	1.39
0.03	0.00449 ± 0.00011	0.85 ± 0.12	$1400. \pm 110.$	$-57. \pm 36.$	1.24
0.04	0.003252 ± 0.000098	0.8 ± 0.13	$2120. \pm 220.$	$-140. \pm 52.$	1.17

TABLE A.7. Knot 6_3

r_{ex}	C_K	$m(K)$	N_K	b_K	χ^2/DF
0	0.2342 ± 0.0057	1.721 ± 0.024	262.6 ± 1.5	-61.3 ± 4.9	0.62
0.005	0.2543 ± 0.0064	1.800 ± 0.027	381.8 ± 3.0	-42.7 ± 6.6	1.15
0.01	0.2948 ± 0.0061	1.778 ± 0.025	$525. \pm 4.4$	$-56.5 \pm 7.$	1.22
0.02	0.3347 ± 0.0032	1.830 ± 0.012	$876. \pm 4.3$	-54.3 ± 4.7	0.5
0.03	0.356 ± 0.0051	1.893 ± 0.019	$1355. \pm 13.$	-44.4 ± 9.2	1.35
0.04	0.3796 ± 0.004	1.896 ± 0.015	$2071. \pm 18.$	-46.7 ± 8.8	0.92
0.05	0.4059 ± 0.0049	1.891 ± 0.016	$3111. \pm 35.$	$-73. \pm 12.$	0.91
0.06	0.4069 ± 0.0073	1.937 ± 0.021	$4344. \pm 78.$	$-67. \pm 18.$	1.32

TABLE A.8. Knot $3_1\#3_1$

r_{ex}	C_K	$m(K)$	N_K	b_K	χ^2/DF
0	0.105 ± 0.0041	1.715 ± 0.038	260.7 ± 2.3	-88.8 ± 7.9	0.61
0.005	0.102 ± 0.0041	1.783 ± 0.043	381.6 ± 4.7	$-67. \pm 11.$	1.04
0.01	0.1051 ± 0.0031	1.786 ± 0.034	519.7 ± 5.9	-73.8 ± 9.7	0.76
0.02	0.1028 ± 0.0017	1.806 ± 0.022	879.6 ± 7.6	-80.2 ± 8.6	0.45
0.03	0.0953 ± 0.0021	1.847 ± 0.030	$1374. \pm 20.$	$-87. \pm 15.$	0.74
0.04	0.0866 ± 0.0022	1.883 ± 0.035	$2075. \pm 43.$	$-76. \pm 22.$	1.17

TABLE A.9. Knot $3_1\#4_1$

r_{ex}	C_K	$m(K)$	N_K	b_K	χ^2/DF
0	0.0418 ± 0.0038	2.797 ± 0.066	258.1 ± 3.1	$-44. \pm 18.$	1.7
0.005	0.0578 ± 0.0037	2.800 ± 0.049	380.3 ± 4.2	$-49. \pm 17.$	1.1
0.01	0.0681 ± 0.0034	2.845 ± 0.039	511.9 ± 5.2	$-53. \pm 16.$	0.73
0.02	0.0873 ± 0.0037	2.864 ± 0.034	860.6 ± 8.9	$-60. \pm 19.$	0.72
0.03	0.0957 ± 0.0056	2.925 ± 0.047	$1337. \pm 24.$	$-46. \pm 33.$	1.17
0.04	0.1105 ± 0.0053	2.900 ± 0.039	$2051. \pm 34.$	$-80. \pm 37.$	0.81
0.05	0.1122 ± 0.0088	2.954 ± 0.059	$2955. \pm 88.$	$-68. \pm 69.$	1.15
0.06	0.143 ± 0.013	2.837 ± 0.065	$4700. \pm 200.$	$-204. \pm 89.$	0.86

TABLE A.10. Knot $3_1\#3_1\#3_1$

r_{ex}	C_K	$m(K)$	N_K	b_K	χ^2/DF
0	0.0311 ± 0.0032	2.727 ± 0.073	259.2 ± 3.5	$-103. \pm 20.$	1.23
0.005	0.0357 ± 0.0028	2.781 ± 0.059	378.1 ± 4.9	$-100. \pm 21.$	0.8
0.01	0.0387 ± 0.0042	2.792 ± 0.086	$516. \pm 11.$	$-94. \pm 35.$	1.68
0.02	0.0385 ± 0.0029	2.878 ± 0.061	$855. \pm 16.$	$-74. \pm 35.$	0.94
0.03	0.0331 ± 0.0021	3.012 ± 0.050	$1299. \pm 23.$	$8. \pm 36.$	0.4
0.04	0.0348 ± 0.0034	2.956 ± 0.078	$1981. \pm 64.$	$-61. \pm 74.$	1.02

TABLE A.11. Knot $3_1\#3_1\#4_1$

Acknowledgments

The authors would like to thank many participants of the AMS special session: "Topology of Biopolymers", at the Spring Eastern Sectional Meeting of AMS, Northeastern Univ., Boston, Apr 21-22, 2018. Furthermore, they are grateful to Yasuyuki Tezuka and Koya Shimokawa for helpful comments on topological polymers and the topology of spatial graphs such as the theta curves. The present research is partially supported by JSPS KAKENHI Grant Number JP17H06463.

References

[1] H. L. Frisch and E. Wasserman, *Chemical Topology,* J. Amer. Chem. Soc. **83** (1961), 3789–3795, DOI 10.1021/ja01479a015.

[2] M. Delbrück, *Knotting problems in biology,* in *Mathematical Problems in the Biological Sciences,* ed. by R.E. Bellman, Proc. Symp. Appl. Math. (1962), **14**, 55.

[3] D. W. Sumners and S. G. Whittington, *Knots in self-avoiding walks,* J. Phys. A **21** (1988), no. 7, 1689–1694, DOI 10.1088/0305-4470/21/7/030. MR951053

[4] N. Pippenger, *Knots in random walks,* Discrete Appl. Math. **25** (1989), no. 3, 273–278, DOI 10.1016/0166-218X(89)90005-X. MR1026336

[5] Y. Diao, *The knotting of equilateral polygons in* $\mathbf{R}^3$, J. Knot Theory Ramifications **4** (1995), no. 2, 189–196, DOI 10.1142/S0218216595000090. MR1331746

[6] A. V. Vologodskiĭ, A. V. Lukashin, M. D. Frank-Kamenetskiĭ, and V. V. Anshelevich, *The knot problem in statistical mechanics of polymer chains* (Russian, with English summary), Ž. Èksper. Teoret. Fiz. **66** (1974), no. 6, 2153–2163; English transl., Soviet Phys. JETP **39** (1974), no. 6, 1059–1063 (1975). MR0376030

[7] J. P. J. Michels and F. W. Wiegel, *On the topology of a polymer ring,* Proc. Roy. Soc. London Ser. A **403** (1986), no. 1825, 269–284, DOI 10.1098/rspa.1986.0012. MR834761

[8] E. J. Janse van Rensburg and S. G. Whittington, *The knot probability in lattice polygons,* J. Phys. A **23** (1990), no. 15, 3573–3590, DOI 10.1088/0305-4470/23/15/028. MR1068243

[9] K. Koniaris and M. Muthukumar, *Knottedness in ring polymers.* Phys. Rev. Lett. (1991) **66**, 2211–2214, DOI 10.1103/PhysRevLett.66.2211.

[10] T. Deguchi and K. Tsurusaki, *Topology of closed random polygons,* J. Phys. Soc. Japan **62** (1993), no. 5, 1411–1414, DOI 10.1143/JPSJ.62.1411. MR1224827

[11] S. G. Whittington, *Knot probabilities for equilateral random polygons in R^3*, Joint work with Alexander Taylor and Mark Dennis, a talk in Topology and Its Applications, Bowing Green, KT, USA, July 17–20, 2018.

[12] T. Deguchi and E. Uehara, *Statistical and dynamical properties of topological polymers with graphs and ring polymers with knots*, Polymers, **9**, 252 (2017), DOI 10.3390/polym9070252.

[13] V. V. Rybenkov, N. R. Cozzarelli and A. V. Vologodskii, *Probability of DNA knotting and the effective diameter of the DNA double helix*, Proc. Natl. Acad. Sci. USA **90**, 5307 (1993), DOI 10.1073/pnas.90.11.5307.

[14] S. Y. Shaw and J. C. Wang, *Knotting of a DNA Chain During Ring Closure*, Science **260**, 533 (1993), DOI 10.1126/science.8475384.

[15] C. Plesa, D. Verschueren, S. Pud, J. van der Torre, J. W. Ruitenberg, M. J. Witteveen, M. P. Johnsson, A. Y. Grosberg, Y. Rabin, C. Dekker, *Direct observation of DNA knots using a solid-state nanopore* Nat. Nanotechnol **11**, 1093 (2016), DOI 10.1038/nnano.2016.153.

[16] D. W. Bowden and R. Calendar, *Maturation of Bacteriophage P2 DNA in Vitro: A Complex, Site-specific System for DNA Cleavage*, J. Mol. Biol. **129** (1979) 1-18, DOI 10.1016/0022-2836(79)90055-X.

[17] A. D. Bates and A. Maxwell, *DNA Topology* (Oxford Univ. Press, Oxford, 2005).

[18] A. Vologodoskii, *Biophysics of DNA*, (Cambridge Univ. Press, Cambridge, 2015).

[19] G. Witz, K. Rechendorff, J. Adamcik, and G. Dietler, *Conformation of Circular DNA in two Dimensions* Phys. Rev. Lett. **101**, 148103 (2008), DOI 10.1103/PhysRevLett.101.148103.

[20] T. Sakaue, G. Witz, G. Dietler, H. Wada, *Universal bond correlation function for two-dimensional polymer rings*, Europhys. Lett. **91**, 68002 (2010), DOI 10.1209/0295-5075/91/68002.

[21] S. Ropelewski, E. Uehara, C. Lehmann, T. Deguchi and G. Dietler, *Two-point correlation function of ring polymers: experiments and numerical simulations for the case of circular DNA in 2 dimensions*, React. Funct. Polym. **133** (2018) 66-70, DOI 10.1016/j.reactfunctpolym.2018.10.001.

[22] E. Uehara and T. Deguchi, *Characteristic length of the knotting probability revisited*, J. Phys. Condens. Matter **27**, 354104 (2015), DOI 10.1088/0953-8984/27/35/354104.

[23] E. Uehara and T. Deguchi, *Knotting probability of self-avoiding polygons under a topological constraint*, J. Chem. Phys. **147**, 094901 (2017), DOI 10.1063/1.4996645.

[24] E. Uehara, L. Coronel, C. Micheletti, and T. Deguchi, *Bimodality in the knotting probability of semiflexible rings suggested by mapping with self-avoiding polygons*, React. Funct. Polym. **134** (2019), 141–149, DOI 10.1016/j.reactfunctpolym.2018.11.008.

[25] F. C. Rieger and P. Virnau, *A Monte Carlo study of knots in long double-stranded DNA chains*, PLoS Comput. Biol. **12**, e1005029 (2016), DOI 10.1371/journal.pcbi.1005029.

[26] L. Coronel, E. Orlandini and C. Micheletti, *Non-monotonic knotting probability and knot length of semiflexible rings: the competing roles of entropy and bending energy*, Soft matter **13** (2017) 4260–4267, DOI 10.1039/c7sm00643h.

[27] T. Deguchi and K. Tsurusaki, *A statistical study of random knotting using the Vassiliev invariants*, J. Knot Theory Ramifications **3** (1994), no. 3, 321–353, DOI 10.1142/S0218216594000241. Random knotting and linking (Vancouver, BC, 1993). MR1291863

[28] E. Orlandini and S. G. Whittington, *Statistical topology of closed curves: some applications in polymer physics*, Rev. Modern Phys. **79** (2007), no. 2, 611–642, DOI 10.1103/RevModPhys.79.611. MR2326799

[29] C. Micheletti, D. Marenduzzo, and E. Orlandini, *Polymers with spatial or topological constraints: theoretical and computational results*, Phys. Rep. **504** (2011), no. 1, 1–73, DOI 10.1016/j.physrep.2011.03.003. MR2800174

[30] K. C. Millett, *Knotting and linking in macromolecules*, React. Funct. Polym. **131** (2018) 181–190, DOI 10.1016/j.reactfunctpolym.2018.07.023.

[31] T. Deguchi and K. Tsurusaki, *Universality of random knotting*, Phys. Rev. E. **55**, 6245 (1997), DOI 10.1103/PhysRevE.55.6245.

[32] K. Murasugi, *Knot theory and its applications*, Birkhäuser Boston, Inc., Boston, MA, 1996. Translated from the 1993 Japanese original by B. Kurpita. MR1391727

[33] M. Le Bret, *Monte Carlo computation of supercoiling energy, the sedimentation constant, and the radius of gyration of unknotted and knotted circular DNA*, Biopolymers, **19** (1980) 619–637, DOI 10.1002/bip.1980.360190312.

[34] K. V. Klenin, A. V. Vologodskii, V. V. Anshelevich, A. M. Dykhne, M. D. Frank-Kamenetskii, *Effect of excluded volume on topological properties of circular DNA*, J. Biomol. Struct. Dyn. **5** (1988) 1173–1185, DOI 10.1080/07391102.1988.10506462.

[35] K. C. Millett, *Knotting of regular polygons in 3-space*, J. Knot Theory Ramifications **3** (1994), no. 3, 263–278, DOI 10.1142/S0218216594000204. Random knotting and linking (Vancouver, BC, 1993). MR1291859

[36] M. Kapovich and J. J. Millson, *The symplectic geometry of polygons in Euclidean space*, J. Differential Geom. **44** (1996), no. 3, 479–513, DOI 10.4310/jdg/1214459218. MR1431002

[37] S. Alvarado, J. A. Calvo, and K. C. Millett, *The generation of random equilateral polygons*, J. Stat. Phys. **143** (2011), no. 1, 102–138, DOI 10.1007/s10955-011-0164-4. Supplementary material available online. MR2787976

[38] T. Deguchi and K. Tsurusaki, *A new algorithm for numerical calculation of link invariants*, Phys. Lett. A **174** (1993), no. 1-2, 29–37, DOI 10.1016/0375-9601(93)90537-A. MR1207758

[39] M. Polyak and O. Viro, IMRN., **No. 11**, 445 (1994), DOI 10.1155/S1073792894000486.

[40] N. Madras and G. Slade, *The self-avoiding walk*, Probability and its Applications, Birkhäuser Boston, Inc., Boston, MA, 1993. MR1197356

[41] M. Baiesi, E. Orlandini, A. L. Stella, *The entropic cost to tie a knot*, J. Stat. Mech., (2010) P06012, DOI 10.1088/1742-5468/2010/06/P06012.

[42] E. J. Janse van Rensburg and A. Rechnitzer, *On the universality of knot probability ratios*, J. Phys. A **44** (2011), no. 16, 162002, 8, DOI 10.1088/1751-8113/44/16/162002. MR2787081

Department of Physics, Faculty of Core Research, Ochanomizu University, 2-1-1 Ohtsuka, Bunkyo-ku, Tokyo 112-8610, Japan

Email address: deguchi@phys.ocha.ac.jp

Department of Physics, Faculty of Science, Ochanomizu University, 2-1-1 Ohtsuka, Bunkyo-ku, Tokyo 112-8610, Japan

Contemporary Mathematics
Volume **746**, 2020
https://doi.org/10.1090/conm/746/15003

Knotting of replication intermediates is narrowly restricted

Dorothy Buck and Danielle O'Donnol

ABSTRACT. The fundamental biological process of DNA replication contains several potential topological pitfalls. These, if not properly managed, lead to cell death. The enzymes (*topoisomerases*) responsible for averting these topological death warrants can inadvertently knot the same DNA they are supposed to be simplifying. These knotted DNA replication intermediates are still poorly understood. Here, we characterize which knotted forms these DNA molecules can take. Surprisingly, these forms are very narrowly restricted, representing a small fraction of possible entangled configurations.

1. Introduction

DNA replication is the process of creating two (daughter) DNA replicas of a given DNA molecule. This is an intricate choreography, and introduces a variety of potential lethal side effects if not managed properly. Potential complications arise because DNA is not a straight ladder but a double helix, with one complete turn of the helix every $\sim$ 10.5 rungs (*basepairs*). So the unzipping of the double helix during DNA replication could create extraordinary torsional strain on the parental DNA (see Figure 1). Additionally the daughter molecules could be linked together forming a $(2, n)$-torus link (with $n \sim$ (# basepairs)/10.5) rather than the required split link that is subsequently compartmentalized into two daughter cells (see Figure 2).

Fortunately specialized proteins have evolved to mitigate these topological disasters. Termed (type II) *topoisomerases*, they regulate the topology (and geometry) of DNA, by performing crossing changes between two arcs of double-stranded DNA (duplex DNA). The result of these crossing changes can be either unknotting/unlinking or the release of torsional strain on DNA molecules (by removing supercoils, aka performing Reidemeister I moves), neatly sidestepping both potentially lethal dangers above.

2010 *Mathematics Subject Classification.* Primary 57M25, 57M15, 92E10; Secondary 05C10, 92C40.

Key words and phrases. theta-graph, theta-curve, replication intermediate, unknotting.

The first author was supported in part by EPSRC grants EP/H0313671, EP/G0395851. She also acknowledges generous funding from The Leverhulme Trust grant RP2013-K-017.

The second author was supported in part by EPSRC grant EP/G0395851, AWM Mentor Travel Grant and the National Science Foundation grant DMS-1406481, DMS-1600365, DMS-1851675.

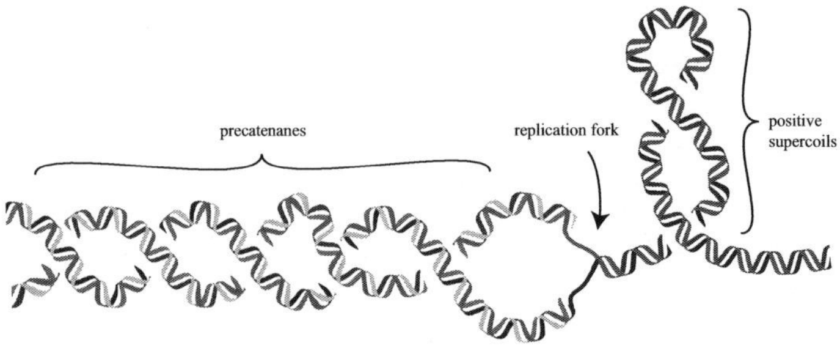

FIGURE 1. As the DNA unzips at what is called, the replication fork, torsional stress leads to positive supercoiling ahead of the replication fork. The freshly replicated duplexes form similar structures (precatenanes) behind the replication fork.

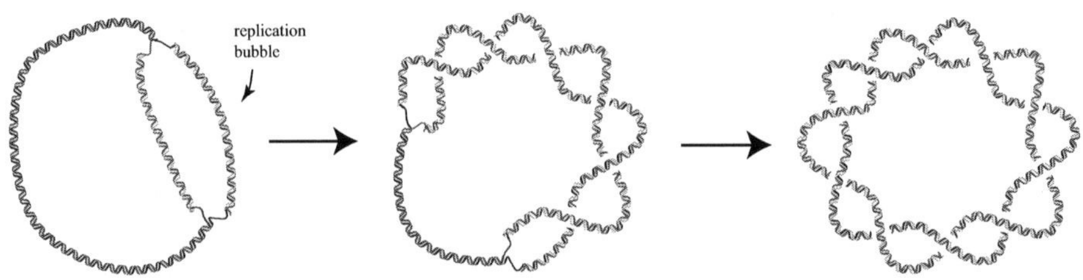

FIGURE 2. The process of DNA replication of a circular molecule. *(Left)* During replication: A DNA molecule, shown in the plane to highlight the replication bubble. *(Middle)* During replication: The freshly replicated duplexes become interwound. *(Right)* After replication: daughter molecules are linked, shown here a $(2, 10)$-torus link (catenane).

Unfortunately however these topoisomerases do not always act perfectly. In particular, when performing crossing changes, they can inadvertently knot (rather than unknot/unlink) DNA molecules. This is seen in the middle of DNA replication, and has been observed experimentally in a series of papers by the Schvartzmann lab [**LMRH$^+$12, SSMR$^+$99**]. In these so-called replication intermediates, knots can clearly be seen in microscopy images.

In related work, the authors, together with Andrzej Stasiak, have characterized the knotting pathways during DNA replication—i.e. *how* knots arise [**OSB18**].

Replication intermediates (shown in Figure 2) are modeled with θ-curves. In the current work, we answer the related important question '*Which* knotted θ-curves arise during DNA replication?'. Our main result is:

THEOREM 1.1. *Based on mathematically modeling biological restrictions, the only knotted DNA replication intermediates that could occur due to a single DNA-DNA passage are:* $\mathbf{3_1}, \mathbf{4_1}, \overline{\mathbf{5_5}}, \overline{\mathbf{5_6}}, \overline{\mathbf{6_5}}, \overline{\mathbf{6_6}}, \overline{\mathbf{6_7}}, \mathbf{7_{28}}, \mathbf{7_{29}}$ *and* $\mathbf{7_{30}}$. *See Figure* 9.

This theorem narrows the potential replication intermediates to a particular class of θ-curves. They are all θ-curves with a single non-trivial cycle that is a right-handed twist knot.

The paper is organized as follows. In Section 2, we introduce the mathematical background needed to model knotting during DNA replication. In particular we introduce θ-curves, which will be used to represent replication intermediates. To model the knotting of DNA, we will consider *un*knotting of θ-curves. So we also introduce the unknotting number of a θ-curve.

In Section 3, we give a brief biology overview of the fundamental process of DNA replication, how topoisomerases control topological complications and how knotting can occur during DNA replication. We summarize the current understanding of knotting during replication, discussing experiments and models [**LMRH**$^+$**12**, **SSMR**$^+$**99**, **OSB18**]. We close this section by translating the biological phenomena (topoisomerase-mediated knotting of DNA replication intermediates) into mathematical formalizm (*un*knotting of θ-curves).

In Section 4, we examine the biological restrictions on the replication intermediates. We translate these biological restrictions into topological constraints in our model. We then eliminate various θ-curves (or classes of θ-curves) as potential DNA replication intermediates. We state our main theorem, and discuss these results in the context of experimentally observed knotted replication intermediates.

2. Mathematical background

In this article we model replication intermediates with embedded θ-graphs equivalent up to ambient isotopy. In this section we will give definitions and outline the pertinent mathematics.

2.1. Equivalence and diagrams. The θ-graph is the graph with two vertices and three edges between them. An **embedding of a θ-graph** into $\mathbb{R}^3$ or S^3 is a homeomorphism from a θ-graph to $\mathbb{R}^3$ or S^3 that is onto its image. We will call an embedded θ-graph, a **θ-curve**. We study embedded graphs up to ambient isotopy. (Roughly speaking, this means that a graph can be moved or stretched and we still consider it the same graph. However, it cannot be cut or moved through itself.) We work with **diagrams of θ-curves**, which are projections to the plane where there are only double points away from vertices, with over and under crossings indicated. There are many different diagrams for a given θ-curve. If two graph diagrams are of the same embedded graph then there exists a finite sequence of Reidemeister moves (and planar isotopy) from one diagram to the other [**Kau89**]. (For general information on the Reidemeister moves for knots see [**Rol90**].) In Figure 3, we show the Reidemeister moves for trivalent graphs. In each move it shows how the diagram changes, the rest of the diagram (which is not pictured) does not change.

A **knot** is an embedded circle, also equivalent up to ambient isotopy. So we can think of knots as subgraphs of our θ-curves. Each pair of edges of the θ-graph form a cycle, these are called **constituent knots**.

Given a knot K, the **mirror image** of K, denoted $\overline{K}$, is the knot whose diagram is obtained by changing every crossing in a diagram of K. Similarly, given a θ-curve θ, the **mirror image** of θ, denoted $\overline{\theta}$, is the graph whose diagram is obtained by changing every crossing in a diagram of θ.

2.2. Prime knots and θ-curves. In order to define a prime θ-curve we will first define some operations. They are two operations that will produce new θ-curves, they are called connect sum and order-3 vertex connect sum. The **connect**

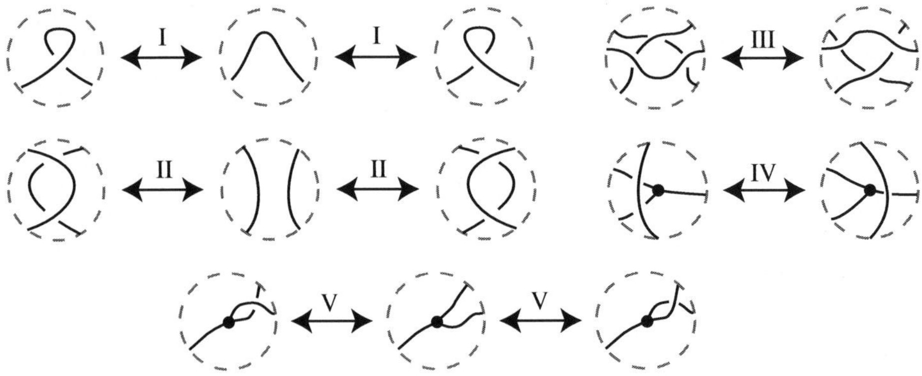

FIGURE 3. The Reidemeister moves for trivalent graph embeddings. Moves I, II, and III are the same as for knots. In move IV an arc passes in front of or behind a vertex. In move V the edges incident to a vertex switch places.

sum of a knot K and θ-curve θ, denoted $K\#\theta$, is a result of removing a 3-ball neighborhood including an arc of K and a 3-ball neighborhood of an arc of one of the edges of θ and gluing the remaining 3-balls together so that the 2 points of the knot and θ-curve on the boundary are matched. This connect sum operation can also be done to two knots to get a new knot. Given two θ-curves in S^3, call them θ_1 and θ_2, both with a chosen vertex, the **order-3 vertex connect sum** of θ_1 and θ_2, denoted $\theta_1\#_3\theta_2$, is the result of removing a 3-ball neighborhood of each of the vertices and gluing the remaining 3-balls together so that the 3 points of θ_1 on the boundary are matched with the 3 points of θ_2 on the boundary of the other ball in the natural way. For examples see Figure 4. (For more information see [**Wol87**].) These definitions are restricted to $\mathbb{R}^3$ in the usual way. A **prime θ-curve** is one that is neither an order-3 vertex sum of two nontrivial θ-curves, nor a connect sum of a nontrivial knot with the edge of a (possibly trivial) θ-curve. Similarly, a **prime knot** is a knot that is not a connect sum of two nontrivial knots.

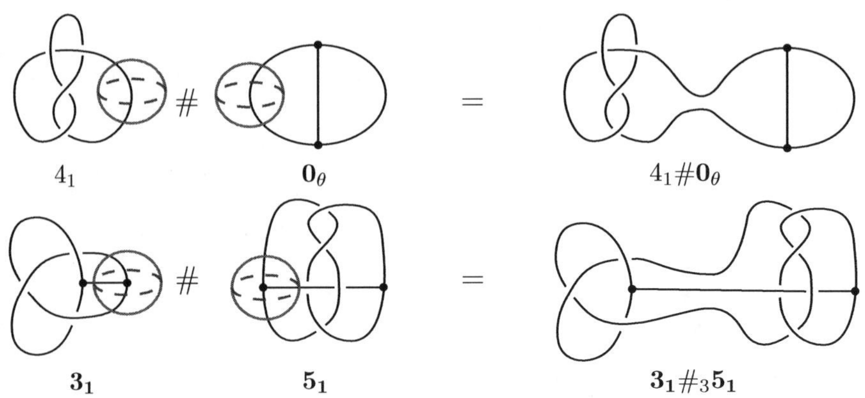

FIGURE 4. *(Top)* A connect sum of 4_1 and $\mathbf{0}_\theta$. *(Bottom)* An order-3 vertex connect sum of $\mathbf{3_1}$ and $\mathbf{5_1}$.

There are tables of all of the prime knots up to 16 crossings [**HT**]. For knots we will use the names given in Rolfsen's tables [**Rol90**]. For convenience, we have included a table of all the prime knots up to seven crossings in Figure 5. The prime θ-curves with up to 7 crossings were originally compiled by Litherland. They were later published and verified by Moriuchi [**Mor09**]. We will use the names given in this table. (See Figure 10.) Though it should be clear from context, to further distinguish θ-curves from knots, the θ-curve names will appear in **bold**.

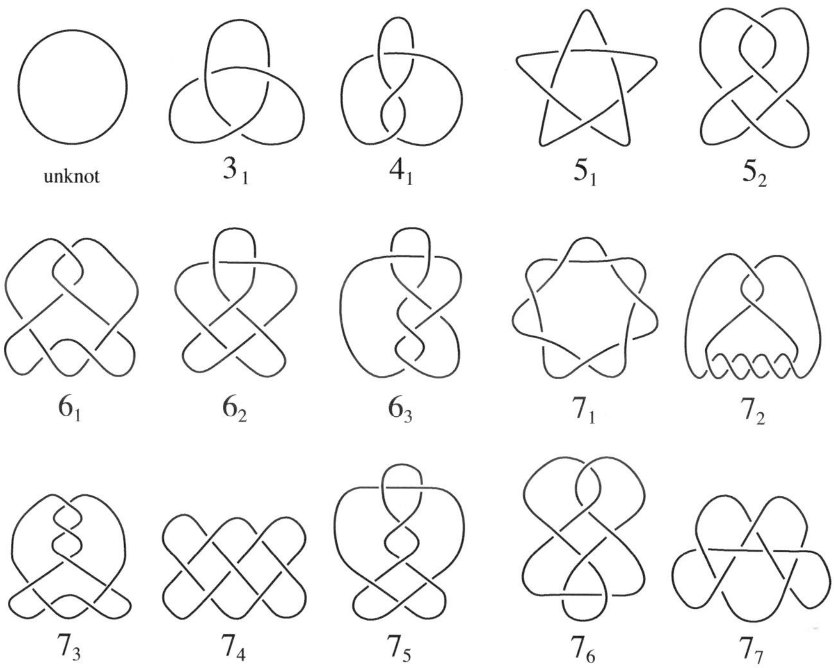

FIGURE 5. The prime knots up to seven crossings

2.3. Unknotting number. We will model the action of topoisomerases in terms of a **crossing move**: in a knot a segment is passed through another segment, or in a graph an edge is passed through itself or another edge. In a diagram a crossing move can be made at a crossing of the diagram. In this case, the segments at that crossing are passed through each other resulting in moving the under-segment to the over-segment and *vice versa.*

The **trivial knot** (also called the **unknot**) is a circle. Similarly, the **trivial θ-curve** is the planar θ-curve (up to equivalence). We will denote the trivial θ-curve as $\mathbf{0}_\theta$. The **unknotting number** of an embedded graph (or knot) g, $u(g)$, is the minimum number of crossing changes needed to obtain a planar embedding, over all possible diagrams of g. A trivial embedding is also called **unknotted.**

All θ-curves with $u(\theta) = 1$ can be created by a single crossing change on $\mathbf{O}_\theta$ but those with higher unknotting numbers require more passages for their formation from $\mathbf{0}_\theta$. The unknotting number for a θ-curve, θ is at least as large as $u(K)$ where θ contains K as a constituent knot. We will use this relationship to build on prior work of the first two authors to rule out many θ-curves as candidates for knotted replication bubbles.

The authors (with an appendix by Ken Baker) have determined $u(\theta)$ for all of the θ-curves in the Litherland-Moriuchi Table [**BO17**]. In Figure 10, we indicate a set of crossing changes that will unknot each θ-curve. The unknotting crossing changes are highlighted in gray (purple). For example, Figure 6 shows how $\mathbf{5_7}$ can be moved to a planar embedding via Reidemeister moves after the indicated crossing change.

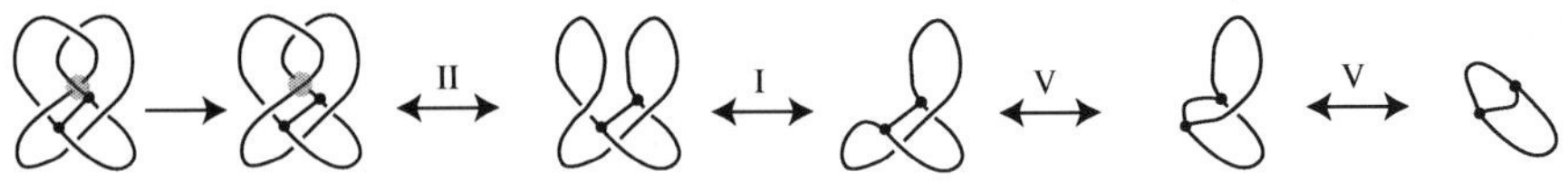

FIGURE 6. The θ-curve $\mathbf{5_7}$ with unknotting crossing change. The arrow indicates the highlighted crossing change, and the double arrows are the indicated Reidemeister moves.

3. Biological Background

3.1. DNA Replication. For a cell to divide, its DNA must first be copied (*i.e. replicated*). This is a complicated process that involves a number of enzymes, and precise coordination. During DNA replication, the double helix is 'unzipped' and each half is paired with a newly synthesised complementary strand; these halves are termed *sister duplexes.* The end result of replication is two new identical DNA molecules, termed *daughter molecules.*

If DNA were a straight ladder, this process would be topologically straightforward. However, there are two complications.

Firstly, the double helix has a natural twist (of approximately 10.5 basepairs/turn), so to unzip this twisted ladder, the natural twist in the DNA must be removed from that region. As the DNA unzips at the *replication fork*, there is formation of nontrivial writhe both in front and behind the fork. Torsional stress leads to positive supercoiling ahead of the replication fork, while the sister duplexes wind around each other in a left-handed fashion behind the replication fork, in so-called *precatenanes.* See Figure 1.

Additionally, in most organisms, DNA molecules is topologically constrained. Bacterial, chloroplast, or mitochondrial DNA are physically circular (covalently closed). In most other organisms, DNA is tethered by large protein complexes that trap DNA entanglement. Additionally, most experimental labs use *plasmid* DNA: small circular molecules. During replication the sister duplexes form a *replication bubble.* After replication the daughter molecules are linked (*catenated*). See Figure 2.

To mitigate these two complications, there is a family of enzymes (*topoisomerases*) whose sole function is to reduce the topological complexity of DNA. In particular these topoisomerases can remove the positive supercoils ahead of the replication fork, and unlink (*decatenate*) daughter molecules. They act by performing crossing changes between arcs of double-stranded DNA.

3.2. Knotting during DNA replication. However, during replication these topoisomerases may also form knots, by performing crossing changes between segments of the replicating DNA. Such knotting seems to be disadvantageous for the molecule. Knotting in these so-called *replication intermediates* is not well-understood, despite the fact that the process by which the daughter plasmids become unlinked/decatenated has been studied in detail.

To better understand how and why knotting occurs, Schvartzman's laboratory produced DNA plasmids that could be stopped in the middle of replication inside bacteria (*E.coli*) cells [**LMRH$^+$12**]. They were able to experimentally determine that there are at least nine different knotted replication intermediates, though the exact θ-curve types remained uncharacterized (and there are likely more that are at very low concentrations). They directly observed (via atomic force microscopy) four types of unknotted replication intermediates, and the single knotted replication intermediate $\mathbf{3_1}$ [**LMRH$^+$12**]. In other experiments the DNA was visualized after cutting the parental edge [**SSMR$^+$99**]. They directly observed (via electron microscopy) knots in the replication bubble (aka freshly replicated knots) formed by the sister duplexes. These freshly replicated knots were shown to be $\overline{3_1}$, 4_1, $\overline{5_2}$ and $\overline{6_2}$.

The toposiomerase Topo IV, which is responsible for the knotting in the replication intermediates in *E. coli* [**LMRH$^+$12**], acts via a DNA-DNA passage. Recently, together with Andrzej Stasiak, we developed a general model of knotting in replication intermediates that explains how Topo IV can produce experimentally observed knotted θ-curves [**OSB18**].

3.3. Modeling the process. We will model a replication intermediate with a θ-curve. Vertices are at the replication forks, and edges represent segments of duplex DNA. See Figure 7. This means that the knotting can be modeled by a crossing change of the corresponding θ-curve. Because of the chirality of the DNA double helix, it is well-defined, and standard in the molecular biology literature, to refer to a 'positive' (in the same sense as the double helix with both backbones oriented parallel) or 'negative' crossing.

4. DNA Replication Intermediate Structure

In this section we examine the biological restrictions on the formation of replication intermediates. We translate these biological restrictions into topological constraints in our model, and subsequently rule out different θ-curves (or classes of θ-curves) from being biologically possible. We will restrict to looking at knotted replication intermediates with seven or fewer crossings, as θ-curves have only been classified up to 7 crossings. We will use names in the Litherland-Moriuchi Table of prime θ-curves. (For more information see Section 2.) All of the possible knotted θ-curves up to seven crossings, are in this set of 90 prime θ-curves, together with any seven crossing or less θ-curve that can be formed via connect sum and order-3 vertex connect sums.

4.1. Ruling out multiple DNA-DNA passages. In this article we will only considering those knotted replication intermediates that could result from one DNA-DNA passage. We make this restriction because experiments seem to suggest, but not definitively show, that significant concentrations of knotted RIs appear as a result of a single round of topoisomerase action [**SSMR$^+$99**]. Thus

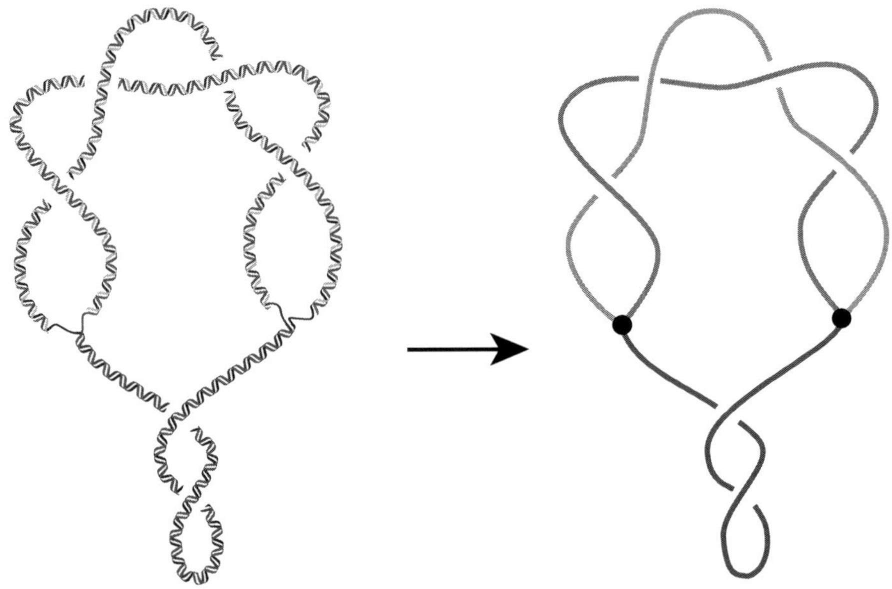

FIGURE 7. The replication intermediate, with negatively supercoiled parental DNA and precatenated sister duplexes in the replication bubble. *(Left)* The replication intermediate is shown with the detail of the helical DNA backbone and the replication forks. *(Right)* The mathematical model of a replication intermediate, with vertices representing replication forks and single edges representing segments of duplex DNA.

we can eliminate all θ-curves that have unknotting number greater than one. In previous work, Buck and O'Donnol (with an Appendix by Baker) determined the unknotting numbers of all of the prime θ-curves in the Litherland-Moriuchi Table [**BO17**]. See Table 1. Using these results, we can eliminate 37 of the prime knotted θ-curves.

4.2. Ruling out θ-curves with multiple constituent knots. The knots experimentally observed in replication intermediates are such that if the replication continues to the end the result is two catenated molecules where each of the two is unknotted [**SSMR$^+$99**]. So we will assume that two of the constituent knots of the θ-curve are trivial. So the only nontrivial constituent is the one that is formed by the sister duplexes. This assumption eliminates an additional 12 of the prime θ-curves. See Table 1. It also eliminates the possibility of a connect sum of a nontrivial knot with any θ-curve, because this will result in at least two nontrivial constituents. This leaves only prime curves and order-3 vertex connect sums of prime curves with non-trivial curves with all trivial constituents.

4.3. Ruling out formation of loops in sister duplexes. A large complex of proteins act at the replication fork, which has recently been shown can rotate [**CCM$^+$15**]. This rotation would rapidly dissipate any loops transiently formed in the sister duplexes. Thus, there should not be any non-transient loops in sister duplexes.

θ	Constituent Knots			u	4.*
3_1	2x0	3_1		1	
4_1	2x0	4_1		1	
5_1	3x0			1	4
5_2	2x0	3_1		1	4
5_3	2x0	5_1		2	1
5_4	0	3_1	5_1	2	1, 2
5_5	2x0	5_2		1	
5_6	2x0	5_2		1	
5_7	0	3_1	5_2	1	2
6_1	3x0			1	4
6_2	2x0	3_1		1	4
6_3	0	3_1	4_1	1	2, 4
6_4	0	3_1	4_1	1	2, 4
6_5	2x0	6_1		1	
6_6	2x0	6_1		1	
6_7	2x0	6_1		1	
6_8	0	4_1	6_1	1	2
6_9	2x0	6_2		1	5
6_{10}	2x0	6_2		1	5
6_{11}	2x0	6_2		1	5
6_{12}	0	3_1	6_2	2	1, 2, 5
6_{13}	0	4_1	6_2	1	2, 5
6_{14}	2x0	6_3		1	5
6_{15}	2x0	6_3		1	5
6_{16}	0	3_1	6_3	1	2, 5
7_1	3x0			1	4
7_2	3x0			1	4
7_3	3x0			1	4
7_4	3x0			1	4
7_5	2x0	3_1		2	1, 4
7_6	2x0	3_1		1	4
7_7	2x0	3_1		1	4
7_8	0	2x3_1		2	1, 2, 4
7_9	0	2x3_1		1	2, 4
7_{10}	0	2x3_1		2	1, 2, 4
7_{11}	2x0	5_2		1	4
7_{12}	2x0	4_1		1	4
7_{13}	2x0	4_1		1	4
7_{14}	0	2x4_1		1	2, 4
7_{15}	2x0	5_1		2	1, 4
7_{16}	2x0	5_1		2	1, 4
7_{17}	2x0	5_1		2	1, 4
7_{18}	0	5_1	5_2	2	1, 2, 4
7_{19}	2x0	5_2		1	4
7_{20}	2x0	5_2		1	4
7_{21}	2x0	5_2		1	4
7_{22}	0	3_1	5_2	2	1, 2, 4
7_{23}	0	4_1	5_2	2	1, 2, 4
7_{24}	0	4_1	5_2	2	1, 2, 4
7_{25}	2x0	7_1		3	1
7_{26}	0	3_1	7_1	3	1, 2
7_{27}	0	5_1	7_1	3	1, 2
7_{28}	2x0	7_2		1	
7_{29}	2x0	7_2		1	
7_{30}	2x0	7_2		1	
7_{31}	0	3_1	7_2	1	2
7_{32}	0	5_2	7_2	1	2
7_{33}	2x0	7_3		2	1
7_{34}	2x0	7_3		2	1
7_{35}	0	3_1	7_3	2	1, 2
7_{36}	0	5_1	7_3	2	1, 2
7_{37}	0	5_2	7_3	2	1, 2
7_{38}	2x0	7_4		2	1
7_{39}	2x0	7_4		2	1
7_{40}	0	3_1	7_4	2	1, 2
7_{41}	0	3_1	7_4	2	1, 2
7_{42}	0	5_2	7_4	2	1, 2
7_{43}	2x0	7_5		2	1
7_{44}	2x0	7_5		2	1
7_{45}	0	3_1	7_5	2	1, 2
7_{46}	0	3_1	7_5	2	1, 2
7_{47}	0	3_1	7_5	2	1, 2
7_{48}	0	5_1	7_5	2	1, 2
7_{49}	0	5_2	7_5	2	1, 2
7_{50}	2x0	7_6		1	5
7_{51}	2x0	7_6		1	5
7_{52}	2x0	7_6		1	5
7_{53}	2x0	7_6		1	5
7_{54}	2x0	7_6		1	5
7_{55}	0	3_1	7_6	2	1, 2, 5
7_{56}	0	3_1	7_6	1	2, 5
7_{57}	0	4_1	7_6	2	1, 2, 5
7_{58}	0	5_2	7_6	2	1, 2, 5
7_{59}	2x0	7_7		1	5
7_{60}	2x0	7_7		1	5
7_{61}	2x0	7_7		1	5
7_{62}	2x0	7_7		1	5
7_{63}	2x0	7_7		1	5
7_{64}	0	3_1	7_7	2	1, 2, 5
7_{65}	0	4_1	7_7	1	2, 5

TABLE 1. Theta curves, their constituent knots, their unknotting numbers, and the subsection of section 4 which rules it out as a potential replication intermediate.

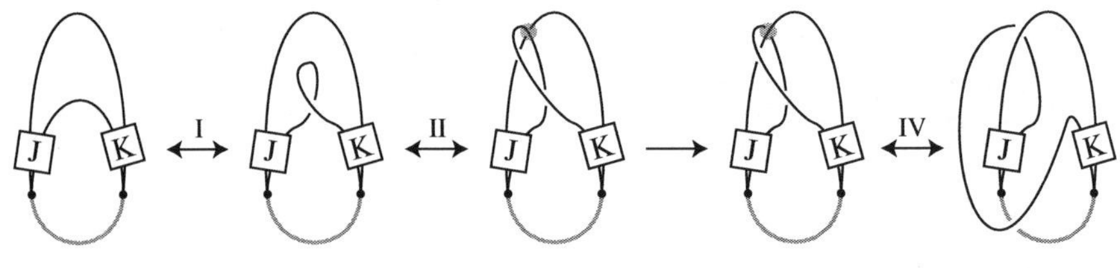

FIGURE 8. This shows how if a loop is first formed in one edge, followed by a crossing change, it can result in a knotted θ-curve where the minimal diagram involves crossings with the edge that was not in the crossing change. The boxes with J and K indicate a pattern that the two edges are making. Reidemeister moves are indicated with double arrows, and the crossing change is indicated with a single arrow at the highlighted crossing. The edge that is not involved in the crossing change is thickened and shown in blue (gray).

This is important when considering mathematically modelling the replication intermediates, as we explain in the following. In Figure 8, we see that if a loop is introduced in one of the sister duplexes, and is followed by a DNA-DNA passage, then it could be that a Reidemeister IV move is needed to move to the minimal diagram. This move introduces a crossing between one of the sister duplexes and the parental duplex.

4.4. Ruling out θ-curves with parent duplex interwoven with the sister duplexes. During replication the intact supercoiled parent duplex is sequestered from the sister duplexes [**PPC99**]. As shown in Figure 1, both the parental DNA (ahead of the replication fork) and sister duplexes (behind the replication fork) exhibit supercoiling (the latter in the form of precatenanes). Experiments have shown that this supercoiled DNA is typically close together, which would preclude the invasion of another duplex. (For a more extensive discussion of this, see [**BF07**].) It is thus reasonable to assume that neither the supercoiled parental segment nor interwound sister segments could pierce through the precatenanes or supercoils, respectively. In particular, although there may be thermal fluctuations, there will not be a DNA-DNA passage between the parental duplex and a sister duplex. This together with the absence of additional topology like a loop (see Section 4.3) in a sister duplex, means that crossing between the parental duplex and the sister duplex will not be introduced. See Figure 8.

This eliminates any θ-curve that has higher crossing number than its constituents. So 16 more of the prime θ-curves are eliminated. This also eliminates the remaining non-prime θ-curves which are each a order-3 vertex connect sum of a prime θ-curve with a non-trivial θ-curve with all trivial constituents.

4.5. Ruling out particular freshly replicated knots. Several experimental works have discussed that topoisomerase-mediated DNA-DNA passages introduce either two negative crossings or two positive crossings [**WC91, SSMR$^+$99**]. Of course if additional crossings (via twisting or other more complicated geometric formations) are introduced into the DNA before the passage occurs more crossings can be trapped. Trapped crossings will both be non-transient and contribute to the underlying knotted structure. For example if a loop is formed, followed by an

DNA-DNA passage then three crossings are introduced. The crossings could be: three positive, two positive and one negative, two negative and one positive, or three negative.

As discussed in Section 4.4 there is a formation of precatenanes between the two daughter duplexes. Consider now the freshly replicated knot composed of the two daughter duplexes within the replication bubble: if we orient the daughter strands consistently in an unknotted replication intermediate the crossings between them will be positive. (See Figure 7.) So the only way to obtain a negative crossing in a diagram of the freshly replicated knot is via the crossing change or the introduction of a loop. In Section 4.3 we ruled out the introduction of a loop that could result in trapping an additional negative crossing. Thus in the resulting diagram there will be at most two negative crossings.

Starting from a minimal diagram of a knot, using Reidemeister moves to change to a different diagram, cannot decrease the number of positive crossing and similarly cannot decrease the number of negative crossings. Thus the freshly replicated knots cannot be knots with more than two negative crossings in their minimal diagram. This eliminates knots 6_3, 7_7 and their reflections as possible freshly replicated knots because they each have three or more negative crossings. This in turn eliminates an additional 7 of the prime θ-curves.

With the exception of 4_1 all other knots have at least three positive crossing or at least three negative crossings. So either knot or its reflection will not be a possible freshly replicated knot. We can thus eliminate each of the remaining θ-curves or its reflection, whichever has three or more negative crossings in its non-trivial constituent.

This also means that the negative crossings must be the unknotting crossing.

Conjecture: The unknotting number one knots $\overline{6_2}$ and $\overline{7_6}$ cannot be unknotted with a crossing change at a negative crossing.
Assuming the Conjecture, we remove $\overline{6_2}$ and $\overline{7_6}$ as freshly replicated knots. This removes another 8 of the prime θ-curves.

4.6. Candidates for knotted Replication Intermediates. In summary, we have employed the following biological restrictions.

Biological Restrictions:

- The knotted replication intermediates arise from a single round of topoisomerase action.
- The unknotted daughter plasmids result from unknotted sister duplexes.
- There are no persistent loops in the sister duplexes.
- In the replication intermediates the sister duplexes wind around each other in a left-handed fashion.
- Freshly replicated knots have at most two negative crossings.
- Neither the supercoiled parental segment nor interwound sister segments pierce through the precatenanes or supercoils, respectively.

The arguments (and Conjecture) in the preceding five sections together leave 10 prime θ-curves as candidates for knotted replication intermediates (see Figure 9).

THEOREM 4.1. *Based on mathematically modelling biological restrictions, the only knotted DNA replication intermediates that could occur due to a single DNA-DNA passage are:* $\mathbf{3_1}, \mathbf{4_1}, \overline{\mathbf{5_5}}, \overline{\mathbf{5_6}}, \overline{\mathbf{6_5}}, \overline{\mathbf{6_6}}, \overline{\mathbf{6_7}}, \mathbf{7_{28}}, \mathbf{7_{29}}$ *and* $\mathbf{7_{30}}$. *See Figure* 9.

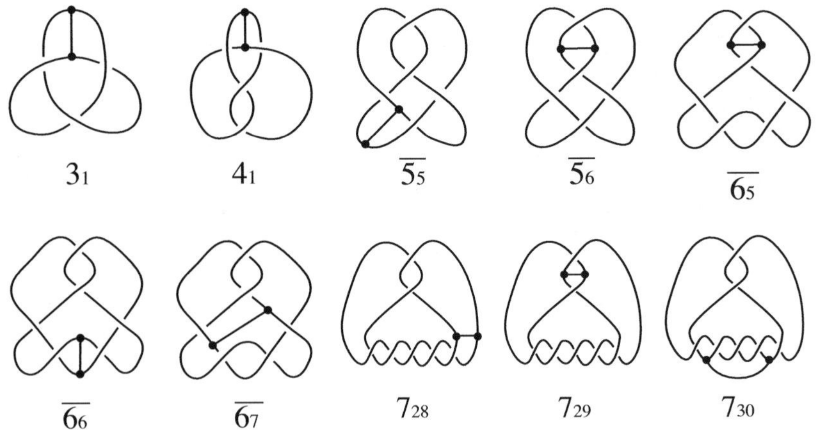

FIGURE 9. The θ-curves that are candidates for knotted replication intermediates.

Breaking down the biological restrictions piece by piece we see that the freshly replicated knots will in fact always be a twist knot (for seven crossings or less). Notice for all of the remaining θ-curves that the non-trivial constituent is a right-handed twist knot.

We will conclude by contexualizing our results within the experimental results. The current model of a single DNA-DNA passage predicts the θ-curves $\mathbf{3_1}, \mathbf{4_1}, \overline{\mathbf{5_6}}, \overline{\mathbf{6_5}}$ and $\mathbf{7_{29}}$ [**OSB18**]. The θ-curve $\mathbf{3_1}$ has been experimentally visualized as a replication intermediate [**LMRH⁺12**]. Freshly replicated knots that have been observed visually are the right-hand knots $\overline{3_1}$, 4_1, $\overline{5_6}$, and $\overline{6_2}$ [**SSMR⁺99**]. The knot $\overline{3_1}$ is the non-trivial constituent of $\mathbf{3_1}$, 4_1 is the non-trivial constituent of $\mathbf{4_1}$, and $\overline{5_2}$ is the non trivial constituent of $\overline{\mathbf{5_5}}$ and $\overline{\mathbf{5_6}}$. The freshly replicated knot $\overline{6_2}$ is believed to be the result of two DNA-DNA passages. In the θ-curves $\overline{\mathbf{6_5}}, \overline{\mathbf{6_6}}$ and $\overline{\mathbf{6_7}}$ the non-trivial constituent is the knot $\overline{6_1}$. In the curves $\mathbf{7_{28}}, \mathbf{7_{29}}$ and $\mathbf{7_{30}}$ the non-trivial constituent is $\overline{7_2}$. We predict with large enough circular DNA, the freshly replicated knots $\overline{6_1}$ and $\overline{7_2}$ will be formed.

FIGURE 10. Theta curves with their unknotting crossing changes shown in purple (gray).

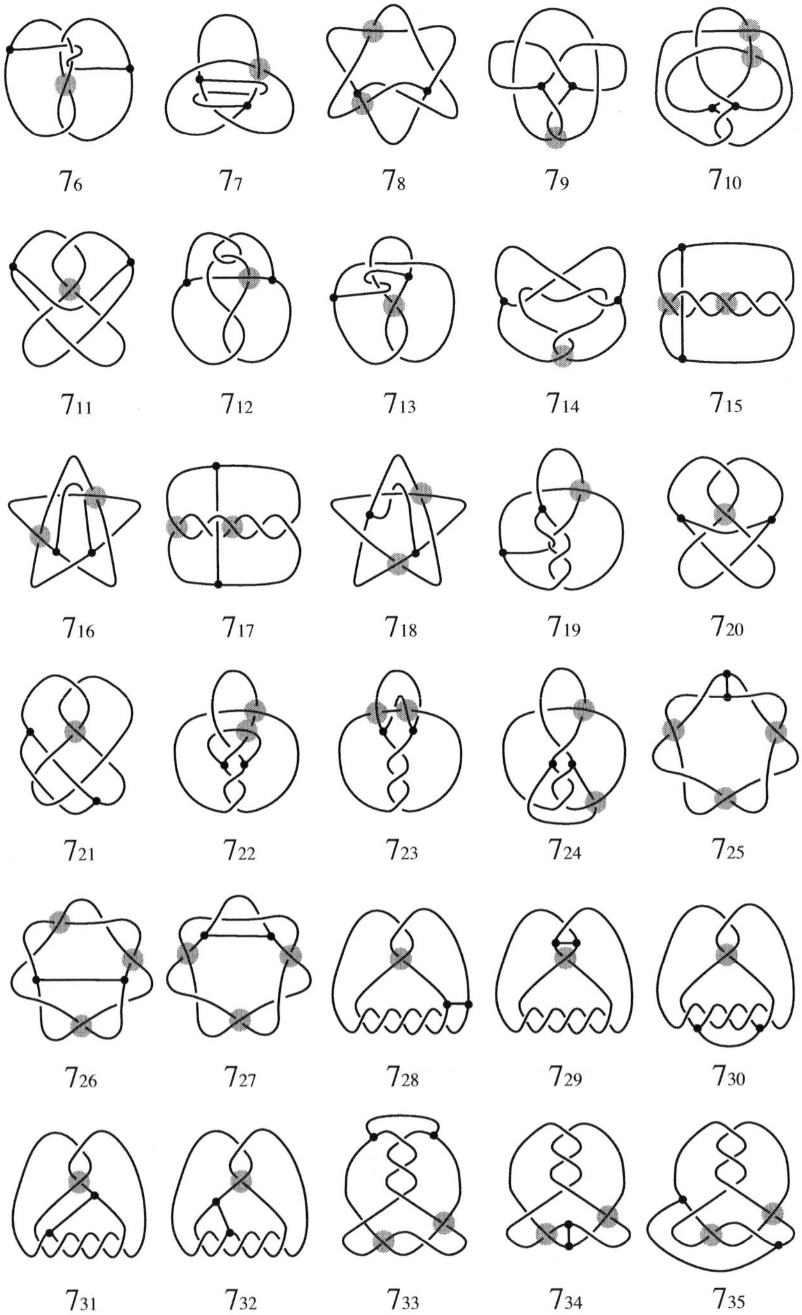

FIGURE 10. CONTINUED. Theta curves with their unknotting crossing changes shown in purple (gray).

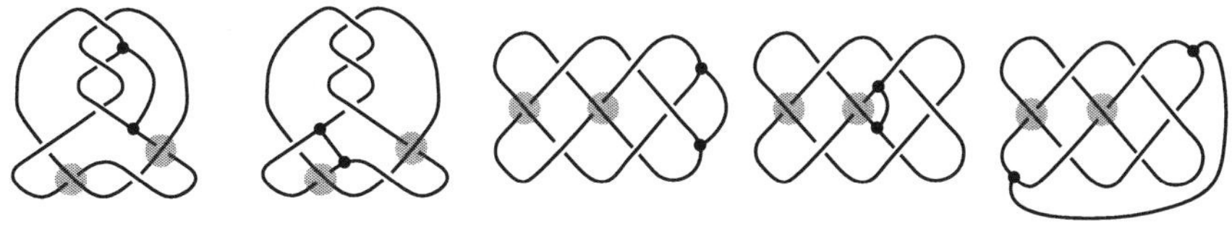

7_{36} 7_{37} 7_{38} 7_{39} 7_{40}

7_{41} 7_{42} 7_{43} 7_{44} 7_{45}

7_{46} 7_{47} 7_{48} 7_{49} 7_{50}

7_{51} 7_{52} 7_{53} 7_{54} 7_{55}

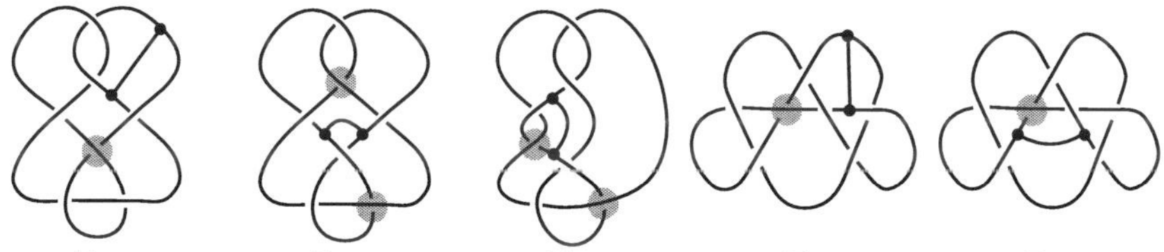

7_{56} 7_{57} 7_{58} 7_{59} 7_{60}

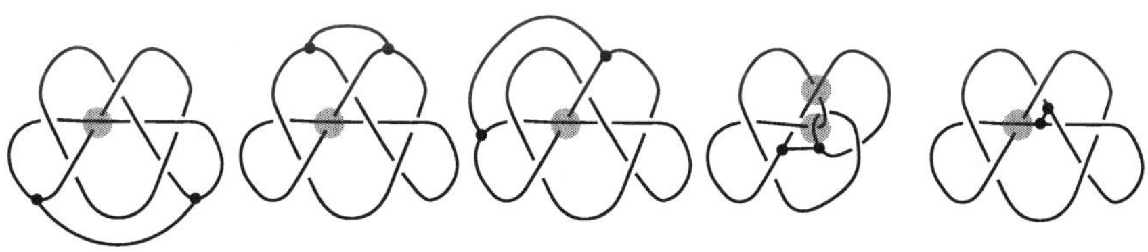

7_{61} 7_{62} 7_{63} 7_{64} 7_{65}

FIGURE 10. CONTINUED. Theta curves with their unknotting crossing changes shown in purple (gray).

Acknowledgments. The authors would like to thank Bernardo Schvartzman and Andrzej Stasiak for interesting and useful conversations. We also thank the Isaac Newton Institute of Mathematical Sciences for hosting us during our early conversations.

References

[BF07] D. Buck and E. Flapan, *Predicting knot or catenane type of site-specific recombination products*, J. Mol. Biol. **374** (2007), no. 5, 1186–1199.

[BO17] D. Buck and D. O'Donnol, *Unknotting numbers for θ-curves, with an appendix by K. Baker*, ArXiv:1710.05237 [math.GT] (2017).

[CCM+15] J. Cebrian, A. Castan, V. Martinez, M. J. Kadomatsu-Hermosa, C. Parra, M. J. Fernandez-Nestosa, C. Schaerer, P. Hernandez, D. B. Krimer, and J. B. Schvartzman, *Direct Evidence for the Formation of Precatenanes during DNA Replication*, J. Biol. Chem. **290** (2015), no. 22, 13725–13735.

[HT] J. Hoste and M. B. Thistlethwaite, *Knotscape*.

[Kau89] Louis H. Kauffman, *Invariants of graphs in three-space*, Trans. Amer. Math. Soc. **311** (1989), no. 2, 697–710, DOI 10.2307/2001147. MR946218

[LMRH+12] V. Lopez, M. L. Martinez-Robles, P. Hernandez, D. B. Krimer, and J. B. Schvartzman, *Topo IV is the topoisomerase that knots and unknots sister duplexes during DNA replication*, Nucleic Acids Res. **40** (2012), no. 8, 3563–3573.

[Mor09] Hiromasa Moriuchi, *An enumeration of theta-curves with up to seven crossings*, J. Knot Theory Ramifications **18** (2009), no. 2, 167–197, DOI 10.1142/S0218216509006884. MR2507922

[OSB18] D. O'Donnol, A. Stasiak, and D. Buck, *Two convergent pathways of DNA knotting in replicating DNA molecules as revealed by theta-curve analysis*, Nucleic Acids Res. **46** (2018), no. 17, 9181–9188.

[PPC99] L. Postow, B. J. Peter, and N. R. Cozzarelli, *Knot what we thought before: the twisted story of replication*, Bioessays **21** (1999), no. 10, 805–808.

[Rol90] Dale Rolfsen, *Knots and links*, Mathematics Lecture Series, vol. 7, Publish or Perish, Inc., Houston, TX, 1990. Corrected reprint of the 1976 original. MR1277811

[SSMR+99] J. M. Sogo, A. Stasiak, M. L. Martinez-Robles, D. B. Krimer, P. Hernandez, and J. B. Schvartzman, *Formation of knots in partially replicated DNA molecules*, J. Mol. Biol. **286** (1999), no. 3, 637–643.

[WC91] S. A. Wasserman and N. R. Cozzarelli, *Supercoiled DNA-directed knotting by T4 topoisomerase*, J. Biol. Chem. **266** (1991), no. 30, 20567–20573.

[Wol87] Keith Wolcott, *The knotting of theta curves and other graphs in* S^3, Geometry and topology (Athens, Ga., 1985), Lecture Notes in Pure and Appl. Math., vol. 105, Dekker, New York, 1987, pp. 325–346. MR873302

Department of Mathematics, University of Bath, Claverton Down, Bath, England BA2 7AY

Email address: d.buck@bath.ac.uk

Department of Mathematics, Marymount University, 2807 N Glebe Rd, Arlington, VA 22207

Email address: dodonnol@marymount.edu

Contemporary Mathematics
Volume **746**, 2020
https://doi.org/10.1090/conm/746/15004

Recent advances on the non-coherent band surgery model for site-specific recombination

Allison H. Moore and Mariel Vazquez

ABSTRACT. Site-specific recombination is an enzymatic process where two sites of precise sequence and orientation along a circle come together, are cleaved, and the ends are recombined. Site-specific recombination on a knotted substrate produces another knot or a two-component link depending on the relative orientation of the sites prior to recombination. Mathematically, site-specific recombination is modeled as coherent (knot to link) or non-coherent (knot to knot) banding. We here survey recent developments in the study of non-coherent bandings on knots and discuss biological implications.

1. Introduction

Local events where two linear segments along a curve in three-dimensional space come together and undergo cleavage followed by reconnection have been observed in a surprising variety of natural settings and at widely different scales. Examples range from microscopic site-specific recombination on circular DNA molecules, to reconnection of fluid vortices and magnetic reconnection in solar coronal loops [**52**, **58**, **85**].

Enzymes in the family of site-specific recombinases mediate local reconnection on DNA molecules in order to address a variety of biological problems. We refer to the DNA molecule as a polymer chain, or simply a *chain*. Because local reconnection reactions introduce two breaks and rearrange the strands of the chains involved, when acting on circular chains they can induce topological changes. In particular, they can resolve problematic knotting and linking that threatens the health of the cell. Experimental data show striking similarities between the pathways of topology simplification of newly replicated circular DNA plasmids by recombination [**85**, **87**] and those of interlinked fluid vortices [**53**–**55**], thus pointing to a universal process of unlinking by local reconnection. This process has been the focus of numerous experimental and theoretical studies, e.g. [**11**, **12**, **33**, **42**, **85**, **87**].

Here, we are interested in *site-specific recombination* on circular DNA molecules, where site-specific recombinases perform local reconnection on two short DNA segments with identical nucleotide sequences, the *recombination sites* (reviewed in Section 2). We model knotted and linked DNA molecules as topological knots and

2010 *Mathematics Subject Classification.* Primary 92E10, 57M25, 57M27.

AM and MV were partially supported by DMS-1716987. MV was also partially supported by CAREER Grant DMS-1519375 and DMS-1817156.

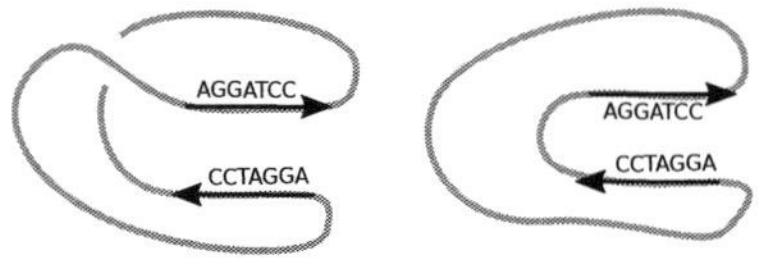

FIGURE 1. Each figure illustrates a circular DNA molecule with two recombination sites. The DNA is modeled as the curve drawn by the axis of the double helix (see Section 2). The two sites have the same nucleotide sequence, AGGATCC. Since the sequence is non-palindromic, one can assign an orientation to it. (Left) A pair of sites in inverted repeats, i.e. the sites induce opposite orientations into the circle (head-to-head). (Right) A pair of sites in direct repeats (head-to-tail).

links in the three-sphere, and the recombination sites as two oriented segments on the knot or link (one in each component). Two recombination sites on a DNA knot are in direct repeat or in inverted repeat depending on whether their DNA sequences induce the same or opposite orientations into the knot, respectively (Figure 1).

Site-specific recombination on a knot is modeled by *coherent* band surgery if the recombination sites are in direct repeats, and as *non-coherent* band surgery if the sites are in inverted repeats. Non-coherent band surgery on a knot yields another knot, while coherent band surgery on a knot yields a link of two components. Band surgery along knots and links is defined in Section 3. Both operations have biological relevance. For example in *Escherichia coli*, site-specific recombination is used to monomerize dimers that arise as products of homologous recombination on damaged replication links (reviewed in [**57**]). Mathematically, in its simplest form, this is a coherent band surgery taking an unknot to an unlink. However, monomerizing DNA dimers can also refer to splitting a knot of length l into a two-component link where each component is of length $\frac{l}{2}$. Coherent banding is also used by enzymes to integrate viral DNA into the genomes of their hosts, and by transposons, in a two-step reaction, to move pieces of DNA from one region of the genome to another. Non-coherent banding is observed in nature in processes that require inversion of a DNA segment. For example, recombinases Gin of bacteriophage Mu and Hin of *Salmonella* recognize two specific sites in inverted repeats along a DNA molecule, and invert the segment of DNA between the sites. The action of these enzymes is reviewed in section 2.3.

While the coherent band surgery model for site-specific recombination has received much attention in the literature, (e.g. [**11, 12, 18, 44, 85**]), non-coherent band surgery is less-understood and is intrinsically difficult from an analytical perspective. This is due to the implicit existence of a non-orientable surface induced by the banding. In Section 3 we give an overview of non-coherent band surgery and mention several topological criteria for the existence of a banding relating a pair of knots.

Our own approach to the study of non-coherent band surgery and its applications to site-specific recombination is two-fold: analytical, described in Section 4; and numerical, described in Section 5. Analytically, a band surgery operation relating a pair of knots is understood as a certain tangle replacement which may be "lifted" to rational Dehn surgery in the branched double covers of the knots involved. The theory of Dehn surgery and knot invariants, together with some heavy machinery from the realm of Floer homology can then be used to study such operations. The goal in this approach is to produce theoretical results that can be used to obstruct, classify, or characterize band surgery operations. We refer the reader to Section 4.

The second facet of our approach, described in Section 5, is computational. A band surgery operation relating a pair of knots can be realized as an operation on knotted polygonal chains embedded in three-dimensional space. Here we use BFACF, a well-known Markov Chain Monte Carlo algorithm, to uniformly sample random conformations of knotted self-avoiding polygons (SAPs) in the simple cubic lattice (reviewed in [**65**]). Using *Recombo*, our own software package, we algorithmically identify binding sites, perform recombination and identify the resulting product knots [**87**]. A robust statistical analysis of the resulting transition probability networks gives us insight into the quantitative behavior of topological simplification via recombination.

In sum, we here focus on the non-coherent band surgery model for site-specific recombination on circular DNA molecules at sites in inverted repeat. In Section 2, we provide biological background, review site-specific recombination on circular DNA, and provide examples of these events in the biological literature. In Section 3, we define band surgery, and survey the current state of non-coherent band surgery results in low dimensional topology. In Section 4 we discuss recent progress by the authors in non-coherent band surgery resulting from related work in three-manifolds. In Section 5, we give an overview of our recent numerical studies of non-coherent band surgery along knots in the simple cubic lattice.

2. DNA recombination

2.1. DNA, damage, and recombination. The nucleic acids DNA and RNA are molecules that carry the genetic code of every organism. DNA is composed of two sugar-phosphate backbones flanked by nitrogenous bases A, G, T and C (adenine, guanine, thymine and cytosine). The backbones are held together by hydrogen bonds between each pair of complementary bases (A with T; G with C). The two backbones wrap around each other, forming the familiar right-handed double-helix. Words in the nucleotide sequence determine the genetic code. The genetic code is unique to each individual and can be used to differentiate a bacterium from a human, or two humans from each other. Double-stranded DNA molecules may be linear or circular. Most bacterial organisms have circular DNA. Length is measured by the number of base pairs (bp) of nucleotides, and varies widely across different organisms. For example, the chromosome of the bacterium *Escherichia coli* is circular, with length ranging from 4.5 to 5.5 million bp. The human genome is comprised of 23 chromosomal pairs of linear DNA molecules, and is approximately 3 billion bp long. Circular DNA molecules may trap interesting geometrical and topological information. For example, links naturally arise as products of DNA replication on circular DNA molecules. Replication is the process by which a cell copies its genetic code in preparation for cell division. Interlinking of chromosomes prevents the newly replicated DNA molecules from properly segregating at cell division. Enzymes that mediate local changes can correct for topological problems; local changes include strand-passage and reconnection, and have the potential to change the global topological type of a knotted or linked DNA chain. The action of these enzymes often leads to topological simplification (e.g. a reduction in minimal crossing number) of the substrate DNA knots and links [**33**, **57**, **81**, **99**].

When exposed to sparsely ionizing radiation, or to some chemotoxic agents such as chemotherapy drugs, the DNA in cells is subjected to double-stranded breaks (DSBs). If not repaired properly, the DNA damage may result in misrejoinings that lead to large-scale genome rearrangements (reviewed in [**40**]). Depending on the extent of the damage and on the phase of the cell cycle when it happens, DSBs are repaired by non-homologous end-joining (NHEJ) or by homologous recombination (HR). The two repair pathways have widely different levels of accuracy, HR being the most faithful. However, in both cases the process of DNA damage and repair starts with two DNA sites in close proximity that undergo cleavage. The loose ends are ideally put back together without any DNA loss or rearrangement. Occasionally the ends are resealed with the wrong partner thus leading to a chromosomal rearrangement. If the strands of DNA are swapped during HR, the resulting local reconnection event is referred to as a *cross-over*. Chromosomal rearrangements induced by DNA damage may lead to chromosomal instability and cell death. Chromosomal rearrangements, in particular DNA inversions, are also considered as drivers of genome evolution [**94**, **101**].

Reconnection events mediated by site-specific recombinases are common in nature, and have been extensively studied from the knot theoretic perspective. During a site-specific recombination event the enzymes recognize and bind two recombination sites. The nucleotide sequence along the two sites is identical. The two sites may be found along one DNA molecule or two distinct molecules. Site-specific recombinases act by cleaving both segments, and then recombining and joining the ends. Because the sites targeted by site-specific recombinases are short (typically 5-50 bp) and the mechanism is highly specific, these enzymes are often used for genetic engineering. They also are ubiquitous in the natural environment, and are involved in the integration or excision of genetic material [**32**], dimer resolution [**86**], or the inversion of genetic material to alter gene expression [**37**].

A recombination site is usually a short, non-palindromic word composed of letters A, T, C, G corresponding to its nucleotide sequence. Hence a given site can be assigned an unambiguous orientation (e.g., think of a 'word' as being read left-to-right). If two sites appear along a single circular chain, they will induce either the same or different orientations along the chain. In the first case they are said to be in *direct repeats*. In the second case, they are said to be in *inverted*

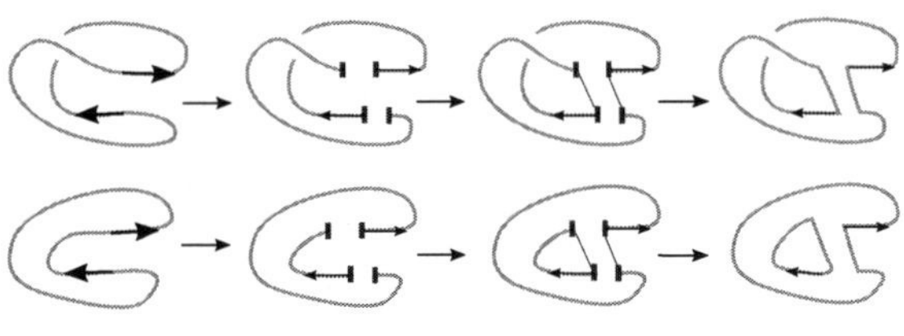

FIGURE 2. (Top) The product of site-specific recombination at inversely repeated sites along a knot is a knot. (Bottom) The product of site-specific recombination at directly repeated sites along a knot is a two-component link.

repeats. Figure 1 shows a schematic of two sites in inverted repeats and direct repeats, respectively, along a single circular molecule.

We make the following simple combinatorial observation. Given a single component knot containing two sites in direct repeats, recombination will yield a two-component link, perhaps non-trivially linked. Given a single component knot containing two inversely repeated sites, the product will yield another single component knot, perhaps of a different knot type. See Figure 2. Conversely, given a two-component link with one site on each component, the product is necessarily a knot, now containing two directly repeated sites. These distinctions will become quite important when we discuss the different band surgery models in Section 3. In the biological literature, when two distinct molecules are fused into one molecule in the recombination process, the product is called a *dimer*. Likewise, in biology two circular molecules linked together are called *catenanes*.

2.2. The relevance of $T(2,n)$ torus knots and links in biology. Certain classes of knots and links are especially relevant in the biological context for mechanistic reasons, for example two-bridge links and torus links. Recall that the torus link $T(2,n)$ has two components when n is even and one component (i.e. is a knot) when n is odd. Consider the two interlinked backbones of a DNA double-helix. If the DNA is circular, the two backbones naturally form a $T(2,2k)$ right-handed torus link, where k corresponds to the number of turns of the helix. See Figure 3 for an example. During DNA replication, enzymes called helicases unwind and 'unzip' the double-helix at the replication fork. The replication process yields two interlinked daughter chromosomes. Each of these daughter chromosomes is itself a circular molecule. It has been experimentally confirmed that the respective axes of the two interlinked daughter molecules naturally realize the two components of a $T(2,2k)$ torus link [**2**]. Notice that the link type is a direct consequence of the right-handed double-helical structure of DNA. To ensure the survival of the cell line, these two components must become unlinked. DNA unlinking mediated by enzymes is reviewed in the next subsection

Knotting in long circular DNA molecules may arise as a result of enzymatic reactions, or as products of a random knotting process. The Frisch-Wasserman-Delbrück Conjecture [**27**,**64**] asserts that in long polymer chains knotting will occur

FIGURE 3. The image on the left shows an unoriented $T(2,21)$ torus knot and the image on the right shows an oriented $T(2,6) = 2^2_6$ torus link. The two sites indicated by arrows are said to be in parallel alignment. This link illustrates one possible product of DNA replication on an unknotted substrate. In the context of DNA replication, $T(2,2k)$ torus links are sometimes called $2k$-catenanes, e.g. see [**33**].

with near certainty. The conjecture is known to hold for various polymer models [**19**, **20**, **88**] and has been experimentally verified on randomly circularized DNA chains [**4**, **62**, **80**, **84**]. Furthermore, when polymer chains are subjected to confinement, for example in a cell nucleus or viral capsid, the knotting probability and knot complexity appears to increase (e.g. [**4**]). The most probable knot observed in random knotting processes is the trefoil knot $\pm T(2,3)$ [**80**, **84**].

2.3. Enzymes that change the topology of DNA. This is a good time to interrupt the exposition in order to clarify some biological naming conventions for the mathematical reader. General classes of enzymes are usually suffixed by '-ase' (e.g. helicase, topoisomerase, recombinase). The name of a particular enzyme is capitalized (e.g. Gin, Hin, Xer, or λ-Int), and the name of a recombination site is indicated in italics (e.g. *att, dif, psi, res*).

In E. coli, topological simplification of replication links is mediated by type II topoisomerases. These are enzymes that cut one segment of double-stranded DNA, pass another segment of double-stranded DNA through the break, and then reseal the broken ends. Strand-passage mediated by type II topoisomerases is typically modeled by crossing changes on curves, linear or circular, representing the DNA. Type II topoisomerases are ubiquitous and essential to life. They are believed to be the predominant decatenases in the cell (i.e. enzymes in charge of unlinking interlinked DNA molecules) [**99**].

Site-specific recombinases can also mediate topological simplification. In a series of *in vitro* experiments Ip *et al.* demonstrated that the recombinases XerC/ XerD, acting at *dif*-sites, and coupled with an auxiliary protein FtsK, were capable of resolving linked plasmid substrates of type $T(2,2k)$ that had been linked by λ-Int, a site-specific recombinase from bacteriophage λ [**42**]. The same research group later showed that the XerCD-*dif*-FtsK enzymatic complex could unlink replication links tied *in vivo* in E. coli cells deficient in the decatenase topoIV [**33**]. Several groups of authors, including our own, studied mathematically the mechanism and pathways of DNA unlinking by site-specific recombination [**11**, **12**, **85**, **87**]. In [**85**] we showed that at least $2k$ steps are needed to unlink a replication link of type $T(2,2k)$. Each component of the replication link contains one *dif* recombination site, in the orientation indicated in Figure 3. All the intermediate knots and links are also in the family of $T(2,2k+1)$ and $T(2,2k)$ knots and links. Under the assumption that each step strictly reduces the minimal crossing number of the substrate, we concluded that there is a unique shortest pathway of unlinking a $T(2,2k)$ replication link. This minimal pathway corresponds to the one proposed by experimentalists [**33**, **42**]. In [**87**] we extended the mathematical model of [**85**] and also analyzed recombination pathways numerically. We identified minimal pathways of unlinking DNA replication links under a variety of biologically-motivated assumptions and used the transition probabilities obtained numerically to identify a most probable minimal pathway of DNA unlinking of replication links. In [**11**], the authors characterized coherent band pathways between knots and two-component links of small crossing number, and in [**12**], they characterized band surgeries between fibered links in terms of arcs on the fiber surfaces of such links.

The above-mentioned mathematical studies [**11**, **12**, **85**, **87**] were all concerned with site-specific recombination modeled as coherent band surgery, and point to the $T(2,n)$ torus knots and links as having heightened importance in pathways of unlinking of replication links. In the biological literature there are numerous other studies of site-specific recombinases acting at two sites in direct repeats along a single circle, or at two sites on separate circles. Several instances specific to reactions involving the trefoil knot and other $T(2,n)$ torus knots and links are mentioned in [**61**, Section 5]. Here, we review several more specific examples of site-specific recombination on DNA knots at sites in inverted repeats. Recall that in this case, the product of recombination is also a single component knot.

An *inversion* is a chromosomal rearrangement in which a segment of a DNA molecule is reversed, so that the corresponding portion of the nucleotide sequence is read backwards. In addition to potential topological changes (i.e. a change in the isotopy type of a DNA knot), site-specific recombination at sites in inverted repeat will induce an inversion. In some organisms, inversions are potentially harmful genetic abnormalities. However, in some cases, inversions may present as advantageous mutations. They are also considered to be the most common rearrangement in evolution.

Gin and Hin are examples of enzymes in the family of serine site-specific recombinases that mediate the inversion of a DNA segment by acting at sites in inverted repeats. Gin is a site-specific recombinase from bacteriophage Mu. Bacteriophages are viruses that infect bacteria. Bacteriophage Mu infects a wide range of bacterial strains. After finding its target cell, a bacteriophage injects its DNA into the host, and the phage DNA becomes circular. Gin recombination inverts the G-segment, a 3kb portion of the viral genome flanked by the *gix* recombination sites. As a result of G-segment inversion, a gene encoding for a new set of tail fibers is expressed. Tail fibers are used by bacteriophages to recognize their host. In this way Gin site specific recombination determines the family of bacteria that will be infected by the new generation of bacteriophages. Gin is known to act processively, meaning it performs recombination several times successively. During the first round of recombination, it converts an unknotted DNA circle with sites in inverted repeat into another unknot, now with an inverted G-segment. In the second round, the unknot is converted into a trefoil, and the G-segment is flipped back to its original orientation. As the reaction proceeds, the family of twist knots is generated $(3_1, 4_1, 5_2, 6_1, 7_2)$ [**46**]. From a topological perspective, the second step is a non-coherent banding relating an unknot to a trefoil knot, and a third step is a non-coherent banding relating the trefoil with the figure 8 knot (4_1). In Section 3 we discuss this specific instance of non-coherent band surgery.

Another notable example is the Hin site-specific recombination system from *Salmonella*. *Salmonella* sp. are bacteria in the family Enterobacteriaceae. The cells are rod-shaped with lengths $2 - 5\mu m$. A few flagella used for locomotion are arranged on the cell surface. Salmonella cells express two different flagellar proteins in a process called flagellar phase variation. Expression of one protein over the other results in a change of phenotype. Diphasic agglutination was first observed in Salmonella in 1922 by F.W. Andrewes [**3**]. Half a century later it was shown that the observed phenotypic changes were due to an inversion of a genomic region affecting flagellin genes. The site-specific recombinase Hin inverts a 1kb DNA segment in the *Salmonella* sp. chromosome. The invertible segment includes a promoter region for one flagellin gene and a repressor for another flagellin gene. Inversion by Hin selectively determines the expression of one gene and the repression of the other. In [**66**] experiments were conducted *in vivo* and *in vitro* to study the topological mechanism of action of Hin recombinase acting at *hix* sites.

Like Gin, when acting on unknotted DNA circles with two sites in inverted repeats, Hin recombination produces twist knots. Experiments pointed to a mechanism analogous to that of Gin. See [**38**] for a review of bacterial phase variation and [**46**] for a review of inversion by serine recombinases. Mathematical models of Gin and Hin recombination have been reported in [**13, 15, 16, 91, 92**].

3. The band-surgery model for site-specific recombination

3.1. Knot theory background. A *knot* K is an embedding of the circle into $\mathbb{R}^3$ or its compactification, the three-sphere S^3. A *link* L is a disjoint union of knots, possibly linked together. A knot is a link with one component. Two links are considered to be equivalent if one can be continuously deformed into the other via an ambient isotopy of the surrounding space. We will further assume that all links are tame, meaning each component can be represented by piecewise linear curves with finitely many segments. It is common to illustrate links with link diagrams, which are planar projections containing the data of over-crossings and under-crossings. A link which is isotopic to its mirror image is called an *achiral* link, otherwise it is *chiral.* Chirality and orientation are important in biology. When distinguishing between different symmetries of links we follow the nomenclature proposed in [**10, 95**]. Because we will mainly deal with non-coherent band surgery, unless otherwise noted, here we take knots and links as unoriented objects.

3.2. Band surgery. Site-specific recombination on circular DNA will be modeled by a topological operation on knots and links which we now describe. Let L denote a link in S^3. A band b is an embedding of the unit square, $b : I \times I \hookrightarrow S^3$, such that two opposing edges lie on L, i.e. $L \cap b(I \times I) = b(I \times \partial I)$. We say that two links L and L' are related by a *band surgery* if

$$L' = (L - b(I \times \partial I)) \cup b(\partial I \times I).$$

In other words, given a link L and a band b, we replace the two opposing edges of the band on L with the remaining two edges to obtain a new link L'.

If L and L' are oriented links, and the orientation of $L - b(I \times \partial I)$ agrees with the specified orientations on L and L', then the band surgery is called *coherent* (meaning it is coherent with respect to orientation). Otherwise, the band surgery is *non-coherent.* Elsewhere in the literature, coherent band surgery is sometimes called nullification [**25**], and non-coherent band surgery is called an H(2)-move, incoherent, or unoriented surgery [**1, 48, 49**]. A band surgery relating a pair of links induces a surface cobordism between them. A surface cobordism from L to L' is a surface F with $\partial F = L \sqcup L'$ properly embedded in $S^3 \times I$. Band surgery induces a cobordism by attaching a single 1-handle to the surface $L \times I$ in $S^3 \times I$. In the case of coherent band surgery, the surface is oriented and the restriction to the boundary respects the given orientations on L and L'. In the case of non-coherent band surgery, the addition of the band makes the surface non-orientable, as it contains an embedded Möbius strip.

Let us make a similar combinatorial observation to the one we made in the case of site-specific recombination. Suppose that a band surgery along a knot yields another single component knot. This implies that the surface cobordism

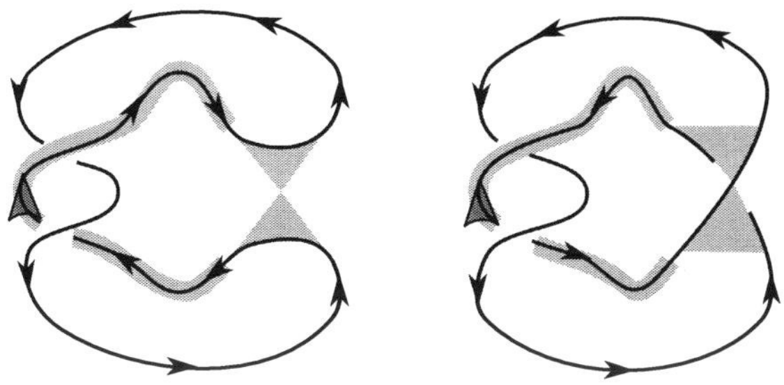

FIGURE 4. Notice that when a banding takes a knot to another knot, it is not possible to orient the knots in such a way that the orientations are consistent with the band surgery. Observe that the non-coherent band surgery is an inversion of the inner-most arc indicated in grey.

induced by the banding contains only two boundary components. In particular, it is non-orientable and therefore the knots cannot both be oriented in a way that is compatible with the banding. See Figure 4. Thus a band surgery which takes a knot to another knot is necessarily non-coherent. Similarly, a coherent band surgery operation along a knot will always produce a two-component link. The local action of site-specific recombinases is thought of as a simple reconnection. This is modeled mathematically as a band surgery. The appropriate model for site-specific recombination is coherent band surgery when the sites are directly repeated, and is non-coherent band surgery when the sites are inversely repeated.

3.3. Band surgery via tangle replacements. It is often convenient to view a band surgery between a pair of knots or links in terms of a standard diagrammatic operation. More precisely, after isotoping the links L and L', perhaps at the expense of complicating their embeddings, we may assume that a coherent or non-coherent band surgery is obtained from a particular two-string tangle replacement.

A tangle is a pair (B, t), where B is a three-ball and t is a properly embedded 1-submanifold with nonempty boundary, that is possibly disconnected and possibly contains closed components. In particular, (B, t) is a *two-string tangle* when t consists of a pair of properly embedded arcs and no closed components. Since S^3 is the union of two 3-balls, every link in S^3 can be written as the union of two complementary tangles. In particular, we may describe the links L and L' in S^3 that are related by band surgery via the following tangle decompositions

$$(S^3, L) = (B_o, t_o) \cup (B, t) \text{ and } (S^3, L') = (B_o, t_o) \cup (B, t'), \tag{3.1}$$

where $S^3 = B_o \cup B$ and the sphere $\partial B = \partial B_o$ intersects L and L' transversely in four points. Here, $t = (B \cap L)$, $t' = (B \cap L')$ are pairs of properly embedded arcs and $t_o = (B_o \cap L) = (B_o \cap L')$ is a 1-submanifold with at least two components. To define a band surgery operation, the tangle (B, t) is replaced with (B, t').

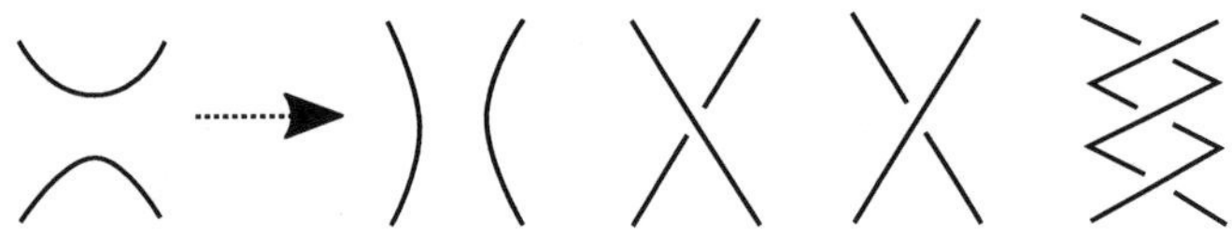

FIGURE 5. Some of the standard tangle replacements used to model reconnection. From left to right we see the following rational tangles: (0), (∞), $(+1)$, (-1), $(-1/3)$.

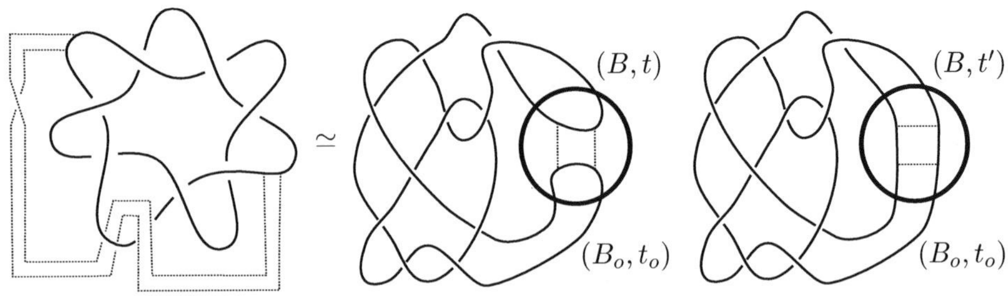

FIGURE 6. A non-coherent band surgery, indicated by the dashed band in the left image, relates the knots $7_1 = T(2,7)$ and 5_2^*, the mirror of 5_2. An isotopy that simplifies the embedding of the band takes the left diagram of 7_1 to the center diagram, also of 7_1. After isotopy, the outside tangle (B_o, t_o) becomes complicated. The outside tangle (B_o, t_o) is shared by both the center and right diagrams. The tangles (B,t) and (B,t') are inside the solid black circles. The right-most knot is 5_2^*.

Here a trivial two-string tangle is one that is homeomorphic to the tangle labeled by (0) in Figure 5. Trivial two-string tangles parameterized by the extended rational numbers are called rational tangles[1]. By convention, we may assume that (B,t) is the trivial two-string tangle labeled by (0) and that (B,t') is either the (∞) or $(\pm 1/n)$ tangle, as shown in Figure 5. This convention is based on the assumption that the arcs t in the tangle (B,t) represent only the cleavage region of each recombination site. Since the arcs are very short DNA segments, and since DNA is a stiff biopolymer, then (B,t) and (B,t') can be assumed to be very simple rational tangles, in particular one of the tangles shown in Figure 5 (see also [**85**,**90**]). The tangle (B_o,t_o) is fixed in the operation, and is shared by both of L and L'. Notice that subscript 'o' stands for 'outside' tangle, and this tangle may become arbitrarily complicated in the isotopy that simplifies the embedding of the band. In particular, (B_o,t_o) need not be rational or trivial. See Figure 6 for an example.

3.4. Dehn surgery and branched double covers. The perspective that band surgery is a two-string tangle replacement adheres to the now classical notion that the action of enzymatic complexes on circular DNA is appropriately modeled by tangle surgery, due to Ernst and Sumners in 1990 [**23**], reviewed in [**70**]. When topoisomerases or recombinases act locally, they essentially split the DNA molecule

[1]Here we follow the definition of trivial and rational tangles as in [**50**], where all rational tangles are considered trivial. Usually in the context of tangle calculus, only the rational tangles (0), (∞), and (± 1) are called trivial, e.g. [**23**,**92**].

into two complementary tangles, replacing one tangle with another. Viewing this operation as a topological tangle surgery opens the door to a broad range of tools in low-dimensional topology that can be used to understand these enzymatic actions. Many of the most successful topological approaches in the study of band surgery go by way of the double cover branched along the knots or links involved in the banding, starting with work of Murasugi [**69**] and Lickorish in the case of bandings relating knots [**59**]. Here, the branched double cover of the pair (S^3, L) denoted by $\Sigma_2(S^3, L)$, is the unique closed, connected, oriented 3-manifold associated with the representation of the knot group onto $\mathbb{Z}/2\mathbb{Z}$ (c.f. [**79**]). Similarly, the branched double cover of a tangle (B, t) is a compact, connected, oriented three-manifold with torus boundary (see, for example [**14**] or [**31**, Lecture 4]).

The classic approach in modeling enzymatic actions as tangle replacements relies on combining important theorems about Dehn surgery with the characterization of rational tangles as those with branched double covers that are solid tori. *Dehn surgery* refers to a surgical operation that transforms a 3-manifold M into another 3-manifold, in which an open tubular neighborhood of a knot or link is removed, and solid tori are glued along the boundary components [**31**]. More precisely, let M be a compact, connected, irreducible, oriented 3-manifold with ∂M a torus. A *slope* is the isotopy class of an unoriented simple closed curve on the bounding torus. Given a pair of slopes r, s the *distance* $\Delta(r, s)$ refers to their minimal geometric intersection number. For any slope r, a closed 3-manifold $M(r)$, called a *Dehn filling* of M, can be constructed by gluing a solid torus to M by a homeomorphism which identifies a meridional curve of the solid torus to the slope r on ∂M. See [**31**] for a review of Dehn surgery.

In a processive recombination reaction the enzymatic complex binds the two sites, and may perform one or more rounds of recombination before releasing the sites. In the particular setup presented by Ernst and Sumners [**23**] the authors used three or more tangle equations to model processive recombination reactions. The Cyclic Surgery Theorem [**17**], applied in the branched double cover, was used to determine the types of tangles that satisfy the equations. This approach to studying the mechanism of an enzymatic action yields the *tangle calculus*, which effectively reduces the problem to the simple combinatorics of rational tangles (see [**21–24**] for an extended discussion of rational tangles and the tangle method). However, one major limitation of tangle calculus is that one must assume, or be able to show, that the tangles involved are rational tangles.

Another topological strategy in the study of band surgery involves analyzing surfaces bounded by the links involved. This is done in [**11, 12, 18, 43, 85**], for example. The essence of the approach is as follows: suppose L and L' are related by coherent band surgery. Let $\chi(L)$ denote $\max\{\chi(S)\}$, where $\chi(S)$ denotes the Euler characteristic and the maximum is taken over Seifert surfaces S for L, i.e. oriented surfaces without closed components bounded by the link. If a band surgery decreases complexity, meaning $\chi(L') \geq \chi(L) + 1$, then the band can be isotoped to lie on a minimal genus Seifert surface for the link L. See [**82**] or [**39**, Theorem 1.6] for precise statements. If the genus minimizing Seifert surfaces are known, these can be used to obstruct the existence of coherent bandings, analyze nullification pathways, or characterize the tangle decompositions that yield such transitions. Here too, there are several limitations to the strategy. The first limitation is that the surface cobordism induced by the banding must be orientable, making this strategy only applicable in the case of coherent band surgery. The second limitation is that genus minimizing Seifert surfaces can be difficult to determine. Thus, many studies are restricted to special cases, e.g. between two-bridge knots and $T(2,n)$ torus knots [**18**] amongst fibered knots [**12**], low crossing knots [**11, 43**], or make strong assumptions on topological complexity, as in [**85**].

3.5. Knot invariants and non-coherent band surgery. Several groups of authors have looked at the behavior of knot invariants under non-coherent band surgery. The study of band surgery as an operation transforming a knot into other knot goes back to the 1980s when Lickorish asked whether a knot could be unknotted by adding a twisted band, and gave a criterion for this in terms of the linking form of the branched double cover [**59**]. His approach suffices to show, for example, that the figure eight knot cannot be unknotted with a single twisted band. In fact, many of the results on non-coherent band surgery involve the branched double cover of the knot. Let e_p denote the minimum number of generators of the homology group $H_1(\Sigma_p(L);\mathbb{Z})$, where $\Sigma_p(L)$ is the p-fold branched cyclic cover. In [**1**], the authors observed that work of Hoste, Nakanishi, and Taniyama [**41**, Theorem 4] shows the following.

PROPOSITION 3.1. [**1**, Lemma 5.1] If two knots K and K' are related by a single non-coherent band surgery, then $|e_p(K) - e_p(K')| \leq p-1$.

Abe and Kanenobu also showed that certain evaluations of the Jones polynomial and Q-polynomial may be used to determine whether knots or links are related by a band surgery (either coherent or non-coherent). In the following theorem, $\omega = e^{i\pi/3}$ and $V(L;\omega)$ denotes the evaluation of the Jones polynomial $V(L;t)$ at $t^{1/2} = e^{i\pi/6}$.

THEOREM 3.2. [**1**, Theorem 5.5] *If two links L and L' are related by a single coherent or non-coherent band surgery, then*

$$|V(L;\omega)/V(L';\omega)| \in \{1, \sqrt{3}^{\pm 1}\}.$$

The appearance of the Jones polynomial in Theorem 3.2 is intrinsically related to the role of the branched double cover of L in band surgery. To see this, consider the following statement of Lickorish and Millet [**60**, Theorem 3], which states that this particular evaluation of the Jones polynomial is in fact related to the dimension of $H_1(\Sigma_2(S^3,L);\mathbb{Z}/3\mathbb{Z})$:

$$V(L;\omega) = \pm i^{c-1}(i\sqrt{3})^{\delta(L)}.$$

Here, c is the number of link components and $\delta(L) = \dim H_1(\Sigma_2(S^3,K);\mathbb{Z}/3\mathbb{Z})$. A similar statement also holds for evaluations of the Q-polynomial at $\bar{\phi}$, where $\bar{\phi}$ is the conjugate of the golden ratio, $\bar{\phi} = (1-\sqrt{5})/2$.

THEOREM 3.3. [**1**, Theorem 5.5] *If two links L and L' are related by a single coherent or non-coherent band surgery, then*

$$|Q(L;-\bar{\phi})/Q(L';-\bar{\phi})| \in \{\pm 1, \sqrt{5}^{\pm 1}\}.$$

Notably, there is also a relationship between the first homology of the branched double cover with coefficients in $\mathbb{Z}/5\mathbb{Z}$ and evaluations of the Q-polynomial, due to Jones [**47**]. Indeed, $Q(L;-\bar{\phi}) = \pm\sqrt{5}^{r}$, where $r = \dim H_1(\Sigma_2(S^3,L);\mathbb{Z}/5\mathbb{Z})$. An additional theorem of Kanenobu [**48**, Theorem 2.2] implies that if a knot or link L

is obtained from an unknotting number one knot K by a coherent or non-coherent band surgery, then either $2\det(L)$ or $-2\det(L)$ is a quadratic residue of $\det(K)$. The proof of this statement also uses the Jones polynomial.

3.6. Unknotting via non-coherent band surgery and crossing changes. Recall that the classical unknotting number $u(K)$ of a knot is defined to be the minimal number of crossing changes required to convert K into an unknot. The analogously defined number in the case of non-coherent band surgery operations is denoted by $u_2(K)$. The notation here is meant to indicate $H(2)$-unknotting number, where an $H(2)$-move is equivalent to a non-coherent band surgery. Several authors have investigated the relationship between $u(K)$ and $u_2(K)$.

Nakajima [**71**] has shown that unknotting number provides an upper bound on non-coherent unknotting as follows.

THEOREM 3.4. [**71**, Theorem 3.2.3] *In general, $u_2(K) \leq u(K) + 1$. Moreover, if both signed unknotting numbers $u_+(K)$ and $u_-(K)$ are even, then $u_2(K) \leq u(K)$.*

A very clear pictorial proof of Nakajima's theorem is given in [**49**, Theorem 3.1]. It is worth noting that this bound is certainly not sharp. For example, the $T(2, 2k+1)$ torus knots have unknotting number k, but because they bound a Möbius strip, they can be unknotted by the addition of a single band. Kanenobu and Miyazawa have directly computed or bounded $u_2(K)$ for most knots of up to 10 crossings [**49**] . Their table indicates that all knots of up to nine crossings have $u_2(K) \leq 2$, with the exception of 9_{49}, for which $u(9_{49}) = u_2(9_{49}) = 3$.

The determinant $\det(K)$, the signature $\sigma(K)$ and the Arf invariant $\mathrm{Arf}(K)$ are integer valued invariants of knots and links defined using the Seifert form [**14**]. The signature of a link is equivalently the signature of the intersection form of the four-manifold obtained by taking the double cover of the four-ball branched over a Seifert surface for the link. For knots the determinant is odd and the signature is even. The Arf invariant takes values in $\mathbb{Z}/2\mathbb{Z}$. In [**98**], Yasuhara gave a criterion on unknotting via banding in terms of the signature and Arf invariant.

THEOREM 3.5. [**98**, Proposition 5.1] *If a knot K bounds a Möbius band in the four-ball, then there exists an integer x such that*

$$|8x + 4\,\mathrm{Arf}(K) - \sigma(K)| \leq 2.$$

Note that slice knots are those which bound a disk smoothly embedded in the four ball, e.g. the unknot is slice. The addition of a twisted band produces a knot K which bounds a Möbius band in the four-ball. Therefore Yasuhara's criterion implies that for a knot K with $u_2(K) = 1$, if $\sigma(K) \equiv 0 \pmod 8$, then $\mathrm{Arf}(K) = 0$ and if $\sigma(K) \equiv 4 \pmod 8$, then $\mathrm{Arf}(K) = 1$.

As mentioned above, using the Jones polynomial, Kanenobu [**48**, Theorem 2.2] proved that if a knot or link L is obtained by a coherent or non-coherent band surgery from a knot K with unknotting number one, then $2\det(L) \equiv \pm s^2 \pmod{\det(K)}$ for some integer s. Kanenobu and Miyazawa also gave several criteria for a knot to have non-coherent band unknotting number one via the signature, the Arf invariant, and certain evaluations of the Jones polynomial or Q-polynomial [**49**]. Their work both recovers Theorem 3.5 and establishes the following criterion.

THEOREM 3.6. [**49**, Theorem 4.5] *Let K be a knot with $u_2(K) = 1$. Suppose that* $\det(K) \equiv 0 \pmod 3$ *and* $\sigma(K) \equiv 2\epsilon \pmod 8$, $\epsilon = \pm 1$. *Then*

$$V'(K; -1) \equiv (-1)^{\mathrm{Arf}(K)} 8\epsilon \pmod{24}.$$

The proof of Kanenobu and Miyazawa's criterion is centered on polynomial invariants, whereas the proof of Yasuhara's criterion is an application of his work on the Euler numbers of surfaces in simply connected four-manifolds. The Jones polynomial and Q-polynomial encode topological data about abelian knot invariants in a mysterious way. Indeed, the Jones polynomial may be defined using methods from many different areas of mathematical physics. The full relationship between the Jones polynomial, Q-polynomial, and other related knot polynomials with traditional geometric objects in knot theory (the knot group, cyclic branched covers, etc) is a topic of immense interest that is central to modern knot theory, topology, and mathematical physics (see for example [**96, 97**]). Any further discussion is beyond the scope of this exposition.

4. Band surgery obstructions via Heegaard Floer homology

4.1. Bandings along the trefoil. Motivated by our interest in site-specific recombination and by the biological significance of the class of $T(2,n)$ torus knots and links, and in particular of the trefoil knot $T(2,3)$, in a joint work with Lidman [**61**] we proved the following classification theorem. Note that $T(2,3)$ is the right-handed trefoil; an analogous statement holds for the left-handed trefoil after mirroring.

THEOREM 4.1. [**61**, Corollary 1.2] *The torus knot $T(2,n)$ is obtained from $T(2,3)$ by a non-coherent banding if and only if n is ± 1, 3 or 7. The torus link $T(2,n)$ is obtained from $T(2,3)$ by a coherent banding if and only if n is ± 2, 4 or -6.*

The main effort of Theorem 4.1 is of course to show that the trefoil is not related to $T(2,n)$ for values of n other than those listed. This statement follows as a corollary of a stronger statement, Theorem 4.2 below.

Let K be a knot in a three-manifold Y with neighborhood $N(K)$ a solid torus. A meridian of K is an essential, simple closed curve on $\partial N(K)$ which bounds a disk in $N(K)$. There is always a trivial filling along the meridian of K for which Dehn surgery returns Y [**31**]. Recall from Section 3 that the *distance* between two surgery slopes is their minimal geometric intersection number. By a *distance one surgery* we mean a surgery slope that intersects the meridian of K exactly once. Elsewhere these are called integral surgeries. We have:

THEOREM 4.2. [**61**, Theorem 1.1] *The lens space $L(n,1)$ is obtained by a distance one surgery along a knot in the lens space $L(3,1)$ if and only if n is one of $\pm 1, \pm 2, 3, 4, -6$ or 7.*

Taking Theorem 4.2 at face value, we can now prove Theorem 4.1.

PROOF OF THEOREM 4.1. Suppose that $n = \pm 1, \pm 2, 3, 4, -6$ or 7. Band moves illustrating these cases are shown in Figures 7 and 8. Conversely, suppose that there exists a non-coherent band surgery relating $T(2,3)$ and $T(2,n)$. Using the tangle decomposition in line (3.1), we may write

$$(S^3, T(2,3)) = (B_o, t_o) \cup (B, t) \text{ and } (S^3, T(2,n)) = (B_o, t_o) \cup (B, t'),$$

where (B,t) is the (0) tangle, (B,t') is the (∞) tangle, and (B_o, t_o) is some outside tangle. The zero and infinity tangles are the first two tangles shown in Figure 5.

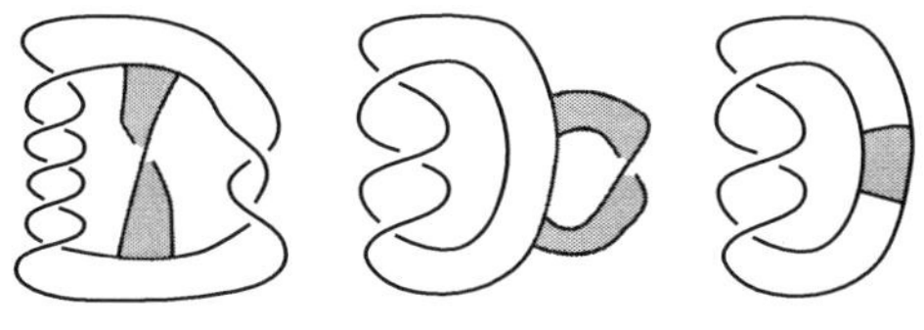

FIGURE 7. Non-coherent bandings: (Left) $T(2,3)$ to $T(2,7)$. (Center) $T(2,3)$ to $T(2,3)$. (Right) $T(2,3)$ to the unknot.

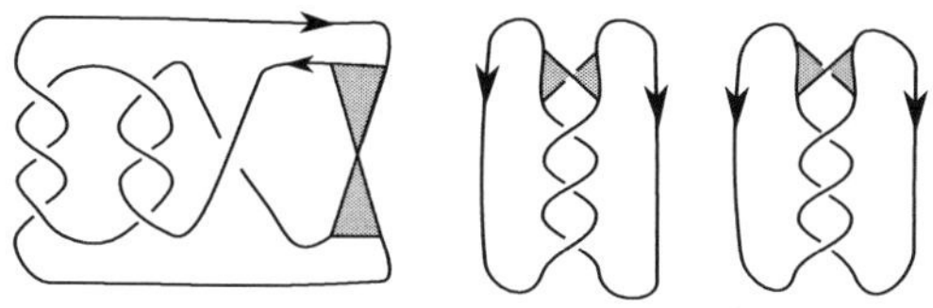

FIGURE 8. Coherent bandings: (Left) $T(2,-6)$ to $T(2,3)$. (Center) $T(2,3)$ to $T(2,2)$. (Right) $T(2,3)$ to $T(2,4)$.

The double cover of S^3 branched over $T(2,n)$ is the lens space $L(n,1)$. We may also write $L(n,1)$ in terms of the tangle decomposition (3.1) as

$$L(3,1) = \Sigma_2(B_o,t_o) \cup \Sigma_2(B,t) \text{ and } L(n,1) = \Sigma_2(B_o,t_o) \cup \Sigma_2(B,t'),$$

where each union of branched double covers of tangles is taken along the bounding tori. The branched double cover of the outside tangle, $\Sigma_2(B_o,t_o)$, is a compact, connected, oriented 3-manifold with torus boundary that may be alternatively described as the complement of a knot J in $L(3,1)$. This knot J is the lift of a properly embedded arc arising as the core of the band from the band surgery that relates $T(2,3)$ and $T(2,n)$. Both $\Sigma_2(S^3,T(2,3)) = L(3,1)$ and $\Sigma_2(S^3,T(2,n)) = L(n,1)$ are obtained by Dehn surgery along J. The slope yielding $L(3,1)$ is just the meridian of J, and the Montesinos trick [**67**] implies that the slope yielding $L(n,1)$ is distance one from the meridian.[2] Theorem 4.2 implies that n cannot fall out of the set of integers $\{\pm1,\pm2,3,4,-6,7\}$. □

If one is only interested in coherent band surgery, Theorem 4.1 can be obtained more readily. The banding is coherent when n is even, in which case $T(2,n)$ is a link of linking number $\pm n/2$, depending on the orientation. When the linking number is positive, the signature of $T(2,n)$ is $1-n$. The statement follows an easy consequence of Murasugi's criterion from 1965 for coherent banding:

THEOREM 4.3. [**69**, Lemma 7.1] *It if two links L and L' are related by a coherent band surgery, then* $|\sigma(L) - \sigma(L')| \leq 1$.

When the linking number is negative, the signature of $T(2,n)$ is 1. In this case, the classification of coherent band surgeries relating $T(2,3)$ and $T(2,n)$ for n even follows from [**18**, Theorem 3.1], in which they characterize coherent band surgeries relating $T(2,n)$ torus links and certain two-bridge knots.

Our proof of Theorem 4.2 is rather technical. In brief, the theorem is proved by analyzing the behavior of the Heegaard Floer homology d-invariants under distance one surgeries. The *d-invariant* $d(Y,\mathfrak{t})$ is a rational number that comes from the

[2] A very informative exposition of the Montesinos trick can also be found in [**31**, Lecture 4].

Heegaard Floer homology $HF^+(Y, \mathfrak{t})$ of a rational homology sphere Y equipped with a Spin^c structure $\mathfrak{t}$. The d-invariants, and the Heegaard Floer package in general, are due to Ozsváth and Szabó [**73**], and analogues of them exist in other Floer theories. Each d-invariant $d(Y, \mathfrak{t})$ is defined to be the minimal grading of a non-torsion class in $HF^+(Y, \mathfrak{t})$ [**73**], and is in fact a Spin^c rational homology cobordism invariant. The most technical aspect of Theorem 4.2 required us to prove series of formulas (see [**61**]) which relate the d-invariants under surgery along knots in the lens space $L(3,1)$ to certain integer-valued knot invariants that are related to Rasmussen's local h-invariant [**78**]. These formulas generalize work of Ni and Wu [**72**].

Theorem 4.2 can be viewed as a partial generalization of the *lens space realization problem*, which was proved by Greene [**34**]. The lens space realization problem asks for a characterization of the lens spaces that arise from integral surgery along nontrivial knots in S^3 (note that distance one surgeries are integral surgeries). This type of problem falls into a broader class of questions involving the characterization of three-manifolds arising by integral surgery along knots, and the characterization of knots that admit lens space surgeries. The Berge conjecture is one of the most prominent of these questions, having survived substantial attacks over several decades [**5,6,8,9,29,35,74,77**]. The problem was posed by Gordon and attributed to Berge in [**30**, Conjecture 4.5] and appears as Problem 1.78 in the Kirby list [**51**].

CONJECTURE 4.4. [**30**, Conjecture 4.5] *A hyperbolic knot K in S^3 has a lens space surgery if and only if K is doubly primitive, and the surgery is the corresponding integral surgery.*

A hyperbolic knot is one whose complement may be given a metric of constant curvature -1. A doubly primitive knot K is one that lies on the surface F of a genus two Heegaard splitting $X \cup_F X'$ of S^3, where $[K]$ belongs to bases for the free groups $\pi_1(X)$ and $\pi_1(X')$. That the surgery coefficient is integral is due to the Cyclic Surgery Theorem [**17**]. A discussion of this conjecture can be found in [**31**]. Another nice discussion of some related problems in knot theory and three-manifold topology can be found in Hedden and Watson's article [**36**, Section 6].

4.2. A criterion from determinant and signature.

Using a similar strategy as that in [**61**], that is, by studying the d-invariants under surgery in the branched double cover, in [**68**] we established a new criterion on banding via the determinant and the signature $\sigma(K)$ of a knot. Before stating our result, we need to recall the following definition of Ozsváth and Szabó [**75**]. Let L_h and L_v denote diagrams obtained by the horizontal and vertical smoothings of crossing in a fixed diagram of L, as in Figure 9.

DEFINITION 4.5. The set QA of quasi-alternating links is the smallest set of links satisfying the following properties:

(1) The unknot is in QA.
(2) The set is closed under the following operation: If L is a link which admits a diagram with a crossing such that
  (a) both resolutions L_h and L_v are in QA
  (b) $\det(L_v) \neq 0$, $\det(L_h) \neq 0$, and
  (c) $\det(L) = \det(L_v) + \det(L_h)$,
then L is in QA.

FIGURE 9. The horizontal and vertical smoothings of a crossing in a link diagram.

By definition, any alternating knot is quasi-alternating.

THEOREM 4.6. [**68**, Theorem 1.1] *Let K and K' be a pair of quasi-alternating knots and suppose that* $\det(K) = m = \det(K')$ *for some square-free integer m. If there exists a band surgery relating K and K', then $|\sigma(K) - \sigma(K')|$ is* 0 *or* 8.

Our criterion for non-coherent banding (Theorem 4.6) is reminiscent of Murasugi's criterion for coherent banding (Theorem 4.3 in section 4.1). It is also similar in spirit to the work of Kanenobu and Kanenobu-Moriuchi, described in Section 3, although our statement is not focused on the case of bandings relating a knot to the unknot.

We remark that given K and K' quasi-alternating such that $\det(K) \neq \det(K')$ (i.e. the conditions of Theorem 4.6 do not hold), a banding may change the signature of a knot by arbitrary amounts. For example, take the $T(2,n)$ torus knot, which has signature $1 - n$, and can be unknotted by one smoothing of a crossing.

The following statement is a direct corollary to the proof of Theorem 4.6.

COROLLARY 4.7. *Excluding* 8_{19}*, let K and K' be knots of eight or fewer crossings with* $\det(K) = m = \det(K')$ *for m a square-free integer. If there exists a banding from K to K', then* $|\sigma(K) - \sigma(K')| = 8$ *or* 0.

One special case where the condition that $\det(K) = \det(K')$ is met is the case of a *chirally cosmetic banding.*

DEFINITION 4.8. A *chirally cosmetic banding* is a non-coherent band surgery relating a knot K with its mirror image K^*.

We then asked how Theorem 4.6 applies to the class of $T(2,n)$ torus knots.

COROLLARY 4.9. *The only nontrivial torus knot $T(2,n)$, with n square free, admitting a chirally cosmetic banding is $T(2,5)$.*

The banding that relates $T(2,5)$ with its mirror $T(2,-5)$ was first found by Zeković [**100**]. Livingston has recently informed us that that the square-free condition of Corollary 4.9 can be removed, with a possible exception when $n = 9$ [**63**]. Indeed, in a recent preprint [**63**], Livingston observes that if K admits a chirally cosmetic banding, then $K \# K$ bounds a ribbon Möbius band in the four-ball. By applying Casson-Gordon theory for non-orientable surfaces in the four-ball [**28**], one may then achieve an improved obstruction for the $T(2,n)$ torus knots, amongst some other knots. The status of the knot $T(2,9)$ remains unknown.

QUESTION 4.10. Does the torus knot $T(2,9)$ admit a chirally cosmetic banding?

Another possible avenue for approaching non-coherent band surgery is to consider the bounds on nonorientable 3- and 4-genus coming from Heegaard Floer homology. The *nonorientable 3-genus* or *3-dimensional crosscap number* $\gamma_3(K)$ of K in S^3 is the minimal first Betti number $b_1(F)$ of any surface F, possibly non-orientable, bounded by K. Likewise the *nonorientable 4-genus* or *smooth 4-dimensional crosscap number* $\gamma_4(K)$ is the minimal $b_1(F)$ of any surface F, possibly non-orientable, properly and smoothly embedded in the four-ball and bounded by K. Obviously $\gamma_4(K) \leq \gamma_3(K)$. The *upsilon invariant* $\upsilon(K)$ is a Heegaard Floer theoretic concordance invariant defined by Ozsváth, Stipsicz, and Szabø [**76**] that gives a lower bound on the nonorientable 4-genus.

THEOREM 4.11. [**76**, Theorem 1.2] *Let K be a knot in S^3. Then*

$$|\upsilon(K) - \frac{\sigma(K)}{2}| \leq \gamma_4(K).$$

Livingston's observation, when combined with the lower bound from υ, implies the following.

THEOREM 4.12. [**63**, Theorem 19] *If K admits a chirally cosmetic banding, then $\upsilon(K) = \sigma(K)/2$.*

We note that for alternating and quasi-alternating knots, it is always true that $\upsilon(K) = \sigma(K)/2$. However, for non-quasi-alternating knots, this provides a good obstruction to chirally cosmetic bandings. The literature on nonorientable 3- and 4-genus is relatively sparse, compared with its orientable counterpart. Additional work in this direction (particularly in dimension four) can be found in [**7, 28, 45, 76, 98**].

5. The frequency of chirally cosmetic bandings, as assessed by numerical simulations

The primary outcome of our analytical work in Theorem 4.1, Theorem 4.6, and their corollaries is to establish mathematical obstructions to the existence of a banding relating a given pair of knots. That is to say, we aim to determine whether a banding cannot possibly occur, and if possible, find examples of bands when such obstructions are not available. We complement the analytical work with numerical simulations of non-coherent banding on cubic lattice knots for which the focus is to determine whether a band surgery relating two knots exists, and if so, determine the relative frequency of such a transition. This study is an ongoing effort. We will discuss here the preliminary findings pertaining to chirally cosmetic bandings, as reported in [**68**].

5.1. Cubic lattice knots. A *cubic lattice knot* is an embedding of a knot in $\mathbb{R}^3$ contained in the simple cubic lattice, $(\mathbb{R}\times\mathbb{Z}\times\mathbb{Z})\cup(\mathbb{Z}\times\mathbb{R}\times\mathbb{Z})\cup(\mathbb{Z}\times\mathbb{Z}\times\mathbb{R})$. It is the union of finitely many unit length line segments with the end points contained on the integer lattice $\mathbb{Z}^3$. Any tame knot K admits a cubic lattice representative, i.e. a self-avoiding polygon (SAP) in the simple cubic lattice with knot type K, called a *lattice conformation* of K. The length of a fixed conformation is the number of unit steps in the embedding.

In computational knot theory, particularly in numerical studies of DNA topology, it is common practice to model knots and links as random SAPs in $\mathbb{R}^3$ (e.g. freely jointed chains, uniform random polygons, wormlike chains) or using lattice models (e.g. knots and links in the face-centered, hexagonal, or simple cubic lattice, as above). While polymer models in $\mathbb{R}^3$ better approximate the physical properties of DNA, modeling DNA as lattice polygons presents a number of analytical advantages. There is extensive analytical work on knotting and linking in the cubic

TABLE 1. **Chirally cosmetic bandings are rare.** The relative likelihood of chirally cosmetic bandings for knots of up to eight crossings is given. The probabilities were computed out of 3×10^6 total band moves performed on each knot type.

Knot	$P(K - K^*) \times 10^{-5}$	*Number observed*
5_1	3.467	104
5_1^*	2.800	84
8_8	0	0
8_8^*	0.003	1
8_{20}	42.833	1285
8_{20}^*	46.400	1392

lattice, and a solid mathematical foundation for the development of algorithms that generate equilibrium distributions of SAPs given by sets of vertices in $\mathbb{Z}^3$ [**65, 93**]. Also, lattice polygons incorporate inherently a volume exclusion, which can be used to model the effective diameter of DNA. In collaboration with C. Soteros' group at the University of Saskatchewan [**83**], we have work in progress in this direction based on the work of [**89**].

5.2. Chirally cosmetic bandings are rare. Our analytical work suggests that chirally cosmetic bandings are unusual events. This is consistent with the behavior observed in our simulations of non-coherent banding in the cubic lattice. For each chiral knot type K of up to eight crossings, we sampled three million reconnection events and produced a transition probability network to summarize the outcome. Each reconnection represented a non-coherent banding. The implementation was adapted from our published work in [**87**] (see Section 5.3 for more details).

In our preliminary report, we found chirally cosmetic bandings to be exceedingly rare. Chirally cosmetic bandings were observed, with transition probabilities given in Table 1, for only three knots: 5_1, 8_8, and 8_{20}. The band move relating 5_1 and its mirror image was known to exist, and had been previously reported in [**100**]. The band move for 8_{20} had not, to our knowledge, been explicitly reported in the literature but could be easily ascertained from the symmetric union presentation given in [**56**]. The banding exhibited for 8_8^* was newly discovered via our computer simulations.

5.3. Methods. In the experiment that we designed, we considered the set of 63 non-trivial prime knots of up to eight crossings, where we included both knots from each chiral pair. For each substrate knot K, we used the BFACF algorithm to generate large ensembles of cubic lattice representatives of K. The BFACF algorithm is a Markov chain Monte Carlo algorithm frequently used in the study of self-avoiding polygons [**65**]. It is ergodic in the state space of all cubic lattice representatives of a self-avoiding polygon. The algorithm acts by perturbing the chain via a set of small moves, called the BFACF moves. These moves preserve the isotopy type of the knotted chain, but may change the length. Our particular implementation of BFACF employed a composite Markov chain (CMC) Monte Carlo process in order to more effectively randomize the samples.

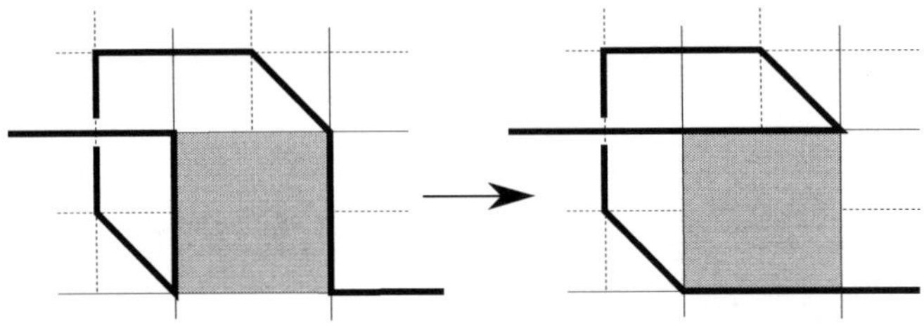

FIGURE 10. An example of reconnection sites in the cubic lattice.

Having generated large, independent ensembles of cubic lattice representatives of a fixed knot type, we then searched for reconnection sites, performed non-coherent banding, and identified the resulting product knots. In the cubic lattice, a reconnection site is a pair of edges along the chain that are of a unit distance from each other and occupy the opposite edges of a square in the lattice. See Figure 10. This was accomplished by extending our software package, originally developed for the case of coherent band surgery simulations in [**87**], to the non-coherent case. We also similarly use batch mean analysis and ratio estimation techniques to ensure statistical robustness. For knot identification, we primarily used the HOMFLY-PT polynomial, with a second identification routine applied in the few cases when identification via the HOMFLY-PT polynomial was inconclusive. Extended information on the BFACF algorithm and our implementation can be found in [**87**, Supplementary Materials], [**65**].

5.4. Future studies. The difference in chirality between a chiral knot K and its mirror image K^* is a well-defined concept. Thus, in [**68**], we focused on bandings that relate a chiral knot K with K^*. For an arbitrary pair of knots K and K', the relative difference in chirality is a quantity that must formally be defined and estimated by numerical simulations. A more extensive study of the chirality trends apparent in the transition probability networks associated to non-coherent banding and its biological relevance is the focus of a forthcoming report [**26**].

Acknowledgments

We are especially grateful to our collaborator Tye Lidman. We also thank Michelle Flanner for her sustained efforts in carrying out the numerical studies described here and Chuck Livingston for sharing with us his new work on chirally cosmetic bandings. The authors thank the anonymous referee for many helpful improvements in the exposition.

References

[1] T. Abe and T. Kanenobu, *Unoriented band surgery on knots and links*, Kobe J. Math. **31** (2014), no. 1-2, 21–44. MR3307286

[2] D. E. Adams, E. M. Shekhtman, E. L. Zechiedrich, M. B. Schmid, and N. R. Cozzarelli, *The role of topoisomerase IV in partitioning bacterial replicons and the structure of catenated intermediates in DNA replication*, Cell **71** (1992), no. 2, 277–288.

[3] F. W. Andrewes, *Studies in group-agglutination i. the salmonella group and its antigenic structure*, The Journal of Pathology and Bacteriology **25** (1922), no. 4, 505–521.

[4] J. Arsuaga, M. Vazquez, S. Trigueros, D. Sumners, and J. Roca, *Knotting probability of DNA molecules confined in restricted volumes: DNA knotting in phage capsids*, Proc. Natl. Acad. Sci. U.S.A. **99** (2002), no. 8, 5373–5377.

[5] K. L. Baker, *Surgery descriptions and volumes of Berge knots. I. Large volume Berge knots*, J. Knot Theory Ramifications **17** (2008), no. 9, 1077–1097, DOI 10.1142/S0218216508006518. MR2457837

[6] K. L. Baker, *Surgery descriptions and volumes of Berge knots. II. Descriptions on the minimally twisted five chain link*, J. Knot Theory Ramifications **17** (2008), no. 9, 1099–1120, DOI 10.1142/S021821650800652X. MR2457838

[7] J. Batson, *Nonorientable slice genus can be arbitrarily large*, Math. Res. Lett. **21** (2014), no. 3, 423–436, DOI 10.4310/MRL.2014.v21.n3.a1. MR3272020

[8] J. Berge, *Some knots with surgeries yielding lens spaces*, Unpublished Manuscript, 1990.

[9] S. A. Bleiler and R. A. Litherland, *Lens spaces and Dehn surgery*, Proc. Amer. Math. Soc. **107** (1989), no. 4, 1127–1131, DOI 10.2307/2047677. MR984783

[10] R. Brasher, R. G. Scharein, and M. Vazquez, *New biologically motivated knot table*, Biochemical Society Transactions **41** (2013), no. 2, 606–611.

[11] D. Buck and K. Ishihara, *Coherent band pathways between knots and links*, J. Knot Theory Ramifications **24** (2015), no. 2, 1550006, 21, DOI 10.1142/S0218216515500066. MR3334659

[12] D. Buck, K. Ishihara, M. Rathbun, and K. Shimokawa, *Band surgeries and crossing changes between fibered links*, J. Lond. Math. Soc. (2) **94** (2016), no. 2, 557–582, DOI 10.1112/jlms/jdw049. MR3556454

[13] D. Buck and M. Mauricio, *Connect sum of lens spaces surgeries: application to Hin recombination*, Math. Proc. Cambridge Philos. Soc. **150** (2011), no. 3, 505–525, DOI 10.1017/S0305004111000090. MR2784772

[14] G. Burde, H. Zieschang, and M. Heusener, *Knots*, Third, fully revised and extended edition, De Gruyter Studies in Mathematics, vol. 5, De Gruyter, Berlin, 2014. MR3156509

[15] H. Cabrera Ibarra and D. A. Lizárraga Navarro, *An algorithm based on 3-braids to solve tangle equations arising in the action of Gin DNA invertase*, Appl. Math. Comput. **216** (2010), no. 1, 95–106, DOI 10.1016/j.amc.2010.01.007. MR2596136

[16] H. Cabrera-Ibarra and D. A. Lizárraga-Navarro, *Braid solutions to the action of the Gin enzyme*, J. Knot Theory Ramifications **19** (2010), no. 8, 1051–1074, DOI 10.1142/S0218216510008327. MR2718627

[17] M. Culler, C. McA. Gordon, J. Luecke, and P. B. Shalen, *Dehn surgery on knots*, Ann. of Math. (2) **125** (1987), no. 2, 237–300, DOI 10.2307/1971311. MR881270

[18] I. K. Darcy, K. Ishihara, R. K. Medikonduri, and K. Shimokawa, *Rational tangle surgery and Xer recombination on catenanes*, Algebr. Geom. Topol. **12** (2012), no. 2, 1183–1210, DOI 10.2140/agt.2012.12.1183. MR2928910

[19] Y. Diao, *The knotting of equilateral polygons in* $\mathbf{R}^3$, J. Knot Theory Ramifications **4** (1995), no. 2, 189–196, DOI 10.1142/S0218216595000090. MR1331746

[20] Y. Diao, N. Pippenger, and D. W. Sumners, *On random knots*, J. Knot Theory Ramifications **3** (1994), no. 3, 419–429, DOI 10.1142/S0218216594000307. Random knotting and linking (Vancouver, BC, 1993). MR1291869

[21] C. Ernst, *Tangle equations*, J. Knot Theory Ramifications **5** (1996), no. 2, 145–159, DOI 10.1142/S0218216596000114. MR1395775

[22] C. Ernst, *Tangle equations. II*, J. Knot Theory Ramifications **6** (1997), no. 1, 1–11, DOI 10.1142/S0218216597000029. MR1442176

[23] C. Ernst and D. W. Sumners, *A calculus for rational tangles: applications to DNA recombination*, Math. Proc. Cambridge Philos. Soc. **108** (1990), no. 3, 489–515, DOI 10.1017/S0305004100069383. MR1068451

[24] C. Ernst and D. W. Sumners, *Solving tangle equations arising in a DNA recombination model*, Math. Proc. Cambridge Philos. Soc. **126** (1999), no. 1, 23–36, DOI 10.1017/S0305004198002989. MR1681651

[25] C. Ernst and A. Montemayor, *Nullification numbers of knots with up to 10 crossings*, J. Knot Theory Ramifications **25** (2016), no. 7, 1650037, 20, DOI 10.1142/S0218216516500371. MR3513937

[26] M. Flanner, A. H. Moore, and M. Vazquez, *Reconnection and chirality in cubic lattice knots*, Preprint, 2018.

[27] H. L. Frisch and E. Wasserman, *Chemical topology*, Journal of the American Chemical Society **83** (1961), no. 18, 3789–3795.

[28] P. M. Gilmer and C. Livingston, *The nonorientable 4-genus of knots*, J. Lond. Math. Soc. (2) **84** (2011), no. 3, 559–577, DOI 10.1112/jlms/jdr024. MR2855790

[29] H. Goda and M. Teragaito, *Dehn surgeries on knots which yield lens spaces and genera of knots*, Math. Proc. Cambridge Philos. Soc. **129** (2000), no. 3, 501–515, DOI 10.1017/S0305004100004692. MR1780501

[30] C. McA. Gordon, *Dehn filling: a survey*, Knot theory (Warsaw, 1995), Banach Center Publ., vol. 42, Polish Acad. Sci. Inst. Math., Warsaw, 1998, pp. 129–144. MR1634453

[31] C. Gordon, *Dehn surgery and 3-manifolds*, Low dimensional topology, IAS/Park City Math. Ser., vol. 15, Amer. Math. Soc., Providence, RI, 2009, pp. 21–71. MR2503492

[32] M. E. Gottesman and R. A. Weisberg, *Prophage insertion and excision*, Cold Spring Harbor Monograph Archive **2** (1971), 113–138.

[33] I. Grainge, M. Bregu, M. Vazquez, V. Sivanathan, S. C. Ip, and D. J. Sherratt, *Unlinking chromosome catenanes in vivo by site-specific recombination*, EMBO J. **26** (2007), no. 19, 4228–4238.

[34] J. E. Greene, *The lens space realization problem*, Ann. of Math. (2) **177** (2013), no. 2, 449–511, DOI 10.4007/annals.2013.177.2.3. MR3010805

[35] M. Hedden, *On Floer homology and the Berge conjecture on knots admitting lens space surgeries*, Trans. Amer. Math. Soc. **363** (2011), no. 2, 949–968, DOI 10.1090/S0002-9947-2010-05117-7. MR2728591

[36] M. Hedden and L. Watson, *On the geography and botany of knot Floer homology*, Selecta Math. (N.S.) **24** (2018), no. 2, 997–1037, DOI 10.1007/s00029-017-0351-5. MR3782416

[37] K. A. Heichman and R. C. Johnson, *The Hin invertasome: Protein-mediated joining of distant recombination sites at the enhancer*, Science **249** (1990), no. 4968, 511–517.

[38] I. Henderson, P. Owen, and J. P. Nataro, *Molecular switches — the on and of of bacterial phase variation*, Molecular Microbiology **33** (1999), no. 5, 919 – 932.

[39] M. Hirasawa and K. Shimokawa, *Dehn surgeries on strongly invertible knots which yield lens spaces*, Proc. Amer. Math. Soc. **128** (2000), no. 11, 3445–3451, DOI 10.1090/S0002-9939-00-05417-4. MR1676336

[40] L. Hlatky, R. K. Sachs, M. Vazquez, and M. N. Cornforth, *Radiation-induced chromosome aberrations: Insights gained from biophysical modeling*, BioEssays **24** (2002), no. 8, 714–723.

[41] J. Hoste, Y. Nakanishi, and K. Taniyama, *Unknotting operations involving trivial tangles*, Osaka J. Math. **27** (1990), no. 3, 555–566. MR1075165

[42] S. C. Y. Ip, M. Bregu, F.-X. Barre, and D. J. Sherratt, *Decatenation of DNA circles by FtsK-dependent Xer site-specific recombination*, The EMBO Journal **22** (2003), no. 23, 6399–6407.

[43] K. Ishihara and K. Shimokawa, *Band surgeries between knots and links with small crossing numbers*, Prog Theor Phys Supplement **191** (2011), 245–255.

[44] K. Ishihara, K. Shimokawa, and M. Vazquez, *Site-specific recombination modeled as a band surgery: applications to Xer recombination*, Discrete and topological models in molecular biology, Nat. Comput. Ser., Springer, Heidelberg, 2014, pp. 387–401, DOI 10.1007/978-3-642-40193-0_18. MR3220014

[45] S. Jabuka and T. Kelly, *The nonorientable 4-genus for knots with 8 or 9 crossings*, Algebr. Geom. Topol. **18** (2018), no. 3, 1823–1856, DOI 10.2140/agt.2018.18.1823. MR3784020

[46] R. C. Johnson, *Site-specific DNA inversion by serine recombinases*, Microbiology Spectrum **3** (2015), no. 1.

[47] V. F. R. Jones, *On a certain value of the Kauffman polynomial*, Comm. Math. Phys. **125** (1989), no. 3, 459–467. MR1022523

[48] T. Kanenobu, *Band surgery on knots and links, III*, J. Knot Theory Ramifications **25** (2016), no. 10, 1650056, 12, DOI 10.1142/S0218216516500565. MR3548474

[49] T. Kanenobu and Y. Miyazawa, *$H(2)$-unknotting number of a knot*, Commun. Math. Res. **25** (2009), no. 5, 433–460. MR2573402

[50] A. Kawauchi, *A survey of knot theory*, Birkhäuser Verlag, Basel, 1996. Translated and revised from the 1990 Japanese original by the author. MR1417494

[51] R. Kirby, *Problems in low dimensional manifold theory*, Algebraic and geometric topology (Proc. Sympos. Pure Math., Stanford Univ., Stanford, Calif., 1976), Proc. Sympos. Pure Math., XXXII, Amer. Math. Soc., Providence, R.I., 1978, pp. 273–312. MR520548

[52] D. Kleckner and W. Irvine, *Creation and dynamics of knotted vortices*, Nature Physics **9** (2013), 253–258.

[53] D. Kleckner and W. T. M. Irvine, *Creation and dynamics of knotted vortices*, Nature Physics **9** (2013), 253 EP –.

[54] D. Kleckner, L. Kauffman, and W. Irvine, *How superfluid vortex knots untie*, Nature Physics **12** (2015).

[55] C. E. Laing, R. L. Ricca, and D. L. Sumners, *Conservation of writhe helicity under anti-parallel reconnection*, Scientific Reports **5** (2015), 9224 EP –.

[56] C. Lamm, *The search for non-symmetric ribbon knots*, arXiv:1710.06909 [math.GT], 2017.

[57] P. Le Bourgeois and F. Cornet, *Chromosome dimer resolution by site-specific recombination*, Brenner's Encyclopedia of Genetics (2013), 551–554.

[58] L. Li, J. Zheng, H. Peter, E. Priest, H. Cheng, L. Guo, F. Chen, and D. Mackay, *Magnetic reconnection between a solar filament and nearby coronal loops*, Nat. Phys. **12** (2016), 847–851.

[59] W. B. R. Lickorish, *Unknotting by adding a twisted band*, Bull. London Math. Soc. **18** (1986), no. 6, 613–615, DOI 10.1112/blms/18.6.613. MR859958

[60] W. B. R. Lickorish and K. C. Millett, *Some evaluations of link polynomials*, Comment. Math. Helv. **61** (1986), no. 3, 349–359, DOI 10.1007/BF02621920. MR860127

[61] T. Lidman, A. H. Moore, and M. Vazquez, *Distance one lens space fillings and band surgery on the trefoil knot*, Algebr. Geom. Topol. **19** (2019), no. 5, 2439–2484, DOI 10.2140/agt.2019.19.2439. MR4023320

[62] L. F. Liu, J. L. Davis, and R. Calendar, *Novel topologically knotted DNA from bacteriophage P4 capsids: studies with DNA topoisomerases*, Nucleic Acids Res **16** (1981), no. 9, 3979–3989.

[63] C. Livingston, *Chiral smoothings of knots*, arXiv:1809.07619 [math.GT]], 2018.

[64] M. Delbrück, *Knotting problems in biology*, Proceedings of Symposia in Applied Mathematics **14** (1962), 55–63.

[65] N. Madras and G. Slade, *The self-avoiding walk*, Probability and its Applications, Birkhäuser Boston, Inc., Boston, MA, 1993. MR1197356

[66] S. K. Merickel and R. C. Johnson, *Topological analysis of Hin-catalysed DNA recombination in vivo and in vitro*, Molecular Microbiology **51** (2004), no. 4, 1143–1154.

[67] J. M. Montesinos, *Three-manifolds as 3-fold branched covers of S^3*, Quart. J. Math. Oxford Ser. (2) **27** (1976), no. 105, 85–94, DOI 10.1093/qmath/27.1.85. MR0394630

[68] A. H. Moore and M. Vazquez, *A note on band surgery and the signature of a knot*, arXiv:1806.02440 [math.GT], 2018.

[69] K. Murasugi, *On a certain numerical invariant of link types*, Trans. Amer. Math. Soc. **117** (1965), 387–422, DOI 10.2307/1994215. MR171275

[70] K. Murasugi, *Knot theory and its applications*, Birkhäuser Boston, Inc., Boston, MA, 1996. Translated from the 1993 Japanese original by Bohdan Kurpita. MR1391727

[71] K. Nakajima, *Calculation of the H(2)-unknotting numbers of knots to 10 crossings*, Master's thesis, Yamaguchi University, Japan, 1997.

[72] Y. Ni and Z. Wu, *Cosmetic surgeries on knots in S^3*, J. Reine Angew. Math. **706** (2015), 1–17, DOI 10.1515/crelle-2013-0067. MR3393360

[73] P. Ozsváth and Z. Szabó, *Absolutely graded Floer homologies and intersection forms for four-manifolds with boundary*, Adv. Math. **173** (2003), no. 2, 179–261, DOI 10.1016/S0001-8708(02)00030-0. MR1957829

[74] P. Ozsváth and Z. Szabó, *On knot Floer homology and lens space surgeries*, Topology **44** (2005), no. 6, 1281–1300, DOI 10.1016/j.top.2005.05.001. MR2168576

[75] P. Ozsváth and Z. Szabó, *On the Heegaard Floer homology of branched double-covers*, Adv. Math. **194** (2005), no. 1, 1–33, DOI 10.1016/j.aim.2004.05.008. MR2141852

[76] P. S. Ozsváth, A. I. Stipsicz, and Z. Szabó, *Unoriented knot Floer homology and the unoriented four-ball genus*, Int. Math. Res. Not. IMRN **17** (2017), 5137–5181, DOI 10.1093/imrn/rnw143. MR3694597

[77] J. Rasmussen, *Lens space surgeries and a conjecture of Goda and Teragaito*, Geom. Topol. **8** (2004), 1013–1031, DOI 10.2140/gt.2004.8.1013. MR2087076

[78] J. A. Rasmussen, *Floer homology and knot complements*, ProQuest LLC, Ann Arbor, MI, 2003. Thesis (Ph.D.)–Harvard University. MR2704683

[79] D. Rolfsen, *Knots and links*, Mathematics Lecture Series, vol. 7, Publish or Perish, Inc., Houston, TX, 1990. Corrected reprint of the 1976 original. MR1277811

[80] V. V. Rybenkov, N. R. Cozzarelli, and A. V. Vologodskii, *Probability of DNA knotting and the effective diameter of the DNA double helix*, Proceedings of the National Academy of Sciences **90** (1993), no. 11, 5307–5311.

[81] V. V. Rybenkov, C. Ullsperger, A. V. Vologodskii, and N. R. Cozzarelli, *Simplification of DNA topology below equilibrium values by type II topoisomerases*, Science **277** (1997), no. 5326, 690–693.

[82] M. Scharlemann and A. Thompson, *Link genus and the Conway moves*, Comment. Math. Helv. **64** (1989), no. 4, 527–535, DOI 10.1007/BF02564693. MR1022995

[83] M. Schmirler, *Strand passage and knotting probabilities in an interacting self-avoiding polygon model*, 2012, Masters Thesis – University of Saskatchewan.

[84] S. Y. Shaw and J. C. Wang, *Knotting of a DNA chain during ring closure*, Science **260** (1993), no. 5107, 533–536.

[85] K. Shimokawa, K. Ishihara, I. Grainge, D. J. Sherratt, and M. Vazquez, *FtsK-dependent XerCD-dif recombination unlinks replication catenanes in a stepwise manner*, Proc Natl Acad Sci USA **110** (2013), no. 52, 20906–20911.

[86] C.J. Stirling, G. Stewart, and D.J. Sherratt, *Multicopy plasmid stability in escherichia coli requires host-encoded functions that lead to plasmid site-specific recombination*, Molecular and General Genetics MGG **214** (1988), no. 1, 80–84.

[87] R. Stolz, M. Yoshida, R. Brasher, M. Flanner, K. Ishihara, D. Sherratt, K. Shimokawa, and M. Vazquez, *Pathways of DNA unlinking: a story of stepwise simplification*, Scientific Reports **7** (2017), no. 1, 12420.

[88] D. W. Sumners and S. G. Whittington, *Knots in self-avoiding walks*, J. Phys. A **21** (1988), no. 7, 1689–1694. MR951053

[89] M. C. Tesi, E. J. Janse van Rensburg, E. Orlandini, D. W. Sumners, and S. G. Whittington, *Knotting and supercoiling in circular dna: A model incorporating the effect of added salt*, Phys. Rev. E **49** (1994), 868–872.

[90] M. Vazquez, S. D. Colloms, and D. Sumners, *Tangle analysis of Xer recombination reveals only three solutions, all consistent with a single three-dimensional topological pathway*, J. Mol. Biol. **346** (2005), no. 2, 493–504.

[91] M. E. Vazquez, *Tangle analysis of site-specific recombination: Gin and Xer systems*, ProQuest LLC, Ann Arbor, MI, 2000. Thesis (Ph.D.)–The Florida State University. MR2701776

[92] M. Vazquez and D. W. Sumners, *Tangle analysis of Gin site-specific recombination*, Math. Proc. Cambridge Philos. Soc. **136** (2004), no. 3, 565–582, DOI 10.1017/S0305004103007266. MR2055047

[93] F. Wang and D. P. Landau, *Efficient, multiple-range random walk algorithm to calculate the density of states*, Phys. Rev. Lett. **86** (2001), 2050–2053.

[94] M. Wellenreuther and L. Bernatchez, *Eco-evolutionary genomics of chromosomal inversions*, Trends in Ecology & Evolution **33** (2018), no. 6, 427–440.

[95] S. Witte, M. Flanner, and M. Vazquez, *A symmetry motivated link table*, Preprint, 2018.

[96] E. Witten, *Quantum field theory and the Jones polynomial*, Comm. Math. Phys. **121** (1989), no. 3, 351–399. MR990772

[97] E. Witten, *Two lectures on the Jones polynomial and Khovanov homology*, Lectures on geometry, Clay Lect. Notes, Oxford Univ. Press, Oxford, 2017, pp. 1–27. MR3676591

[98] A. Yasuhara, *Connecting lemmas and representing homology classes of simply connected 4-manifolds*, Tokyo J. Math. **19** (1996), no. 1, 245–261, DOI 10.3836/tjm/1270043232. MR1391941

[99] E. L. Zechiedrich, A. B. Khodursky, and N. R. Cozzarelli, *Topoisomerase IV, not gyrase, decatenates products of site-specific recombination in Escherichia?coli*, Genes & Development **11** (1997), no. 19, 2580–2592.

[100] A. Zeković, *Computation of Gordian distances and H_2-Gordian distances of knots*, Yugosl. J. Oper. Res. **25** (2015), no. 1, 133–152, DOI 10.2298/YJOR131210044Z. MR3331990

[101] C.-Z. Zhang, M. L. Leibowitz, and D. Pellman, *Chromothripsis and beyond: rapid genome evolution from complex chromosomal rearrangements*, Genes & Development **27** (2013), no. 23, 2513–2530.

Department of Mathematics and Applied Mathematics, Virginia Commonwealth University, Richmond, VA 23220
Email address: moorea14@vcu.edu

Department of Microbiology and Molecular Genetics / Department of Mathematics, University of California, Davis, One Shields Avenue, Davis, CA 95616
Email address: mariel@math.ucdavis.edu

Part II

The Topology and Geometry of Proteins

Contemporary Mathematics
Volume **746**, 2020
https://doi.org/10.1090/conm/746/15006

Why are there knots in proteins?

Sophie E. Jackson

Contents

Abstract. There are now more than 1700 protein chains that are known to contain some type of topological knot in the protein structure databank. Although this number is small relative to the total number of protein structures solved, it is remarkably high given the fact that for decades it was thought impossible for a protein chain to fold and thread in such a way as to create a knotted structure. There are four different types of knotted protein structures that contain 3_1, 4_1, 5_2 and 6_1 knots and over the past 15 years there has been an increasing number of experimental and computational studies on these systems. The folding pathways of knotted proteins have been studied in some detail, however, the focus of this review is to address the fundamental question "Why are there knots in proteins?" It is known that once formed, knotted protein structures tend to be conserved by nature. This, in addition to the fact that, at least for some deeply knotted proteins, their folding rates are slow compared with many unknotted proteins, has led to the hypothesis that there are some properties of knotted proteins that are different from unknotted ones, and that this had resulted in some evolutionary advantage over faster folding unknotted structures. In this review, the evidence for and against this theory is discussed. In particular, how a knot within a protein chain may affect the thermodynamic, kinetic, mechanical and cellular (resistance to degradation) stability of the protein is reviewed.

1. Introduction

Proteins are linear polymers comprising of amino-acid monomers that are linked *via* a peptide bond. The polypeptide chain that results has a common backbone but each position along the chain can have a different side chain depending upon the amino acid incorporated, Fig. 1. There are 20 naturally occurring amino acids that can be used by nature and therefore, even for a short polypeptide chain, there are a remarkable number of different possibilities in terms of the primary sequence of a protein. Regular elements of secondary structure such as α-helices and β-strands/β-turns/β-sheets can form and these can pack against each other to form a wide range of different tertiary structures. For most globular proteins, during this process a significant number of hydrophobic side chains are buried and form a hydrophobic core. The folded state of a protein is frequently referred to as the native (N) state whilst the unfolded state is commonly called the denatured (D) state. It is the native state that is active and functional. How a polypeptide chain adopts its final three-dimensional structure is known as the protein folding problem.

Proteins vary in length enormously and can be very short, e.g., peptide hormones, or can be extremely long and comprise of thousands of amino acids such as the giant protein titin. As many proteins are essentially very long, linear, unbranched polymers, this raises the question of whether the polypeptide chain can ever become knotted, either in a programmed way or by accident. For decades there was a general consensus that no protein structure would be knotted. This was mainly because only one knotted structure, carbonic anhydrase, had been identified by Richardson (*1*) and then Mansfield (*2*). In this case, the protein only forms a very shallow trefoil knot with a short region of the polypeptide chain passing through a knotting loop to form the knot. In 1994 Mansfield concluded "The most reasonable interpretation of these results is that the protein folding mechanism only explores unknotted conformations..... schemes to predict the native state by locating the global minimum must therefore be designed to reject knotted conformations" (*2*).

It was not until 2000 that many more, and a number of deeply knotted protein structures, were identified (*3*). Willie Taylor used a computer algorithm to detect knots in open chains (most proteins are linear and open not closed). He applied the algorithm to the protein databank (PDB), which by the year 2000 contained many thousands of protein structures not just the 400 that Mansfield had looked at in 1994. Surprisingly, he found a number of knotted structures including the deeply knotted enzyme acetohydroxy acid isomeroreductase from plants that had a 4_1 (figure-of-eight) knot in its polypeptide chain (*3*). This significant finding made the protein structure and protein folding communities rethink what is possible and how knotted protein structures could form given their topological complexity.

At the time of writing, there are a total of 1700 identified knotted protein chains in protein structures deposited in the PDB, of which 622 are knotted and 1078 have slip-knots (see the KnotProt 2.0 database - `http://knotprot.cent.uw.edu.pl/` (*4*)). Sulkowska and coworkers have established and maintain this excellent database that contains all the knotted and slip-knotted structures, as well as providing important information on knot type, the position of the knotted core within the protein chain and the length of residues N- and C-termini to the knotted core which shows whether the protein has a shallow or deep knot (*4*). This review will only focus on knotted proteins for which there has been considerably more investigation and for which much more is known from both computational and experimental studies, compared with slip-knotted proteins. In addition, a recent publication has established that the slip-knotted protein AFV3-109 can fold fast (folding rate constant $k_F = 180\ s^{-1}$ and $t_{1/2} = 4$ ms) and efficiently, suggesting that slip-knotted proteins may not have the topological barriers to forming their native states that knotted proteins can. Any reader interested in slip-knotted proteins is directed towards some recent papers on this subject (*4-9*).

Four different types of knots are found in protein chains — 3_1 (trefoil), 4_1 (figure-of-eight), 5_2 (Gordian) and 6_1 (stevedore) knots, Fig. 2 (*10, 11*). These are all twist knots where only a single threading event is required in order to form the knotted structure. Other than the fact that they are all twist knots, there are few similarities between the different knotted proteins. They occur in proteins from all kingdoms of life from bacteria to mammals, are present as very shallow (only a few amino-acid residues thread through a loop to form the knot) (*2*) or extremely deep knots (more than 140 amino-acid residues or more has to thread through a loop to create the knot) (*12*), they have different handedness (*11*), the knot can be relatively localised (for example when tightened in a pulling experiment) or spread over the entire native structure and much of the full polypeptide chain and the protein chain can be relatively short or very long (*4*). Although most knotted proteins known are soluble proteins, there are examples of integral membrane proteins that are knotted (*13*). Knotted proteins also have very different functions — many are enzymes, but some are signal-transducing proteins, DNA- or metal-binding proteins, whilst others have yet to have a function assigned (*11*).

Over the past decade, a number of experimental and computational studies have addressed a variety of questions regarding how knotted proteins fold both *in vitro* and *in vivo* (e.g., after their initial synthesis on the ribosome and in the presence of molecular chaperones, which are a large and diverse class of proteins that aid in the folding, assembly and maturation of many cellular proteins and their complexes). As there are now a number of reviews on this subject including some recent reviews (*10, 11, 14-16*), I will not focus on this material here. Instead, I will concentrate on a specific question: **"Why are there knots in proteins?"** which addresses another question **"Are there any evolutionary advantages for a protein chain to be knotted?"** The discovery of a number of knotted protein structures back in 2000 (*3*) and subsequently the finding that once a knotted protein structure is formed then it is conserved by nature (*17*)[1] has led to the proposal that there is some evolutionary advantage for a polypeptide chain to be knotted and, therefore, that some of the properties of knotted proteins must be different from their unknotted counterparts. This is a question that many groups have tried to address over the past decade with different degrees of success. There is still no consensus on this issue, which can be controversial in the field. Here, I first review the hypotheses on how knotted proteins may differ from unknotted ones and then go

[1]It is of interest that evolutionary pathways connecting knotted and unknotted carbamoyltransferases have been identified and these illustrate how, by the insertion of a knot-promoting loop, this class of trefoil-knotted proteins have evolved from unknotted precursors. 18. R. Potestio, C. Micheletti, H. Orland, Knotted vs. Unknotted Proteins: Evidence of Knot-Promoting Loops. *PLoS Computational Biology* **6**, (2010).

on to address each of these in turn, reviewing the experimental and computational evidence for and against. Finally, I provide some thoughts as to why this has turned into such a difficult question to address and what is needed from the experimental and computational communities to definitively answer the question.

2. Stability

Stability is a term used throughout Physical, Chemical and Biological Sciences and its definition is often different dependent on the discipline and the context in which it is being used. This frequently leads to confusion and a misunderstanding of the term and interpretation of experimental and computational results. Here, I consider four different types of stability — thermodynamic, kinetic, mechanical, degradative (cellular). In each case, I first define the stability in question and then consider the evidence from both experimental and computational studies that the stability of knotted proteins is significantly different from unknotted ones.

3. Thermodynamic stability

Definitions of thermodynamic stability. In the case of the simplest system — a protein which populates only two states its folded or native (N) state and the unfolded or denatured (D) state — the thermodynamic stability of the protein is straightforward to define as it is the difference in Gibbs' free energy between N and D, Fig. 3A. It is usually defined as the difference in free energy between the two states in water, ΔG_{D-N} (H_2O) (from herein referred to as simply ΔG_{D-N}), under a specific set of experimental conditions, as it can change with pH, temperature, salt concentration etc. Usually, the Gibbs' free energy of unfolding is used, ΔG_{D-N}, in this case, the values are positive as the native state is more stable (lower in energy) than the denatured state, i.e., the protein is folded, Fig. 3A. Many small, monomeric proteins show two-state behaviour under equilibrium conditions, and in these cases the ΔG_{D-N} (H_2O) has been measured and found to vary from between 1.5 kcal mol^{-1} (very unstable) to 25 kcal mol^{-1} (very stable). The thermodynamic stability of a protein can vary considerably with primary (amino-acid) sequence and a single mutation, i.e., a substitution of one amino acid at one position in the protein chain to a different amino acid, can change the thermodynamic stability significantly - by up to 4-5 kcal mol^{-1} or even more (*19*). In many cases, this is sufficient to make an otherwise stable protein into an unstable protein (*20*).

Many larger proteins including multiple-domain proteins (*21*), oligomeric proteins that adopt dimers, trimers etc, (*22, 23*) and proteins with complex topologies (*24, 25*) do not follow simple two-state behaviour under equilibrium conditions. In these cases, partially structured intermediate states are observed between the native and denatured states, Fig. 3B. The thermodynamic stability of these systems can be defined in a number of different ways including as the difference in free energy between the native and denatured states, ΔG_{D-N}. The free energy differences between all of the individual substates can also be used, ΔG_{I-N} and ΔG_{D-I}, Fig. 3B. In the case of knotted proteins where it is known that the protein first unfolds to a denatured state which contains the native knot, and then only slowly unties to populate the denatured state which has no knot, the thermodynamic stability can be defined in such a way as to refer to the free energy difference between the native state and either the knotted or unknotted denatured state, $\Delta G_{Dknot-N}$ and $\Delta G_{Dunknot-N}$, respectively, or between the different substates, ΔG_{I-N}, $\Delta G_{Dknot-I}$, $\Delta G_{Dunknot-Dknot}$, Fig. 3C. As some knotted proteins are oligomeric, i.e, bacterial trefoil-knotted methyltransferases are often dimers, the situation is even more complex as the free energy differences between states depend upon protein concentration which they do not for a monomeric system (*22*).

What definition of thermodynamic stability should be used when comparing the stabilities of knotted proteins with their unknotted counterparts? This may depend upon context, however, the best measure of the thermodynamic stability of a non-two-state protein may well be the difference in free energy between the native state and the next state that the native protein unfolds to, in Fig. 3B and C this would be the intermediate state and not the fully denatured state either knotted or unknotted, i.e., ΔG_{I-N} . Why is this the case? When considering whether knotted proteins might have evolved and been conserved in order to enhance the thermodynamic stability of the native state, one also has to consider which states have any biological activity. In almost all cases, even if proteins have only partially unfolded, partially unstructured intermediate states have no activity at all and only the native state is functional and active. In this case, ΔG_{I-N} is the crucial value that relates stability and activity *in vivo*.

Why is this important? Several experimental and computational studies have shown that, for proteins that have deep knots in their structures, the native state of the protein unfolds to a knotted denatured state that shows remarkable resilience to untying (*26-28*). It has therefore been proposed that knotted proteins may be a new class of superstable proteins which may potentially have an impact on a number of fields including applications as therapeutics and in biotechnology. In these examples, activity associated with the native state of the protein is key, and therefore, ΔG_{I-N} is the key value that determines the functional stability of the protein.

Results from experimental and computational studies. There are now a number of experimental studies that have measured the thermodynamic stability of knotted proteins and the values are well within the range found for unknotted proteins, Table 1 (*24*, *27-33*). Two experimental studies claim that knotted proteins are more thermodynamically stable than unknotted ones (*34*, *35*). However, these two studies raise two very important issues, the first of which is reversibility.

The discussion above regarding thermodynamic stability relies upon the transitions from native to intermediate to denatured states being fully reversible, i.e, on fully unfolding the protein and on changing the conditions back to those that favour the native state all of the protein molecules refold to the active, native structure. If the unfolding is not fully reversible then the thermodynamic parameters measured are not accurate and should not be used as a measure of stability (*36*). This is because, if not fully reversible, an additional reaction takes place during folding, that of aggregation, which is frequently irreversible, Fig. 3D. In this case, the thermodynamic stability measured is less than the real thermodynamic stability if it could be measured, but by some unknown amount. The more aggregation the lower the ΔG_{D-N} observed. In some studies, such as that on a polymer created of related knotted and unknotted variants of a designed trefoil knotted protein (*34*) and more recently on knotted and unknotted transcarbarmylases (*35*) , "apparent" ΔG_{D-N} values are reported. This is misleading as apparent can mean anything from, the values might be quite close to the real values or very far from them. The problem is that it just isn't possible to know. It cannot be assumed that different proteins with the same/similar structure aggregate to the same extent as it is well established that small differences in sequence can lead to large differences in aggregation behaviour and therefore in how much the aggregation affects the stabilities measured. Therefore, apparent ΔG_{D-N} values should not be used to as a guide of thermodynamic stability.

The recent knotted and unknotted transcarbarmylase study also raises another important issue. That is of undertaking a comparison of the correct things, i.e., not comparing the stability of a knotted apple with that of an unknotted pear! Apart from the issue of irreversibility in the studies of the transcarbamylases, there is also the issue of all the differences between the knotted and unknotted variants. The three transcarbamylases studied all vary with respect to amino-acid sequence, i.e., their primary sequences differ in many positions. It is very well established that a single point mutation in a protein can destabilise its native state considerably — up to and above 4-5 kcal mol^{-1} (*19*). Therefore, it is simply impossible to say whether a difference in stability comes from a structure being knotted *versus* unknotted, or it arises as a simple consequence of one of the many amino acid differences between the two proteins.

Monte Carlo simulations using simple lattice models have also been used and these approaches have shown that a knotted backbone has no effect on the thermodynamic stability of the native state of a protein (*37*). Together with the experimental studies, there is no clear evidence that knotted structures are thermodynamically more stable than their unknotted counterparts. In fact, recently one experimental study established that knotting actually destabilised the native state of a designed protein with a trefoil knot (*38*). In this case, NMR H/D exchange techniques were used to demonstrate that some regions of the designed knotted monomeric protein were up to 5 kcal mol^{-1} less stable than in the unknotted dimeric variant of the protein (*38*).

4. Kinetic stability

The kinetic stability of a protein is simply the rate constant for unfolding which experimentally can be measured directly in high concentrations of a chemical denaturant or which can be extrapolated from measurements at high denaturant concentrations to the value it would be in their absence, i.e., in water. As with thermodynamic stability, the unfolding rate constant can vary considerably with conditions, e.g., pH, temperature etc. For a simple two-state system, there is just a single unfolding event and therefore the rate constant corresponds to the transition from native to denatured states. This is determined by the difference in free energy between the transition state and the native state, $\Delta G_{\ddagger-N}$, Fig. 4A, and the temperature at which the reaction takes place through the Arrhenius equation, Fig. 4D. For non-two state systems, there are additional unfolding events as the protein first unfolds from its native state to an intermediate state, Fig. 4B, and if the protein is initially knotted in the denatured state, D_{knot}, then there is an additional event that corresponds to the untying of the knot to form an unknotted denatured state, D_{unknot}, Fig. 4C. In all cases, the kinetic stability is defined by the initial unfolding event, i.e, k_u or k_{1u} is the critical parameter to consider.

The unfolding rate constants for a number of knotted proteins including 3_1-knotted bacterial methyltransferases that have deep knots (*24, 30, 39, 40*), shallow trefoil-knotted proteins (*33*), designed trefoil-knotted proteins (*32*), 5_2-knotted ubiquitin C-terminal hydrolases (UCHs) (*30-32*) and the 6_1-knotted dehalogenase enzyme (*41*) have been determined experimentally. The values extrapolated to water span many orders of magnitude, from 2 s^{-1} to 6 x 10^{-7} s^{-1}, corresponding to unfolding half lives of 0.35 s to 14 days, Table 1. These values cover the range determined experimentally for unknotted proteins such as 1.3 s^{-1} for the R15 domain of spectrin (*42*) and 3 x 10^{-10} s^{-1} for GFP (*25*).

A number of computational studies have also addressed the issue of whether a knot can increase the kinetic stability of the native state of a protein. In their 2008 study, Sulkowksa, Cieplak and coworkers employed molecular dynamic simulations using coarse-grained structure-based models to probe the thermal and mechanical stability of knotted and unknotted variants of the transcarbamylase family (*43*). In addition to a naturally occurring knotted and unknotted variant, they are also "rewired" the backbone of the knotted variant such that an unknotted structure is created. This was an excellent control as this structure is as similar as it can get to the knotted structure without being knotted. The simulations showed that the knotted proteins had longer unfolding times than the unknotted ones suggesting that, in this case, the knot can confer kinetic stability on the native state of a protein (*43*).

More recently, unbiased atomistic molecular dynamic simulations were employed to study the unfolding of YbeA, a naturally occurring trefoil-knotted methyltransferase from *E. coli*, and also a variant in which again the protein chain was rewired in order to generate a similar unknotted variant *in silico* (*44*). In this study, the knotted variant was found to have higher kinetic stability with respect to unfolding by chemical denaturants, heat or force, than the unknotted one (*44*). These simulations also provided a basis for the increased kinetic stability, showing that opening of the hydrophobic core *via* separation of two α-helices is a crucial step initiating unfolding, which is restrained for the knotted protein by topological and geometrical frustrations (*44*).

Although these is some evidence from computational studies that knots may increase the kinetic stability of the native state of a protein with respect to its unfolding, as yet, there remains no indication from experimental investigations that this is the case.

5. Mechanical stability

The mechanical stability of the native state of proteins can be probed using experimental methods such as atomic force microscopy (AFM) or optical tweezers (OT), as well as *in silico* by the application of force to a model structure, be it a simple lattice model or an atomistic model. In AFM or OT experiments the protein is covalently attached to either a surface or beads using well-characterised linker molecules, frequently DNA, Fig. 5A. The force and extension of the molecule are recorded and distinct unfolding events can be measured.

The first study of the force-induced unfolding of a knotted protein was on the 4_1-knotted chromophore-binding domain of the red/far red photoreceptor phytochrome (*45*). The Bornschogl and Rief groups used atomic force microscopy to unfold both apo (no chromophore) and holo (chromophore bound) states (*45*). The apo-phytochrome unfolded at forces of approx. 47 pN whilst the holo-phytochrome was more resistant to mechanical unfolding requiring forces of approx. 73 pN (as expected as ligand binding increases the stability of the native state of the protein). Subsequent to this, AFM studies were conducted on the slip-knotted protein AFV3-109 which was unfolded to a state in which it contained a tightened 3_1 knot (*9*). In this case, much larger forces in the region of 240 pN were required to unfold the protein. Recently, optical tweezer methods have been used to study the force-induced unfolding of the 5_2-knotted protein UCH-L1 (*46*). In this study, different pulling positions were used such that the protein unfolded and tightened to a denatured state containing a 5_2, 3_1 or the unknot. The forces required to unfold UCH-L1 varied from 18 pN (unfold to the 3_1-knot), 37 pN (to unfold to the 5_2-knot) to greater than 40 pN (unfold to the unknot). In all these three cases, the forces required to unfold the knotted proteins were similar to or lower than those of other unknotted proteins whose mechanical stability has been investigated, which range from 20-50 pN for maltose-binding protein (*47*), 130-200 pN for the Ig domains from titin (*48*) to 100-600 pN for GFP (*49*). It should be noted that both the pulling speed and the pulling direction affects the unfolding force. Together, these studies provide no experimental evidence that knotted proteins are more stable with respect to force-induced unfolding than other classes of unknotted proteins. However, no suitable control, i.e., an unknotted protein with a very similar sequence and structure to a knotted protein but no knot was used.

Although these studies did not show that knotted proteins have high mechanostability, they were important in determining the degree to which different types of knot could be tightened within a polypeptide chain. A 3_1 knot in the unfolded state of either AFV03-109 or UCH-L1 was between 4.7-5.7 nm in diameter, corresponding to approx. 13-16 amino acids residues in the chain (*9, 46*). Exactly where the chain is tightened along the chain cannot be determined by these experiments. A 4_1 knot can be tightened to about 6.2 nm corresponding 17-amino acid residues (*45*), whilst the 5_2 knot in UCH-L1 showed more complex behaviour (*46*). At low forces it tightens to 14.6 nm corresponding to 40 amino acid residues but at higher forces it compacts further to about 23 residues in length (*46*). The tightening of the knot in this case may be initially limited by the steric clash of side chains and not the polypeptide backbone chain itself. It should be noted that, although it is common for polypeptide chains to be represented by simple, smooth and narrow tubes, Fig. 1C, this is not a very accurate depiction of a protein chain, the side chains are frequently bulky, Fig. 1A, and therefore can hinder the formation of a very tight knot.

The question of whether a knot increases the mechanostability of the native state of a protein has also been investigated using simulations. As described above, Sulkowksa, Cieplak and coworkers used computational methods on knotted and unknotted variants of the transcarbamylase family to establish that, in this case, the knotted variant had a longer unfolding time than the unknotted ones when the protein was stretched at a constant velocity and force (*43*). Other simulations on knotted and unknotted proteins have also been undertaken but these have focussed on the probability of untying a protein knot and how this depends upon pulling speed, direction and temperature (*50*), or knot end movement during unfolding and tightening of the knot (*51*) rather than on the effect on mechanostability *per se*. Interestingly, in the latter study, the knot ends were found to suddenly jump to well defined sequences along the chain many of which corresponded to sharp turns in the native state, however, in most of these cases the protein chain did not refold to its native knotted state after release of the force (*51*). In contrast, the experimental studies on knotted and slip-knotted proteins have established full reversibility after force-induced unfolding (*9, 46*).

As with the kinetic stability studies, although there is some evidence from computational studies that knots may increase the stability of the native state of a protein with respect to unfolding by force, as yet, there remains little evidence from experimental investigations that this is the case.

6. Cellular stability: degradation by proteolytic AAA+ machines

The stability of proteins *in vivo* can depend upon a large number of factors, some proteins can have very short cellular half-lives whilst others do not. In many cases, proteins are degraded, i.e., hydrolysed in to small peptide fragments which can be further degraded back to amino acids to enable recycling, by AAA+ degradation machines (*52*). This includes the ClpXP machinery in bacteria (*52*) and the proteasome in eukaryotic cells (*53*). Both have similar structures and mechanisms, however, there are not identical in every respect. They are comprised of two large cylinders, each of which is made from homo-oligomers, through which runs a central pore that is relatively small in size compared to the overall dimensions, Fig. 6A. The two cylinders pack against each other and one has internal proteolytic sites which non-specifically hydrolyse the polypeptide backbone chain at regular intervals. The other has an ATP-binding and hydrolysis site in each of the subunits and also a signal recognition site which specifically binds to either a short ssrA peptide or the protein ubiquitin, Fig. 6A. These are covalently attached to a target protein by various enzymes, in order to tag the protein in question for degradation. Key to these machines is the fact that a protein which otherwise has a compact and stable native state, needs to be unfolded in order for the polypeptide chain to be translocated through the internal pore to the proteolytic sites, Fig. 6A-C. Translocation is aided by rounds of ATP-binding and hydrolysis which drive the chain through the machinery with a ratchet-type mechanism (*52, 53*). Unfolding may be induced by the force applied on the protein chain that occurs as the chain itself is translocated into the machinery (*54*). In contrast to the AFM and OT studies discussed above, cellular degradation machines apply force to one end of the protein chain leaving the other end free.

The central translocation channel of the major prokaryotic and eukaryotic degradation machines, ClpXP and the proteasome, are relatively small (*52, 53*). At one point it was thought that only a fully unfolded polypeptide chain could pass through the channel, however, it is now known that in some cases several chains can be translocated simultaneously (*55*). Even so, the channel is narrow which led to the hypothesis that knotted proteins may be particularly stable with respect to degradation by these cellular machines, a knot being unable to penetrate into the translocation pore (*56*). This was, in part also suggested by computational studies that have shown that the translocation of a knotted protein through a pore can be blocked under specific conditions (*57*), polyelectrolyte chains that are knotted can hinder translocation through a nanopore (*58*) and flexible filaments tied in twist knots have a higher hindrance to translocation than torus knots (all protein knots being twist knots) (*59*). Recently a number of experimental and computational studies have investigated this. The experimental studies have all used the ClpXP machinery from *E. coli*. In the first study of its kind, San Martin and coworkers investigated the folding of a small trefoil-knotted protein MJ036 by ClpXP and found that on its own the 3_1-knotted protein was rapidly degraded if degradation was initiated from its C-terminus (*60*). Interestingly, when MJ0366 was fused with the large, very stable β-barrel protein GFP, degradation was stalled at 47-amino acid residues from the start of the structured region of GFP. The authors concluded that this corresponds to either a tightened 3_1-knot outside of the ClpXP machine and buttressing GFP or a tightened 3_1-knot within the translocation pore of ClpXP (*60*), Fig 6. G.H, respectively. More recently, the degradation of a family of 5_2-knotted UCHs has been reported by the Hsu group (*61*). In contrast to the MJ0366 study, this study found that wild-type UCH-L1 showed remarkably slow degradation kinetics when a degradation tag was attached to its C-terminus, essentially demonstrating that it was not degraded by the ClpXP system (*61*). This result initially leads one to conclude that some knotted proteins may indeed be particularly resistant to unfolding by mechanical force and also cellular degradation, however, wild-type UCH-L1 has already been shown to unfold at very moderate forces using optical tweezers (*46*). So why the apparent discrepancy between these two sets of results? It has long-been established that the ability of cellular degradation machines to unfold, translocate and hydrolyse substrate proteins is related not to the overall global stability (thermodynamic stability) of the native state of the substrate protein but the local stability near the site of the degradation tag (*62, 63*). In the case of UCH-L1, locating the degradation tag at the C-terminus of the protein results in it residing right next to the final β-strand which is a central strand within the β-sheet lying at the core of the protein structure and which is known to be highly stable (*31, 64*). My own group with that of Laura Itzhaki has recently established that it is the local stability at the C-terminus of UCH-L1 and not the knotted structure that makes it so resistant to mechanical unfolding and degradation by ClpXP (*65*). This was shown in two different and independent ways: first, when the degradation tag was located at the N-terminus of the protein, rapid degradation was observed. Second, when the region at the C-terminus of the protein was destabilised using mutation, thereby reducing the local stability near the degradation tag, UCH-L1 was rapidly degraded even when the tag was located at the C-terminus (*65*).

Computational studies have also addressed whether knotted proteins are harder to degrade than unknotted ones. Cieplak and coworkers used several models of a cellular degradation machine which was represented by an effective potential with an added pulling force (*66*). Their studies showed that on some occasions the presence of a knot can hinder and potentially jam the machinery, the probability of it doing so depending upon the substrate protein, the model of the degradation machine, the magnitude of the pulling force, and the choice of the pulled terminus (*66*). Again, why the apparent discrepancy with experimental results that have now clearly established that ClpXP can easily degrade both 3_1- and 5_2-knotted proteins? The first explanation is that in the simulations it is possible to detect single chains of a knotted protein that aren't degraded against a background of many that are, this is just not possible experimentally unless one develops a single-molecule assay of degradation. Second, is that the model of the degradation machine does not incorporate any dynamic motions of the machinery, in particular the translocation pore, which *in vitro*/*in vivo* may have sufficient thermal motions for even a bulky knotted chain to be pulled through the machinery. Given that Matouschek and coworkers showed back in 2003 that three polypeptide chains can pass through the translocation pore and degradation channel of the proteasome, it is perhaps not so surprising that 3_1- and even 5_2-knotted proteins can too (*55*).

This raises the interesting question of how cellular degradation machines unfold, translocate and degrade knotted protein substrates. There are two models of how this might occur. In the first, ClpXP unfolds the protein which initially retains its knotted topology, then as the chain is pulled through the machinery the knot simply moves along the polypeptide chain until it falls off the free end, Fig. 6D. Alternatively, the knot, whether it be tightened or still stretched over a large amount of the protein chain, can translocate through the machinery until it is unknotted by the action of the proteolysis sites where it is simply degraded into small fragments, Fig. 6E. Is there any evidence from the experimental or computational studies as to what might happen? Interestingly, the studies of the Bustamante and Baez groups and that of the Jackson and Itzhaki groups, both included a series of experiments using knotted proteins fused to highly stable protein domains, either GFP and ThiS (*60, 65*) the results of which help to address this question.

Fig. 6F-H shows the basis of the experiments on the knotted fusion proteins. The degradation tag is placed at the C-terminus whilst the stable domain (GFP or ThiS) is at the N-terminus. The knotted fusion protein engages with the degradation machine through the C-terminal tag and then the knotted protein domain can be unfolded whilst retaining the knot within the chain. The knot can then either travel along the chain until it abuts the stable folded GFP or ThiS domain, Fig. 6G, or it can be translocated into the pore of the machinery, Fig. 6H. In either case, the degradation ceases as the machinery cannot easily unfold GFP (*60*) or ThiS (*65*) and a degradation intermediate is produced. It has been suggested that knotted proteins therefore provide an alternative mechanism to the well-established slippery Gly-Ala rich regions in protein chains for a cell to be able to produce partially degraded protein fragments with new activities (*60*). However, this is not unique to knotted proteins as it is well established that unknotted multi-domain proteins which contain a highly degradation-resistant domain also show this effect (*67*). For the GFP-knotted protein fusions there is evidence that the trefoil knot does not enter the translocation pore but buttresses the GFP outside of the degradation machine (*60*). However, for ThiS-knotted fusions, there is evidence that the trefoil knot does indeed enter the translocation pore (*65*).

Most of the experimental and computational evidence suggests that knotted proteins can be degraded by cellular degradation machines and that the knot either moves along the polypeptide chain as it is being translocated and falls off its free end, or it can enter the translocation pore directly.

7. Conclusions

Despite 15 years of experimental and computational research on knotted proteins, we are still some way off addressing the basic question of "Why are there knots in proteins?" Some studies have provided some hints as to what role the knot may play, particularly the possibility of knotted protein structures showing greater kinetic stability than unknotted ones. However, with just a couple of computational studies showing this and no good evidence from experimental studies, even this remains to be proven. So, why has it been so hard to answer this simple question? From a computational point of view, it is relatively straightforward to "rewire" a protein chain and change the connectivities between different elements of secondary structure in order to create a protein which has the same overall structure in terms of the packing of secondary structural elements together to form a tertiary structure but which is unknotted. However, the results of computational studies can be highly dependent upon the structural model used, the potentials employed, the forces applied, the timescales of the simulations etc. In contrast, it is extremely challenging experimentally to find a good unknotted control. Unknotted proteins with a similar overall structure and function may appear on the surface to be good states for a comparative analysis. However, they vary significantly in primary structure (amino-acid sequence) and therefore any differences observed can be due to one of the many sequence differences and not the knotted *versus* unknotted structure. One way of overcoming this, is to make a circular permutant, i.e., experimentally rewire the protein as can be achieved *in silico*. With circular permutants, the N- and C-termini of the protein are fused whilst new N- and C-termini are created elsewhere in the protein structure thus effectively removing the knot. My own group have attempted this and other rewiring exercises with a number of classes of knotted proteins on a number of occasions. Unfortunately, we have not successfully made a soluble unknotted circular permutant or rewired variant, aggregation currently representing an unsurmountable challenge to us characterising such constructs.

At this point it is worth considering what the assumptions were underpinning the question in the first place. The most important assumption is that knotted proteins have been conserved in nature because they have some advantage over unknotted ones. Maybe this is incorrect. In part, this comes from the fact that knotted proteins can be slower to fold than unknotted ones and therefore this may represent an evolutionary disadvantage which has to be balanced in some way for knotted structures to have been conserved by nature. But some knotted proteins, particularly those with shallow knots, may not be particularly slow or difficult to fold. Estimates from optical tweezer experiments on UCH-L1 suggest that the difference between the folding rate from an unknotted and a 3_1-knotted denatured state for a shallow system is only about a factor of three and this is also true comparing the folding rates from a 5_2-knotted and 3_1-knotted unfolded state (*46*). This should be compared with the effect of a single-amino acid substitution (which has no effect on structure) which can easily change folding rates by a factor of 20 - 100 (*68, 69*). In addition, although there is evidence that the folding rate of deeply knotted proteins is greatly reduced by the presence of the knot (*70*), even in these cases, cellular factors such as molecular chaperones can greatly accelerate the rate of folding (*70*). Perhaps, within a cellular environment knotted proteins can fold fast enough and there is no evolutionary disadvantage of their folding rates being slightly slower than unknotted ones. Certainly, nature has some well-established examples of even unknotted proteins that fold very slowly indeed, e.g., GFP (*71*).

Whether or not, knotted proteins possess significantly different properties from unknotted proteins, these topologically complex protein structures are intriguing and raise many important questions on how they form. We are still far from understanding the folding pathways of this class of protein and more computational and experimental studies are required in order to solve this particularly challenging protein folding problem.[2]

[2]Since submission of this manuscript and during publication, a number of articles have been published that directly address some of the issues discussed in this review. This includes several papers by the Hsu group which describe the formation of an unknotted circular variant of the dimeric trefoil knotted methyltransferase YbeA and characterise it using a number of methods (*78, 79*). These studies showed that the permutation resulted in a change in structure such that the unknotted variant adopts a dimer-swapped dimer which was non-functional. This study suggests that the knot may be essential for the formation of a dimeric structure capable of biological activity. However, it should be noted that even circular permutants of unknotted proteins can be unstable and, in some cases, form domain-swapped dimers.

In addition, the Hsu group published a follow on study from their previous work on the degradation of GFP-fusions of UCH-L1 by ClpXP (*80*). In this study, they show an N-terminal truncation of the polypeptide chain which removes the knot significantly its stability against degradation. In addition, they show that UCH-L5 requires a very significant amount of ATP to be degraded by ClpXP. Although in the paper, it is suggested that it is the knotted topology that is important in conferring resistance to degradation, the results are compatible with our study on UCH-L1 which established that stability, in particular, local stability, is important for resistance against degradation and not the knotted topology per se.

For YbeA, a trefoil-knotted bacterial methyl transferase, computational studies have revealed cooperation between the knotted conformation and the dimerization necessary for ligand binding (*81*), providing further evidence that the correct knotted dimeric structure is required for this class of knotted enzyme. Stating that the correct native structure is required for protein function is, though, nothing new. Whether it is possible to create an unknotted MTase still capable of carrying out the same reaction needs to be proved. Certainly, with other classes of knotted proteins, unknotted variants exist which catalyse very similar chemical transformations.

TABLE 1. Thermodynamic and kinetic stabilities as calculated from experimental data

Knotted protein	Knot type	Shallow (< 10 aa) Deep (> 10 aa)	[a]ΔG_{X-Y} kcal mol^{-1}	$k_u(H_2O)$ s^{-1}	Reference
YibK	3_1	deep	[b]4, 4, 4, 14	$5x10^{-7}$	*(39)*
YbeA	3_1	deep	[c] 3, 13	$4x10^{-4}$	*(24)*
MTase (*T. maritima*)	3_1	deep	[d] 11, 25	$5x10^{-5}$	*(27)*
MJ0366	3_1	shallow	[e]10	$2x10^{-3}$	*(33)*
$HP0242_{mono}$	3_1	deep	[f]2, 13	2	*(32)*
UCH-L1	5_2	shallow/deep	[g]5, 4, 5	$9x10^{-4}$	*(30, 64)*
UCH-L3	5_2	shallow/deep	[h] 7	$1x10^{-5}$	*(31)*
Dehl	6_1	deep	[i]7, 10, 6	$6x10^{-7}$	*(41)*

[a]The difference in Gibbs' free energy between two states populated under equilibrium conditions, e.g., native (N), intermediate (I) and denatured (D) states.

[b]Values for ΔG_{D1-I1}, ΔG_{D2-I2}, $\Delta G_{I1 or I2-I3}$ and ΔG_{I3-N}, respectively at pH 7.5 calculated from kinetic data. For YibK, three intermediate states can be observed during refolding kinetics. Intermediates I1 and I2 are intemediates populated early during folding, they both fold further to populate a more structured intermediate state I3, which then converts to the native dimer. Please note that the denatured state remains knotted in this case.

[c]Values for ΔG_{D-I} and ΔG_{I-N}, respectively at pH 7.5 calculated from kinetic data. For YbeA, only a single intermediate state can be observed during refolding kinetics. Please note that the denatured state remains knotted in this case.

[d]Values for ΔG_{D-I} and ΔG_{I-N}, respectively calculated from data acquired under equilibrium conditions. For MTase from *T. maritima* the protein was incubated in chemical denaturant for either 24 hours or four weeks.

[e]No intermediate states are observed during the unfolding of MJ0366 under equilibrium conditions in chemical denaturants, the value shown is for ΔG_{D-N}.

[f]Values for ΔG_{I-N} and ΔG_{2D-I2} respectively are shown. The values are calculated from chemical denaturant unfolding experiments under equilibrium conditions.

[g]Values for ΔG_{D-I}, ΔG_{I1-N} and ΔG_{I2-N} respectively are shown. The value for ΔG_{D-I} is from chemical denaturant unfolding experiments under equilibrium conditions where only one

8. Figure Legends

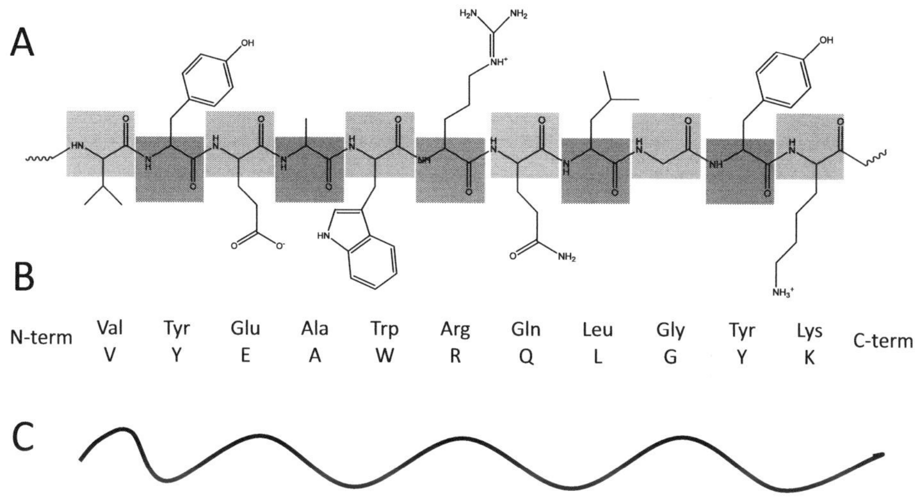

FIGURE 1. Structure of a fragment of a polypeptide chain

A Chemical structure of a fragment of a polypeptide chain from the knotted protein YibK. The main chain (sometimes referred to as the backbone) of the polypeptide is shown by the coloured rectangles. Whilst the side chains (which point away from the main chain) are not coloured. The peptide chain runs from the N-terminal end (left) to the C-terminal end (right). The wiggly lines at the end indicate the peptide chain can continue in both directions.

B Amino acid (primary sequence) of the peptide chain shown using the three-letter and single-letter amino acid code. Also indicated are the N- and C-terminal ends of the chain.

C Simple representation of an unstructured region of polypeptide chain frequently used.

---

intermediate state can be observed. Values for ΔG_{I1-N} and ΔG_{I2-N}, are calculated from kinetic data where the two intermediate states can be distinguished from each other.

[h]The value shown is for ΔG_{D-N}. Although it is known that two intermediate states are populated during unfolding/refolding the data can only be fit to give information on the overall free energy change between the denatured and native states.

[i]Values for ΔG_{D-I}, ΔG_{I-I*} and ΔG_{I-N} respectively are shown. The values are calculated from chemical denaturant unfolding experiments under equilibrium conditions.

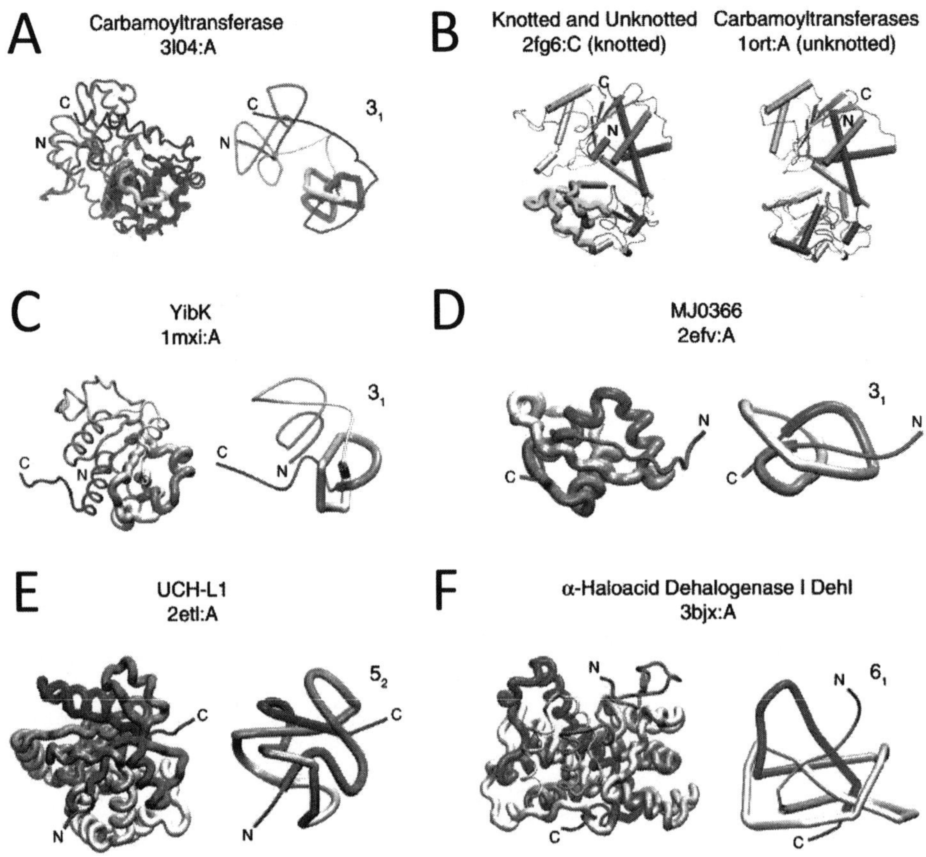

FIGURE 2. Structures of knotted proteins

The structures are all taken from the coordinates deposited in the protein databank (PDB) and the PDB code used in each case is shown. Ribbon diagrams are used which trace the main chain of the protein but which do not show the side chains. The protein chain is coloured from N-terminus (red) to C-terminus (blue) and the letters N and C are used to indicate the N- and C-termini respectively. On the right of the ribbon diagrams is a simplified (smoothed) representation of the polypeptide chain coloured the same as the ribbon diagram to more clearly show the knot and its position within the entire length of the protein chain.

A Structure of a trefoil (3_1) knotted carbamoyl transferase.

B Structures of knotted (left) and unknotted (right) variants of carbamoyltransferases. In the knotted structure the two loop regions, which if either/both were excised would generate a unknotted structure, are shown in yellow and orange.

C Structure of the bacterial trefoil (3_1) knotted methyltransferase YibK that has been studied extensively experimentally (*12, 24, 26, 29, 39, 40, 46, 70, 72, 73*).

D Structure of a small trefoil (3_1) knotted protein MJ0366 (*33*).

E Structure of the 5_2 knotted enzyme UCH-L1 from human neurons that has been studied extensively experimentally (*31, 46*) (*61, 64, 74, 75*).

F Structure of the 6_1-knotted dehalogenase enzyme DehI (*41, 76*). Figure adapted from a recent review (*11*).

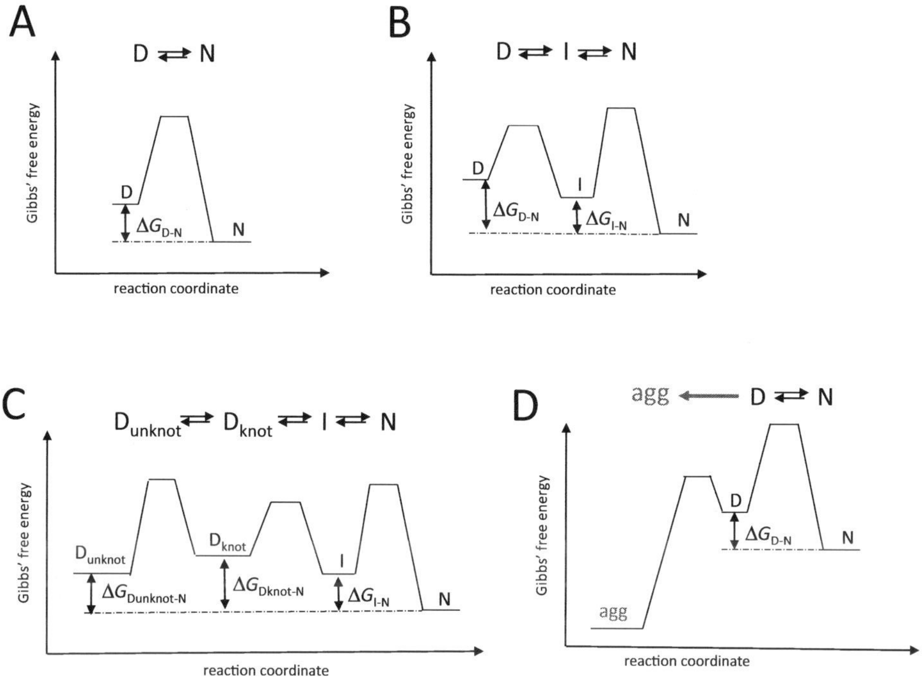

FIGURE 3. Free energy diagrams illustrating the thermodynamic stability of the native state of a protein

A Thermodynamic stability for a two-state system that only populates the folded, native (N) and unfolded, denatured (D) states under equilibrium conditions. The thermodynamic stability is given by ΔG_{D-N}, which under native conditions is positive as the native state is a lower energy than the denatured state. ΔG_{D-N} is usually the difference in free energy between the two states in the absence of denaturant, i.e., in water.

B Thermodynamic stability for a three-state system which also populates a partially unfolded/folded intermediate (I) state under equilibrium conditions. ΔG_{I-N} and ΔG_{D-I} are the differences in free energy between the native and intermediate state, and the intermediate and denatured states, respectively, and where the overall thermodynamic stability is given by ΔG_{D-N}.

C Free energy diagram showing the relative energy levels of the different sub-states that can be populated during the unfolding of a knotted protein that has an intermediate state and that initially unfolds to a denatured state that is knotted, D_{knot}, and then there is a slow conversion to an unknotted state, D_{unknot}. The thermodynamic stabilities, i.e., differences in free energy for the different sub-states are shown. Recently the groups of Bustamante and Baez have used optical tweezer techniques to estimate the difference in free energy of a knotted and unknotted denatured polypeptide chain, $\Delta G_{Dunknot-Dknot}$, and found it to be in the region of 6 kcal mol^{-1} (77). They also used Monte Carlo dynamic simulations to show the energy cost of knot formation is largely entropic in nature (77).

D Free energy diagram illustrating an irreversible reaction such as aggregation linked to the unfolding of a protein, in this case, a simple example of a two-state system. In this case, the apparent ΔG_{D-N} measured is less than the real ΔG_{D-I} as the system is not in equilibrium.

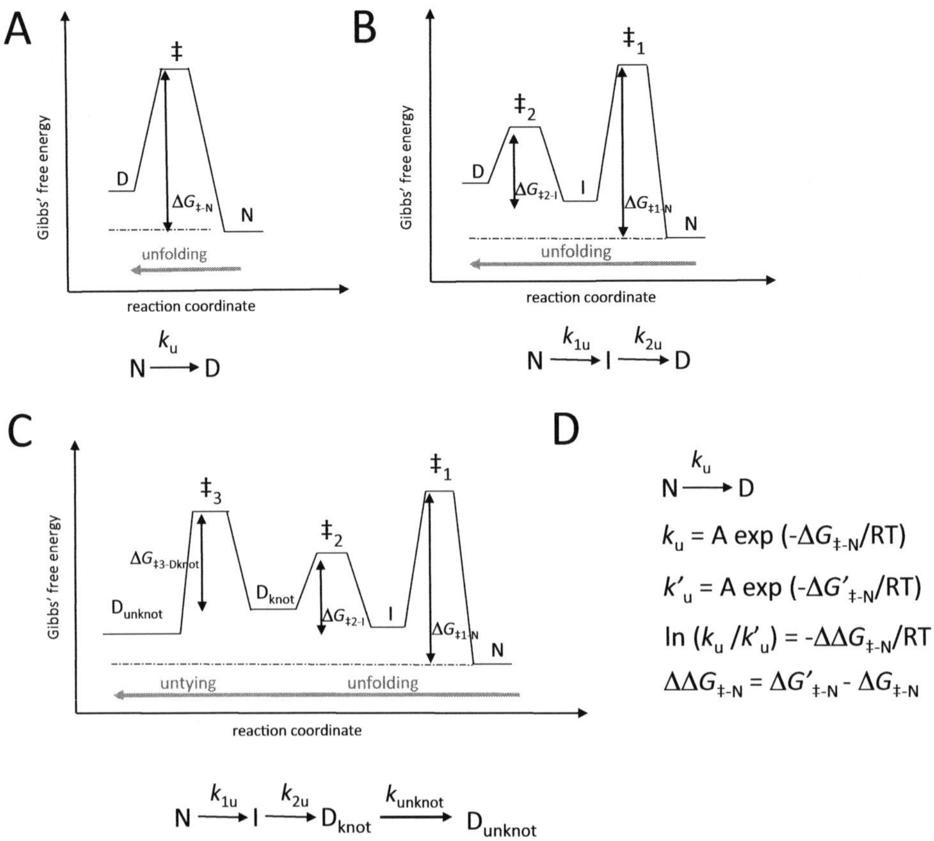

FIGURE 4. Free energy diagrams showing the different kinetic stabilities of proteins

A Free energy diagram showing the kinetic stability of a simple two-state which unfolds from the native (N) state through a transition state (‡) to a denatured (D) state. In this case, the rate constant for unfolding, k_U, is determined by the difference in free energy between the native and transition states, $\Delta G_{\ddagger-N}$.

B Free energy diagram showing the kinetic stability of a three-state which unfolds from the native (N) state through a transition state ($\ddagger_1$) first to an intermediate (I) state, and from there unfolds further through a second transition state ($\ddagger_2$) to the denatured (D) state. In this case, the rate constant for unfolding from native to intermediate state, k_{1U}, is determined by the difference in free energy between the native and transition state 1, $\Delta G_{\ddagger 1-N}$, and the rate constant for unfolding from intermediate to denatured state, k_{2U}, is determined by the difference in free energy between the intermediate and transition state 2, $\Delta G_{\ddagger 2-I}$.

C Free energy diagram showing the kinetic stability of a knotted protein that unfolds via a knotted intermediate state to a denatured state that is initially knotted ($\mathrm{D}_{\mathrm{knot}}$) and which then unties to a fully unstructured unknotted denatured state ($\mathrm{D}_{\mathrm{unknot}}$). In this case, the rate constants k_{1U} and k_{2U} are the same as in (B), however, there is an additional untying reaction which has a rate constant, k_{unknot}, that is determined by the energy barrier, $\Delta G_{\ddagger 3-Dknot}$.

D Arrhenius equation showing the exact relationship between the rate constant for a reaction and the energy barrier for a simple two-state system, where A is the pre-exponential factor, R is the Gas constant, and T the temperature in Kelvin. The pre-exponential factor has been estimated for protein folding, however, remains somewhat controversial. Generally, the ratio of the rate constants for the unfolding of one variant of a protein, k_U, compared with the unfolding of another variant, $k_{U'}$, is taken therefore eliminating the need to know or use the pre-exponential factor.

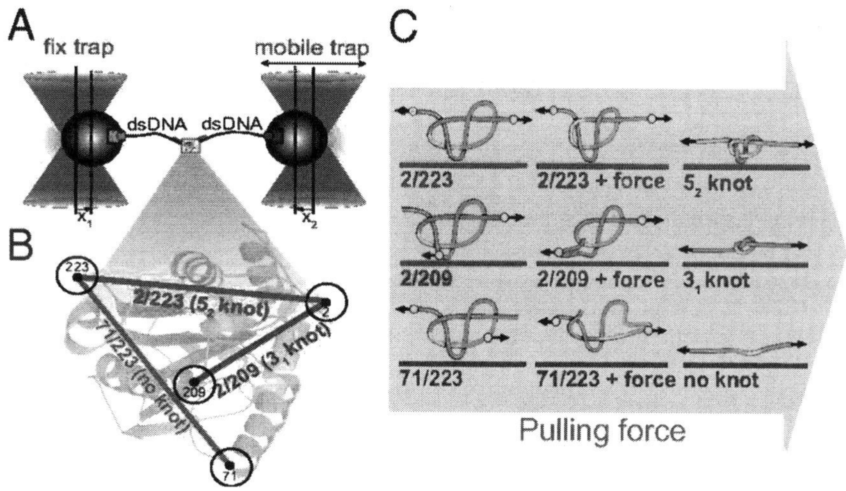

FIGURE 5. Mechanical unfolding of a 5_2-knotted protein using different pulling directions.

A Schematic of the experimental setup of a dual-beam optical tweezers assay. Forces are measured through the deflections of the beads (x_1, x_2) out of the trap centres. The knotted protein, in this case UCH-L1, is shown in the middle of the diagram attached to the beads using double-stranded (ds) DNA.

B Structure of UCH-L1 shown as a grey ribbon with the different attachment points: Pulling at positions 2 and 223 (N and C termini) leads to an unfolded structure with a 5_2-knotted topology (red); pulling at positions 2 and 209 leads to an unfolded structure with a 3_1-knotted topology (violet); and pulling at positions 71 and 223 leads to an unfolded structure without a knot (blue).

C Schematic representation of the unfolding and tightening of the mutant protein structures on application of force, colored as in *B*.

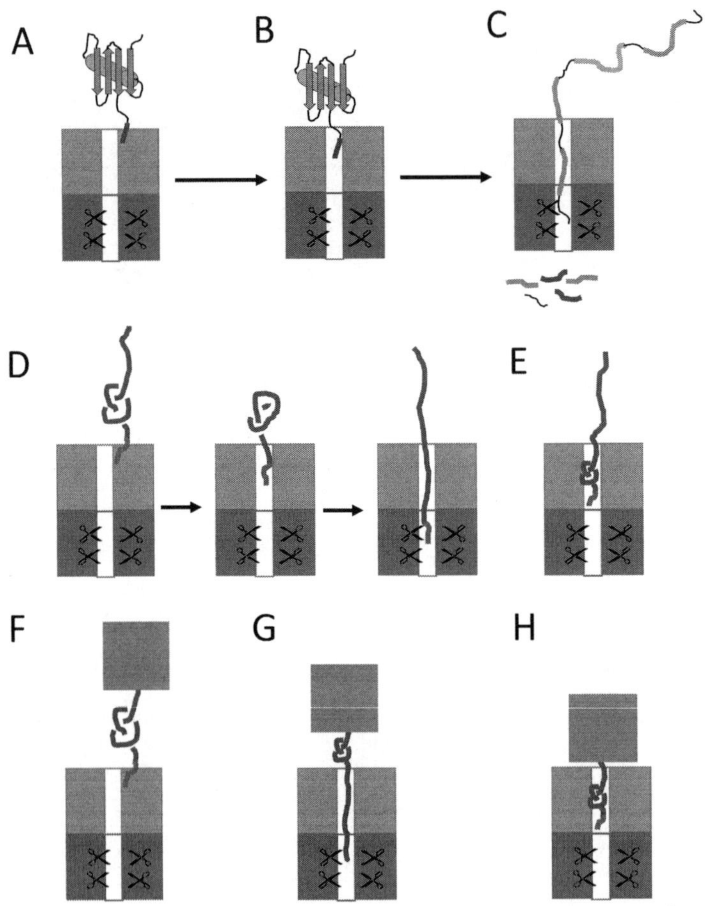

FIGURE 6. Schematic showing the key steps and mechanisms in the unfolding, translocation and degradation of unknotted and knotted proteins by a cellular degradation machine.

In all the panels, the degradation machine, e.g., ClpXP, is shown by a red and blue rectangle representing the cylinders created by oligomers of ClpX (blue) or ClpP (red), which stack on top of each other to create the full active complex. ClpX contains an ATP-binding and hydrolysis site (not shown), a recognition site for the degradation tag — shown in purple. It is the part of the machinery responsible for the initial binding of the target protein and then unfolding and translocating it into the central translocation pore which leads to the degradation sites in ClpP shown by scissors.

A Diagram indicates the initial binding of a protein through its degradation tag (purple). The protein shown is unknotted and contains region of β-sheet (green) and an α-helix (yellow).

B After initial binding through the degradation tag, the machinery binds the protein further within the translocation pore. If the protein remains folded the chain cannot be further translocated.

C After local or global unfolding, the protein chain can be translocated down and into the degradation chamber and undergo hydrolysis. Translocation is linked with rounds of ATP-binding and hydrolysis (not shown).

D A model of how the degradation machine can efficiently degrade knotted protein substrates. In this case, the protein unfolds and is translocated, as translocation of the chain takes place the knot slides along the chain and eventually falls off the free end.

E A model of how the degradation machine can efficiently degrade knotted protein substrates. In this case, the knot can enter the translocation pore and be pulled to the degradation sites where it is unknotted by fragmentation.

F A model of a knotted protein fused to a highly stable folded domain (orange square) which cannot be unfolded by ClpX, e.g., GFP (*60*) or ThiS (*65*).

G Tightened knot buttressing the stable GFP or ThiS domain. In this case, further translocation cannot occur and the partially degraded substrate can be released from the machinery. This would result in an intermediate which was the length of the stable domain (GFP or ThiS) plus the length of the tightened knot plus approx. 30 amino acid residues which is the amount of chain that spans from the opening of the translocation pore to the proteolytic sites.

H The stable GFP or ThiS domain is buttressing the ClpX cylinder, whilst the knot has entered the translocation pore. In this case, further translocation cannot occur and the partially degraded substrate can be released from the machinery. This would result in an intermediate which was the length of the stable domain (GFP or ThiS) plus approx. 30 amino acid residues which is the amount of chain that spans from the opening of the translocation pore to the proteolytic sites.

9. References

1. J. S. Richardson, beta-Sheet topology and the relatedness of proteins. *Nature* **268**, 495-500 (1977).

2. M. L. Mansfield, Are there knots in proteins. *Nature Structural Biology* **1**, 213-214 (1994).

3. W. R. Taylor, A deeply knotted protein structure and how it might fold. *Nature* **406**, 916-919 (2000).

4. M. Jamroz *et al.*, KnotProt: a database of proteins with knots and slipknots. *Nucleic acids research* **43**, D306-314 (2015).

5. W. R. Taylor, Protein knots and fold complexity: Some new twists. *Computational biology and chemistry* **31**, 151-162 (2007).

6. N. P. King, E. O. Yeates, T. O. Yeates, Identification of rare slipknots in proteins and their implications for stability and folding. *Journal of molecular biology* **373**, 153-166 (2007).

7. J. I. Sulkowska, P. Sulkowski, J. N. Onuchic, Jamming Proteins with Slipknots and Their Free Energy Landscape. *Physical review letters* **103**, (2009).

8. C. He, G. Z. Genchev, H. Lu, H. Li, Mechanically untying a protein slipknot: multiple pathways revealed by force spectroscopy and steered molecular dynamics simulations. *Journal of the American Chemical Society* **134**, 10428-10435 (2012).

9. C. He, G. Lamour, A. Xiao, J. Gsponer, H. Li, Mechanically tightening a protein slipknot into a trefoil knot. *Journal of the American Chemical Society* **136**, 11946-11955 (2014).

10. I. Coluzza, S. E. Jackson, C. Micheletti, M. A. Miller, Knots in soft condensed matter. *Journal of physics. Condensed matter : an Institute of Physics journal* **27**, 350301 (2015).

11. S. E. Jackson, A. Suma, C. Micheletti, How to fold intricately: using theory and experiments to unravel the properties of knotted proteins. *Current opinion in structural biology* **42**, 6-14 (2017).

12. A. L. Mallam, S. C. Onuoha, J. G. Grossmann, S. E. Jackson, Knotted fusion proteins reveal unexpected possibilities in protein folding. *Molecular Cell* **30**, 642-648 (2008).

13. M. R. Sanders, H. E. Findlay, P. J. Booth, Lipid bilayer composition modulates the unfolding free energy of a knotted alpha-helical membrane protein. *Proceedings of the National Academy of Sciences of the United States of America* **115**, E1799-e1808 (2018).

14. P. F. N. Faisca, R. D. M. Travasso, T. Charters, A. Nunes, M. Cieplak, The folding of knotted proteins: insights from lattice simulations. *Physical biology* **7**, (2010).

15. J. I. Sulkowska *et al.*, Knotting pathways in proteins. *Biochemical Society transactions* **41**, 523-527 (2013).

16. P. F. Faisca, Knotted proteins: A tangled tale of Structural Biology. *Computational and structural biotechnology journal* **13**, 459-468 (2015).

17. E. J. Rawdon, K. C. Millett, J. I. Sulkowska, A. Stasiak, Knot localization in proteins. *Biochemical Society transactions* **41**, 538-541 (2013).

18. R. Potestio, C. Micheletti, H. Orland, Knotted vs. Unknotted Proteins: Evidence of Knot-Promoting Loops. *PLoS computational biology* **6**, (2010).

19. A. R. Fersht, A. Matouschek, L. Serrano, The folding of an enzyme .1. Theory of protein engineering analysis of stability and pathway of protein folding. *Journal of molecular biology* **224**, 771-782 (1992).

20. A. N. Bullock *et al.*, Thermodynamic stability of wild-type and mutant p53 core domain. *Proceedings of the National Academy of Sciences of the United States of America* **94**, 14338-14342 (1997).

21. J. H. Han, S. Batey, A. A. Nickson, S. A. Teichmann, J. Clarke, The folding and evolution of multidomain proteins. *Nature reviews. Molecular cell biology* **8**, 319-330 (2007).

22. J. A. Rumfeldt, C. Galvagnion, K. A. Vassall, E. M. Meiering, Conformational stability and folding mechanisms of dimeric proteins. *Progress in biophysics and molecular biology* **98**, 61-84 (2008).

23. C. M. Doyle *et al.*, Energetics of oligomeric protein folding and association. *Archives of biochemistry and biophysics* **531**, 44-64 (2013).

24. A. L. Mallam, S. E. Jackson, A comparison of the folding of two knotted proteins: YbeA and YibK. *Journal of molecular biology* **366**, 650-665 (2007).

25. J. R. Huang, T. D. Craggs, J. Christodoulou, S. E. Jackson, Stable intermediate states and high energy barriers in the unfolding of GFP. *Journal of molecular biology* **370**, 356-371 (2007).

26. A. L. Mallam, J. M. Rogers, S. E. Jackson, Experimental detection of knotted conformations in denatured proteins. *Proceedings of the National Academy of Sciences of the United States of America* **107**, 8189-8194 (2010).

27. D. T. Capraro, P. A. Jennings, Untangling the Influence of a Protein Knot on Folding. *Biophysical journal* **110**, 1044-1051 (2016).

28. D. J. Burban, E. Haglund, D. T. Capraro, P. A. Jennings, Heterogeneous side chain conformation highlights a network of interactions implicated in hysteresis of the knotted protein, minimal tied trefoil. *Journal of physics. Condensed matter : an Institute of Physics journal* **27**, 354108 (2015).

29. A. L. Mallam, S. E. Jackson, Folding studies on a knotted protein. *Journal of molecular biology* **346**, 1409-1421 (2005).

30. F. I. Andersson, D. G. Pina, A. L. Mallam, G. Blaser, S. E. Jackson, Untangling the folding mechanism of the 5(2)-knotted protein UCH-L3. *Febs Journal* **276**, 2625-2635 (2009).

31. F. I. Andersson *et al.*, The Effect of Parkinson's-Disease-Associated Mutations on the Deubiquitinating Enzyme UCH-L1 *J. Mol. Biol.* **407**, 261-272. (2011).

32. L. W. Wang, Y. N. Liu, P. C. Lyu, S. E. Jackson, S. T. Hsu, Comparative analysis of the folding dynamics and kinetics of an engineered knotted protein and its variants derived from HP0242 of Helicobacter pylori. *Journal of physics. Condensed matter : an Institute of Physics journal* **27**, 354106 (2015).

33. I. Wang, S. Y. Chen, S. T. Hsu, Unraveling the folding mechanism of the smallest knotted protein, MJ0366. *The journal of physical chemistry. B* **119**, 4359-4370 (2015).

34. T. C. Sayre, T. M. Lee, N. P. King, T. O. Yeates, Protein stabilization in a highly knotted protein polymer. *Protein Engineering Design & Selection* **24**, 627-630 (2011).

35. M. K. Sriramoju, T. J. Yang, S. D. Hsu, Comparative folding analyses of unknotted versus trefoil-knotted ornithine transcarbamylases suggest stabilizing effects of protein knots. *Biochemical and biophysical research communications* **503**, 822-829 (2018).

36. W. Pfeil, The problem of the stability globular proteins. *Molecular and cellular biochemistry* **40**, 3-28 (1981).

37. M. A. Soler, P. F. Faisca, Effects of knots on protein folding properties. *PloS one* **8**, e74755 (2013).

38. S. D. Hsu, Protein knotting through concatenation significantly reduces folding stability. *Scientific reports* **6**, 39357 (2016).

39. A. L. Mallam, S. E. Jackson, Probing Nature's knots: The folding pathway of a knotted homodimeric protein. *Journal of molecular biology* **359**, 1420-1436 (2006).

40. A. L. Mallam, E. R. Morris, S. E. Jackson, Exploring knotting mechanisms in protein folding. *Proceedings of the National Academy of Sciences of the United States of America* **105**, 18740-18745 (2008).

41. I. Wang, S. Y. Chen, S. T. Hsu, Folding analysis of the most complex Stevedore's protein knot. *Scientific reports* **6**, 31514 (2016).

42. K. A. Scott, S. Batey, K. A. Hooton, J. Clarke, The folding of spectrin domains I: wild-type domains have the same stability but very different kinetic properties. *Journal of molecular biology* **344**, 195-205 (2004).

43. J. I. Sulkowska, P. Sulkowski, P. Szymczak, M. Cieplak, Stabilizing effect of knots on proteins. *Proceedings of the National Academy of Sciences of the United States of America* **105**, 19714-19719 (2008).

44. Y. Xu *et al.*, Stabilizing Effect of Inherent Knots on Proteins Revealed by Molecular Dynamics Simulations. *Biophysical journal*, (2018).

45. T. Bornschlogl *et al.*, Tightening the Knot in Phytochrome by Single-Molecule Atomic Force Microscopy. *Biophysical journal* **96**, 1508-1514 (2009).

46. F. Ziegler *et al.*, Knotting and unknotting of a protein in single molecule experiments. *Proceedings of the National Academy of Sciences of the United States of America* **113**, 7533-7538 (2016).

47. M. Bertz, M. Rief, Mechanical unfoldons as building blocks of maltose-binding protein. *Journal of molecular biology* **378**, 447-458 (2008).

48. M. Rief, M. Gautel, F. Oesterhelt, J. M. Fernandez, H. E. Gaub, Reversible unfolding of individual titin immunoglobulin domains by AFM. *Science (New York, N.Y.)* **276**, 1109-1112 (1997).

49. H. Dietz, M. Rief, Exploring the energy landscape of GFP by single-molecule mechanical experiments. *Proceedings of the National Academy of Sciences of the United States of America* **101**, 16192-16197 (2004).

50. J. I. Sulkowska, P. Sulkowski, P. Szymczak, M. Cieplak, Untying Knots in Proteins. *Journal of the American Chemical Society* **132**, 13954-13956 (2010).

51. J. I. Sulkowska, P. Sulkowski, P. Szymczak, M. Cieplak, Tightening of knots in proteins. *Physical review letters* **100**, (2008).

52. A. O. Olivares, T. A. Baker, R. T. Sauer, Mechanistic insights into bacterial AAA+ proteases and protein-remodelling machines. *Nature reviews. Microbiology* **14**, 33-44 (2016).

53. T. Inobe, A. Matouschek, Paradigms of protein degradation by the proteasome. *Current opinion in structural biology* **24**, 156-164 (2014).

54. R. A. Maillard *et al.*, ClpX(P) generates mechanical force to unfold and translocate its protein substrates. *Cell* **145**, 459-469 (2011).

55. C. Lee, S. Prakash, A. Matouschek, Concurrent translocation of multiple polypeptide chains through the proteasomal degradation channel. *The Journal of biological chemistry* **277**, 34760-34765 (2002).

56. P. Virnau, L. A. Mirny, M. Kardar, Intricate knots in proteins: Function and evolution. *PLoS computational biology* **2**, 1074-1079 (2006).

57. P. Szymczak, Tight knots in proteins: can they block the mitochondrial pores? *Biochemical Society transactions* **41**, 620-624 (2013).

58. A. Rosa, M. Di Ventra, C. Micheletti, Topological jamming of spontaneously knotted polyelectrolyte chains driven through a nanopore. *Physical review letters* **109**, 118301 (2012).

59. A. Suma, Rosa, A. & Micheletti, C., Pore Translocation of Knotted Polymer Chains: How Friction Depends on Knot Complexity. *ACS Macro Letters* **4**, 1420-1424 (2015).

60. A. San Martin *et al.*, Knots can impair protein degradation by ATP-dependent proteases. *Proceedings of the National Academy of Sciences of the United States of America* **114**, 9864-9869 (2017).

61. M. K. Sriramoju, Y. Chen, Y. C. Lee, S. D. Hsu, Topologically knotted deubiquitinases exhibit unprecedented mechanostability to withstand the proteolysis by an AAA+ protease. *Scientific reports* **8**, 7076 (2018).

62. A. Matouschek, Protein unfolding--an important process in vivo? *Current opinion in structural biology* **13**, 98-109 (2003).

63. J. A. Kenniston, R. E. Burton, S. M. Siddiqui, T. A. Baker, R. T. Sauer, Effects of local protein stability and the geometric position of the substrate degradation tag on the efficiency of ClpXP denaturation and degradation. *Journal of structural biology* **146**, 130-140 (2004).

64. S. C. Lou *et al.*, The Knotted Protein UCH-L1 Exhibits Partially Unfolded Forms under Native Conditions that Share Common Structural Features with Its Kinetic Folding Intermediates. *Journal of molecular biology* **428**, 2507-2520 (2016).

65. E. M. Sivertsson, Jackson, S.E. & Itzhaki, L.S., The AAA+ protease ClpXP can easily degrade both a 31 and a 52-knotted proteins. *Scientific reports* **in revision**, (2018).

66. M. Wojciechowski, A. Gomez-Sicilia, M. Carrion-Vazquez, M. Cieplak, Unfolding knots by proteasome-like systems: simulations of the behaviour of folded and neurotoxic proteins. *Molecular bioSystems* **12**, 2700-2712 (2016).

67. J. A. Kenniston, T. A. Baker, R. T. Sauer, Partitioning between unfolding and release of native domains during ClpXP degradation determines substrate selectivity and partial processing. *Proceedings of the National Academy of Sciences of the United States of America* **102**, 1390-1395 (2005).

68. S. E. Jackson, How do small single-domain proteins fold? *Folding & design* **3**, R81-91 (1998).

69. H. M. Went, S. E. Jackson, Ubiquitin folds through a highly polarized transition state. *Protein engineering, design & selection : PEDS* **18**, 229-237 (2005).

70. A. L. Mallam, S. E. Jackson, Knot formation in newly translated proteins is spontaneous and accelerated by chaperonins. *Nat. Chem. Biol.* **in press**, (2011).

71. S. E. Jackson, T. D. Craggs, J. R. Huang, Understanding the folding of GFP using biophysical techniques. *Expert review of proteomics* **3**, 545-559 (2006).

72. A. L. Mallam, S. E. Jackson, The dimerization of an alpha/beta-knotted protein is essential for structure and function. *Structure* **15**, 111-122 (2007).

73. N. C. Lim, S. E. Jackson, Mechanistic insights into the folding of knotted proteins in vitro and in vivo. *Journal of molecular biology* **427**, 248-258 (2015).

74. H. Zhang, S. E. Jackson, Characterization of the Folding of a 52-Knotted Protein Using Engineered Single-Tryptophan Variants. *Biophysical journal* **111**, 2587-2599 (2016).

75. Y. C. Lee *et al.*, Entropic stabilization of a deubiquitinase provides conformational plasticity and slow unfolding kinetics beneficial for functioning on the proteasome. *Scientific reports* **7**, 45174 (2017).

76. D. Bolinger *et al.*, A Stevedore's Protein Knot. *PLoS computational biology* **6**, (2010).

77. A. Bustamante *et al.*, The energy cost of polypeptide knot formation and its folding consequences. *Nature communications* **8**, 1581 (2017).

78. Y. C. Chuang *et al.*, Untying a protein knot by circular permutation. Journal of Molecular Biology **431**, 857-863 (2019).

79. K. T. Ko *et al.*, Untying a knotted SPOUT RNA methyltransferase by circular permutation results in a domain-swapped dimer. Structure **27** 1224-1233.e1224 (2019).

80. M. K. Sriramoju *et al.*, Protein knots provide mechano-resilience to an AAA+ protease-mediated proteolysis with profound ATP energy expenses. Biochim Biophys Acta Proteins Proteom **1868**, 140330 (2020).

81. Y. Xu *et al.*, Revealing cooperationl between knotted conformation and dimerization in protein stabilization by molecular dynamics simulations. J. Phys. Chem. Lett. **10**, 5815-5822 (2019).

Department of Chemistry, University of Cambridge, Lensfield Road, Cambridge, CB2 1EW, UK

Email address: sej13@cam.ac.uk

Contemporary Mathematics
Volume **746**, 2020
https://doi.org/10.1090/conm/746/15007

Knotted proteins: Tie etiquette in structural biology

Ana Nunes and Patrícia F. N. Faísca

ABSTRACT. A small fraction of all protein structures characterized so far are entangled. The challenge of understanding the properties of these knotted proteins, and the why and the how of their natural folding process, has been taken up in the past decade with different approaches, such as structural characterization, *in vitro* experiments, and simulations of protein models with varying levels of complexity. The simplest among these are the lattice Gō models, which belong to the class of structure-based models, i.e., models that are biased to the native structure by explicitly including structural data. In this review we highlight the contributions to the field made in the scope of lattice Gō models, putting them into perspective in the context of the main experimental and theoretical results and of other, more realistic, computational approaches.

CONTENTS

1. Introduction

Proteins are chains of amino acids that fold into specific three-dimensional shapes, known as native conformations, to become biologically functional. Finding the native structure of all encoded amino acid sequences is an on-going task, but more than a hundred thousand protein native conformations are already freely available in the Protein Data Bank (PDB) (1), in the form of Cartesian coordinates for each atom in the protein. It has been known for some time that some of these native structures have their backbone knotted (2,3).

A topological knot (4) is a closed curve in space, and all its continuous deformations are by definition equivalent knots. Any such curve may be represented by a planar projection with a finite number of double crossings, together with the specification of the over- and under-strand at each crossing. A reduced knot diagram is one of these planar representations for which the minimum number of crossings is attained. This number, called the crossing number, is therefore a topological invariant. The simplest knot is the circle, with crossing number 0, and the simplest non-trivial knots are the trefoil (3_1 in the Alexander–Briggs notation that will be used throughout) and the figure-eight (4_1) knots, with crossing numbers three and four, respectively. There are two different knots with crossing number five, the cinquefoil (5_1) and the twisted-three (5_2) knot, and three different knots with crossing number six. The number of knots increases very rapidly with crossing number from then on, but there are no known examples of protein folds with crossing number larger than six.

In rigor proteins do not form topological knots because they are open chains. Therefore, it is more accurate to say that protein conformations embed physical (or open) knots, although in many cases the termini can with little ambiguity be connected with an external arc to form a closed loop **(Fig. 1A-B)**. Specific computational methods have been developed to determine if a given protein conformation is knotted, and to make it amenable to knot invariant calculations (in most cases the Alexander polynomial) so as to find its knot type. An important example is the Koniaris–Muthukumar-Taylor (KMT) algorithm developed in 2000 by Taylor (5), which extends to a wide range of protein conformations the application of the Koniaris–Muthukumar method (6) developed for ring polymers **(Fig. 1C-D)**. Alternative methods to determine the knotting state of proteins based on different loop closure procedures have since then been proposed (7-11). More recently, the concept of knotoids introduced by Turaev (12) set the basis for a new method to study the entanglement of open protein chains (13), further refined in (14), and now available as an open access software tool (15).

Independently of these methodological advances, the overall picture of the topological properties of the known native conformations is by now well established (16,17). The systematic application of knot detection methodologies based on the KMT algorithm to all available protein entries in the PDB revealed that about 1% correspond to knotted proteins (10,16). Although the trefoil is by large extent the most common knot type, it is possible to find a few proteins with more complex knots, including the Stevedore knot 6_1 (with six crossings on a planar projection) (18). An interesting variation amongst tangled proteins is that of the slipknot, in which one (or both (19)) protein termini adopts a hairpin-like conformation that threads a loop formed by the remainder of the chain (20). There are ~ 500 slipknotted proteins in the PDB (16) .

These numbers mean that, in a sense to be precised in the following sections, knotted proteins are uncommon, but the reasons for why the protein universe avoids knots have not yet been fully established. It is possible that local geometric features of proteins that do not favor tangling of the backbone were selected for, but it is also likely that knots were partly removed by evolution due to their unfavorable effect on protein folding. To clarify whether knots were selected against as a global

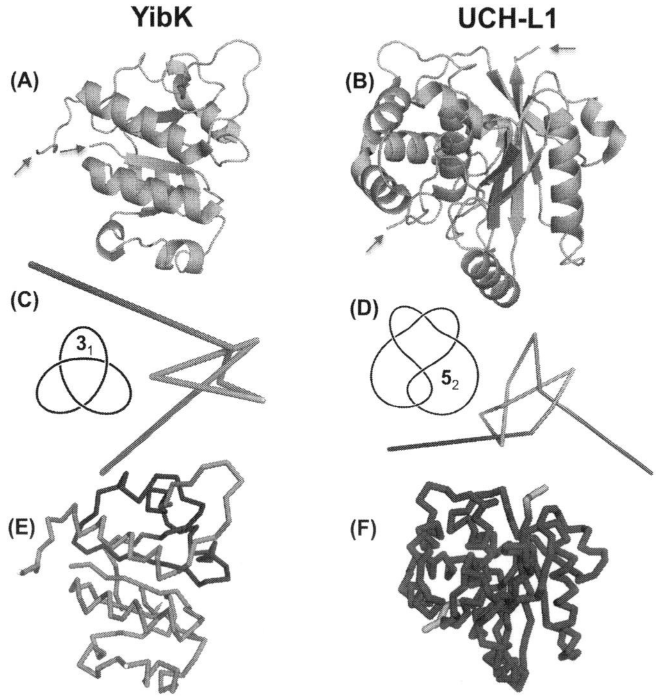

FIGURE 1. Three-dimensional cartoon representation of the native structure of protein YibK (A) and UCH-L1 (B). Note that the termini (denoted by arrows) are located very close together in the case of YibK, and connecting them such that the backbone makes a closed curve in space is also straightforward for UCH-L1. Reduced representation of YibK (C) and UCH-L1 (D) resulting from the application of the KMT algorithm, which highlights the existence of a knot 3_1 in YibK and knot 5_2 in UCH-L1. Ribbon representation with the knotted core (i.e. the minimal part of the chain that contains the knot) highlighted in blue (E) and red (F). In the case of YibK both knot tails are long and the knot classifies as deep, while for UCH-L1 they are short and the knot classifies as shallow.

property or indirectly as a result of selection operating on protein structure at the local level it is necessary to understand how knotted proteins fold.

If it is already challenging to determine how 'regular' proteins fold, it is even more formidable to do it in the case of knotted proteins, which have distinctively complicated native structures. But a complete solution to the protein folding problem must necessarily encompass the solution to this new exciting folding puzzle (21). Therefore, perhaps not surprisingly, during the last decade knotted proteins became a favorite theme within the protein folding community. Several research groups worldwide, both experimental and theoretical have been making significant efforts to solve this tangling puzzle using simulated environments, both in vitro and *in silico* (reviewed in (22)), and first steps have been taken in order to determine how this intricate process occurs inside the living cell (23,24).

Although the overwhelming majority of protein structures in the PDB are unknotted, some proteins are knotted - in contrast with the RNA structures in the PDB, for which the incidence of knots is practically negligible (25). Whatever the selection mechanism that operated, it spared a few structures. The realization that knotted proteins exist motivated the search for the functional advantages that knots may convey to their carriers. The analysis of specific model systems has put forward the idea that knots could enhance thermal (20,26) or mechanical stability (27), just to mention a few examples. However, a universal role for knots in proteins has not yet been identified.

Here we review a set of results, which address the questions outlined above in the scope of molecular simulations. This will not be a comprehensive review article on knotted proteins. We focus on structured-based models (including our own work on lattice models of protein folding) and integrate these views in a broader perspective. There are recent, thorough review articles on the subject that complement the present work (28-32), and the interested reader is invited to consult them for a comprehensive perspective on these interesting matters.

2. Structure-based models

In the 1970s Nobuhiro Gō proposed the now famous Gō potential for protein folding (33). According to the Gō potential, protein folding energetics is driven only by the intramolecular interactions present in the native structure. Therefore, only those interactions can contribute to energetically stabilize a folding chain. The Gō potential was originally developed in the context of a lattice representation **(Fig. 2A-C)**. A lattice representation is one in which the amino acids are reduced to beads of uniform size and placed on the vertices of a two- or three-dimensional regular lattice, with the lattice spacing representing the peptide bond. Lattice models, such as the one used in the original Gō model, retain the fundamental features of polypeptide chains (connectivity, excluded volume etc.) and can be used to explore fundamental aspects of the folding process (e.g. how folding rate depends on the chain length etc.). The combination of the Gō potential with off-lattice representations (including C-alpha and full atomistic resolutions) allows exploring the folding mechanism of specific model systems **(Fig. 2D-F)**.

The Gō model is the archetypal structure-based model (SBM). A SBM is one that imposes a native bias by explicitly including structural data (34). In off-lattice representations the atomic coordinates derived from NMR, X-ray or Cryo-EM must be used to construct the intramolecular potential. SBMs became particularly popular in the beginning of this century after the experimental realization that native 'topology' (geometric properties of the native configuration measured by the contact order (35) or other metrics (36,37)) is a major drive of the two-state folding transition of small, monomeric proteins (see also (38)). On the theoretical side, they are consistent with the energy landscape framework, which envisions proteins as minimally frustrated heteropolymers with a smooth funnel-like energy landscape biased toward the native state (39-41). SBMs can be combined with Monte Carlo methods, and Molecular Dynamics protocols. They offer a clear advantage over more realistic force fields (i.e. those traditionally used in classical Molecular Dynamics such as the GROMOS or AMBER force fields (42)) because of their lower computational cost, especially when combined with replica-exchange simulations (43) and other sampling protocols (44) designed to accelerate conformational search and relaxation

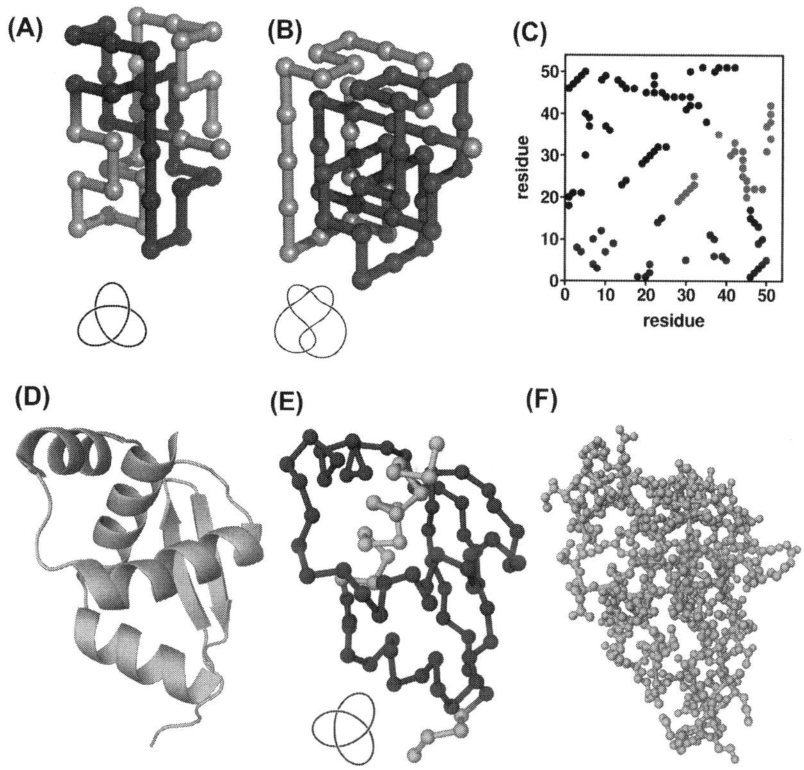

FIGURE 2. Simple lattice representations of a shallow knotted trefoil (A), and of a shallow knotted twisted-three knot (B), both designed by hand, used in Monte Carlo simulations. The lattice representation reduces the amino acids to beads of uniform size, and restricts them to occupy the vertices of a regular lattice. The knotted core is highlighted in blue (knot 3_1) and red (knot 5_2). Contact map (C) representing the set of 52 intra-molecular interactions that exist in the native conformation. On-lattice it is straightforward to define a native interaction: Is one that establishes between beads that are separated by one lattice spacing, but are not linked along the protein's backbone. In a structure-based model, like the Gō model, only native interactions contribute to stabilize the protein as it folds towards the native conformation. In the contact map the red points represent interactions that establish between beads pertaining to the knotted core. Three-dimensional cartoon representation of the native structure of MJ0366, the smallest knotted protein found in the PDB, which embeds a trefoil knot (D). C-alpha continuum representation of MJ0366 (E) where each amino acid is reduced to a bead of uniform size centered at the respective C-alpha atom. Beads are connected along the chain by sticks representing the peptide bond. The knotted core is highlighted in blue in the C-alpha representation. Since one of the knot's tails only spans four residues, the knot classifies as shallow. The most complex representation of the native structure is the full atomistic representation, where all the atoms of the protein (except hydrogen) are explicitly represented (F).

towards equilibrium. Despite their simplicity, SBMs have been successful in the study of protein folding (34,45-47) and, more recently, in exploring other phenomena involving proteins such as aggregation (48-50), protein fold switches (51), and phase separation (52). Clearly, SBMs, and in particular Gō models, will not be able to correctly model misfolding processes leading to compact non-native states and, more generally, regions of the free energy landscape where non-native interactions play an important role (53-55). Interestingly, a seminal study by Shakhnovich and co-workers, which was the first to explore the folding mechanism of knotted trefoil (56), reported that a simple C-alpha representation combined with a Gō potential would not be able to efficiently fold the protein: the process had necessarily to be assisted by non-native interactions. This claim sparked an interesting discussion on the role of non-native interactions in the folding of knotted proteins and triggered a series of studies that contributed to clarify their folding mechanism and stimulate experimental investigations as we outline below.

3. Knotting physics and why knotted proteins are rare

The starting point to understand the role, if any, of topological complexity in biopolymers must be the comparison with the frequency and type of knots we expect to find in homopolymers of comparable length. Before embarking on an overview of the state-of-art on the folding mechanism of knotted proteins we propose to discuss some aspects underlying the physics of knotting in a more general perspective, and then narrowing down our discussion towards the specific case of proteins. This will contribute to clarify the question of rarity of knotted proteins and motivate the need to understand their folding mechanism. Interestingly, many of the results we will overview have been obtained in the scope of simple models including lattice representations.

The self-avoiding random walk (SAW) on a three-dimensional lattice is perhaps the simplest model for a real polymer, one in which excluded volume and stiffness are minimally represented, favoring local entanglement. Indeed, it was conjectured by Frisch, Wassermann and Delbruck that sufficiently long polymers must be knotted (57), and rigorously proved by Sumners and Whittington (58) that almost all sufficiently long self-avoiding walks on the three-dimensional simple cubic lattice contain a knot.

Application to real polymers required some information about the typical length of knotted SAWs, as well as the extension to models including more physical features. This prompted a large body of numerical studies based on self-avoiding polygons, both on-lattice (59,60) and off-lattice with variable thickness (6,61,62). These results, reviewed and expanded in (63), show that probability of the trivial knot in a ring polymer with N segments, is well approximated by an exponential function, $P(N) \sim \exp(-N/N_0)$, so that there exists indeed a model dependent knotting length scale N_0. On lattice, N_0 is of the order of 10^5. Off-lattice models yield values of N_0 that are very sensitive to excluded volume, ranging from ~ 300 for random polygons (no excluded volume) to ~ 10000 when the polymer thickness is about 1% of the edge length.

A study of maximally compact conformations obtained from Hamiltonian walks on cubic lattices found an exponentially decaying probability $P(N)$ with knotting length scale $N_0 \sim 196$ (64), showing that chain compactification favors knot formation even more than ignoring excluded volume. The fact that proteins are compact polymers is therefore a key ingredient to gauge their topological properties through numerical studies of abstract models, and since most proteins in the PDB have a few hundred amino acids, the value found for N_0 in this study already suggested that the low abundance of knotted proteins cannot be ascribed to chance. Moreover, knots in these abstract models should in general be harder to obtain than the knots usually found in proteins, the latter being typically shallow, i.e., they become unknotted by removing a small number of amino acids starting from one of the termini.

More recently, statistical analysis of large samples of random polygons bounded within spheres of variable radius was employed (65-67) to explore in this context the effect of confinement on knotting probability and knot type, as well as on the relation between knot type and average geometrical properties, such as total curvature and total torsion.

Progress towards more realistic physical models was obtained with off-lattice numerical simulations of a simplified model of polyethylene (68), represented by chains of interacting monomers connected by springs, parametrized so as to reproduce experimental data. At high temperatures, the entropic term of the free energy dominates, favoring the swollen or coil state. By contrast, at low temperatures, the energetic contribution of the interaction potential dominates, favoring compact conformations—the globule state. Simulations of the system in these two regimes with chain lengths up $N = 1000$ to recapitulate the two complementary scenarios found in the abstract models mentioned above. Knots are indeed rare in the coil phase, with only around one percent of configurations knotted for $N = 1000$, and common in the globule state, with about half the conformations knotted for $N \sim 600$.

Another way of endowing a linear or a ring polymer with physical features (besides excluded volume) is to introduce its rigidity through a tunable bending energy. A systematic survey of the interplay between chain length, bending rigidity and knotting (69) confirmed the low abundance of knotted conformations for swollen polymers (less than 2% for $N = 500$) and found a surprising effect. As chain flexibility increases, the equilibrium conformations become more compact, but their knotting probability is strongly non-monotonic, exhibiting a pronounced maximum for intermediate bending rigidities. This counterintuitive phenomenon is shown to be due, in a more subtle way than in the previous example, to the competition between the energetic and the entropic contributions.

The studies mentioned above also include estimates for the probability $P_K(N)$ of specific non-trivial knots of type K as a function of chain length and other model parameters, with the trefoil, knot 3_1 being by far always the most abundant, in agreement with PDB data.

Perhaps not surprisingly given the specificity of each protein's amino acid sequence and spatial arrangement, the application these ideas and results to biopolymers has not been as illuminating and conclusive in the case of proteins and RNA as in the case of DNA (29).

Because it is easier to produce and analyze knots in DNA than in other polymers, most experiments on knotting involve DNA (29), and a wealth of experimental data has been collected for the past thirty years. It was shown as early as in 1993 that the description of a random self-avoiding semiflexible chain of a given thickness reproduced both qualitatively and quantitatively the features of free DNA

(70). In this work, the effective diameter of the DNA double helix was determined by comparing experimental results with computer simulations of knotting probability, and the results for the diameter in NaCl solutions at different concentrations were found to be in agreement with independent theoretical predictions based on polyelectrolyte theory.

DNA molecules in the cell are in general in a crowded or confined physical environment that differs drastically from that of free molecules in good solvent conditions. Compactification should increase the knotting probability, and systematic simulations of flexible bead-spring rings confined in a sphere of radius R (71) quantified this effect by finding a scaling function of R and N for the ratio of the probabilities of the trivial knot in the free and in the confined system. Further studies extended the model to include chain thickness and bending rigidity, with realistic parameter choices for DNA, and the equilibrium knot spectrum—the probabilities of formation of the different knot types—was estimated (72). The goal of these theoretical efforts was to reproduce the main features found in a series of experimental data available for bacteriophage DNA packed in the capsids (reviewed in (29) and (73)), but one intriguing aspect of the observed knot spectrum remained unexplained. The bacteriophage genome is very tightly packed in the capsid, displaying an extremely high knotting probability ($\sim$95%) together with a dominance of very complex knots (most of the knots have crossing number larger than 10). Another characteristic of DNA arrangement in the capsid is that the frequency among simple knots of certain knot types, such as 4_1 or 5_2, is much smaller than that of other knot types, such as 3_1 and 5_1. The knot 4_1 is achiral, i.e., equivalent to its own mirror image, while the other three knot types lack this symmetry. The abundant knot types 3_1 and 5_1 are torus knots, i.e., knots that can be drawn on the surface of a torus. This apparent bias of the knot spectrum towards torus knots and against achiral knots could not be reproduced with the inclusion of chain thickness and bending rigidity alone. A successful description was obtained later in (73), by introducing a DNA-DNA interaction potential that translates into a preferred "twist angle" between DNA segments that are close in space.

In contrast with DNA, the incidence of knots in RNA has only recently been systematically explored (25), with a very clear and surprising conclusion: the knotting frequency of RNA in the PDB is virtually zero. Physical and biological causes for the absence of knotted RNA are still being debated (74).

The first systematic comparative study of the geometric and topological properties of natural proteins and compact random homopolymers, as a function of their length, is due to Lua and Grosberg (9). The topological analysis showed that knots in proteins are orders of magnitude less frequent than in random polymers of comparable length, compactness, and bending rigidity. The geometrical analysis showed that, although the overall geometry of the conformations is statistically close to random, there are significant differences at the local (<10 residues) and intermediate (between 15 and 40 residues) levels. On a local scale, due to the formation of secondary structure (α-helices and β-chains), proteins are more stretched than random. On an intermediate scale, protein chains tend to fold back on themselves and crumple, a feature that reflects the most common arrangements of consecutive secondary structure elements and was known to be consistent with the suppression of knots. Thus, a connection was established between knot avoidance, a global property, and the local properties of protein chains.

This explanation of why there are so few knotted proteins leads to the question of why are there knotted proteins at all, and what is special about them. Pioneering experimental work in knotted proteins, Jackson and co-workers found that proteins YibK and YbeA unfold reversibly upon denaturation, and the native knot type appears spontaneously and reproducibly in the same protein location (21,75), inspiring a systematic study of all knotted, versus unknotted, protein structures in the PDB, together with their sequences (76). The analysis revealed in several knotted structures the presence of relatively short segments whose removal results in structurally similar though unknotted configurations, suggesting that there are localized 'knot-promoting' regions within knotted proteins. It also showed that there is little similarity, at the sequence level, between knotted and unknotted proteins.

The relation between knottedness and sequence was explored in the scope of an on-lattice HP model (77). The HP model represents a protein as self-avoiding chain of amino acids of two types only, 'hydrophobic' (H) or 'polar' (P), and the only interaction present is the attraction of neighboring H-H pairs (78). Sampling large ensembles of native conformations for different random and designed sequences showed that the mean unknotting probability, P_0, is about 0.46, slightly larger than for homopolymers, but there is a large variability in sequence space, with P_0 ranging from 0.3 to 0.6 in a set of 100 random sequences and examples of designed sequences with P_0 as high as 0.897 and as low as 0.114.

Taken together, these results strengthen the idea that knot abundance in proteins has to be understood in the light of evolution.

4. First insights into the folding of knotted proteins via molecular simulation

Despite remarkable advances in experimental studies on the folding of knotted proteins (23,79-85), it is not yet possible to obtain a structurally resolved picture of the knotting mechanism. Furthermore, the difficulty in creating unfolded ensembles of unknotted conformations through chemical denaturation (86) creates an additional challenge to explore this process through in vitro studies. A main advantage of molecular simulations is to obtain predictions (often with atomic detail) that can be experimentally tested, leading to a deeper understanding of these intricate processes. Sometimes, the results from simulations also contribute to prompt and design novel experiments, and, on the other hand, experiments can be used to refine simulations (87).

The first computational study addressing the problem of understanding how a knotted protein folds, was carried out by Shakhnovich and co-workers in 2007 (56). They combined Langevin Dynamics simulations with a C-alpha representation to explore the folding mechanism of protein YibK (PDB ID: 1j85), which embeds a trefoil knot. This is a bacterial alpha-beta protein of the methyltransferases (MTases) family, which, together with YbeA (PDB ID: 1vh0), have been extensively studied to shed light on the folding of knotted proteins. YibK comprises 156 residues, and the knotted core, i.e., the minimal segment that contains the knot, is located 77 residues away from the N-terminus and 39 residues away from the C-terminus (Fig. 1 (A-C)). YibK classifies, therefore, as a deeply knotted protein; in contrast, if the removal of a few residues is enough to untie the carrier protein, the corresponding knot classifies as shallow. Shakhnovich's study provided the very first view on the remarkably intricate knotting mechanism, including the structure of the knotting

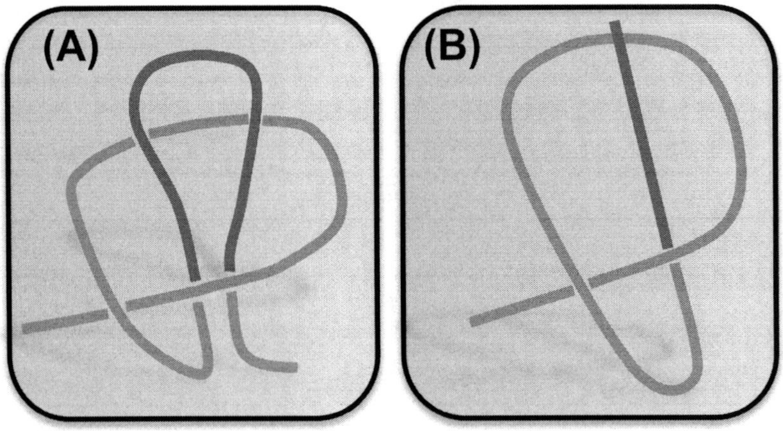

FIGURE 3. Structure of the knotting step as predicted by simulations. In splipknotting (A) one of the chain termini arranges itself in a hairpin like conformation that threads the knotting loop formed by the remainder of the chain. An alternative mechanism involves direct threading of the termini through the knotting loop (B).

step, and timing of knot formation. Indeed, it predicted that to get knotted, part of the polypeptide chain of YibK must first arrange itself in a loop (the so-called knotting loop, which is not necessarily in the native conformation), which is subsequently threaded by the C-terminus **(Fig. 3A)**. The authors pointed out that in most of the cases threading of the C-terminus proceeds directly **(Fig. 3B)**, but in some trajectories the C-terminal region is, instead, arranged in a hairpin-like conformation. Furthermore, knotting in their model system occurs predominantly late, in near-native (and therefore compact) conformations (with fraction of native contacts $Q \sim 0.8$).

The most striking conclusion of this seminal study concerns the energetics of the knotting step. Indeed, it was reported that when only native interactions are considered (as in a pure Gō potential), the conformations are too compact to allow for threading events, and none of the 100 attempted folding runs was successful. Non-native interactions are essential for folding, and in order to observe 100% folding efficiency (i.e. all attempted folding trajectories lead to native conformations) threading must be assisted by a set of specific non-native attractive interactions established between a stretch of residues within the knotting loop (86-93) and the residues pertaining to the C-terminus.

5. Insights into the knotting mechanism from molecular simulations

A follow-up study by Sulkowska *et al.* analysed the folding process of YbiK and YbeA within a similar simulation set-up (88). In line with the results reported by Shakhnovich, only a small ($\sim 1-2\%$) percentage of successful folding trajectories was recorded for a pure Gō potential. It was found that in the folding runs that lead to the native state, the proteins populate a conformation that embeds a slipknot. The slipknot arises when the C-terminal region arranges itself in a hairpin like conformation and threads a partially structured region that contains the knotting loop. This intermediate conformation bears resemblance with that reported by Shakhnovich and co-workers for the less likely folding pathway of the modified Gō potential.

In the wake of these studies several contributions followed, exploring other model systems and using alternative simulation strategies. In particular, inspired by Sulkowska's results, Faísca *et al.*(89) decided to explore in the scope of Monte Carlo simulations the folding of a designed lattice protein embedding a shallow trefoil in its native structure. Like Sulkowska, they also used a pure Gō potential to model intramolecular interactions. All of the thousands attempted folding runs were successful, and the occurrence of both knotting mechanisms was detected. However, direct threading was the most frequent one. The higher efficiency of the lattice model may result from the fact that the analyzed system is considerably smaller (41 residues long), and the knot is shallow, which facilitates direct threading of the shorter knot tail. By measuring the knotting frequency (the fraction of knotted conformations) against the folding probability, P_{fold} (an accurate kinetic estimator of folding progression)(90), it was found that the formation of the knot is mainly a late folding event, occurring with high probability (≥ 0.6) in conformations with $P_{fold} > 0.8$. However, a subsequent study revealed that the folding efficiency decreases dramatically upon tethering the protein to a chemically inert surface through a neutral linker (91). In particular, if surface tethering occurs at the bead that is closer to the knotted core, the folding rate becomes exceedingly (i.e. about two orders of magnitude) slow, and the protein is no longer able to find the native structure in all the attempted folding trajectories. The knotting frequency keeps a negligible value throughout the folding process, increasing abruptly to $\sim$ 1 when the protein is nearly folded, with more than 80% of its native contacts formed. These results predict that the mobility of the terminus closest to the knotted core is critical for efficient folding, which, in turn, highlights the importance of a knotting mechanism that is based on a threading movement of the shorter knot tail. Interestingly, Lim and Jackson provided the first experimental evidence for a knotting mechanism based on the threading of the C-terminus for proteins YbiK and YbeA (24). Indeed, as predicted by the simulations mentioned above, hampering the threading of the C-terminus, i.e., the one closer to the knotted core, also results in a dramatic reduction of the *in vitro* folding rate for these model systems.

Škrbić and co-workers used Monte Carlo simulations of a C-alpha model to investigate the folding process of protein AOTCase (PDB ID: 2g68) (92). This is a 332 amino acid long chain that contains a deep trefoil knot located 173 residues away from the N-terminus and 81 residues away from the C-terminus. When a pure Gō potential is used to model protein energetics, the knotting frequency is always negligible throughout the folding process. However, when non-native interactions are included, the knotting propensity raises considerably. This behavior may result from the fact that non-native interactions energetically stabilize partially folded conformations, therefore lowering the free energy barrier of the entropically costly knotting step (85). Furthermore, for this protein, the knotting step also consists of a direct threading movement of the C-terminus, in line with that observed by Shakhnovich for YibK (56).

A follow-up study looked into the folding mechanism of MJ0366 (PDB ID: 2efv), using an advanced simulation technique known as the dominant reaction pathway, combined with a realistic force field with implicit solvent (93). MJ0366 is the smallest (82 amino acids) knotted protein in the PDB embedding a shallow trefoil knot, located 10 residues away from the C-terminus. The authors reported that knotting typically occurs at a late folding stage, in conformations with $Q = 0.90$, and is almost invariably based on a direct threading mechanism, in line with the observations for the shallow knotted system (89) mentioned above. These results are somehow different from those previously reported by Noel *et al.* for the same model system when analyzed within the scope of a full atomistic model combined with a Gō potential (94). Indeed, in this case, it appears that side-chain packing leads to the population of an intermediate state with about 25% of native contacts formed (i.e. $Q \sim 0.25$), which is structurally characterized for having the native knotting loop formed. Two knotting pathways are observed: one, based on a slipknotting mechanism, dominates below the folding transition temperature, T_f, (i.e. the temperature at which native and unfolded states are equally populated under equilibrium conditions) while the other, based on a direct threading of the C-terminus becomes equally likely at T_f. Interestingly, recent experimental studies based on NMR hydrogen-deuterium analysis also suggested the population of an on-pathway intermediate, although considerably more structurally consolidated than the one predicted by simulations (95). 15 selected conformations containing the slipknot were subsequently used as starting conformations in Molecular Dynamics simulations in explicit solvent with the amber99sb force field (96). These conformations had 10 (out of 15) residues already threaded and had yet to thread another 5 residues to form the native knot. Five of these conformations reached the native state in 450 ns, adding to the plausibility of a knotting mechanism based on a spliknotted conformation in real proteins. Remarkably, it appears that the native contacts that thread the terminus through the knotting loop are conserved in the realistic force field, supporting the use of structure-based models to study complex folding processes.

The results outlined above indicate that in trefoil proteins the knotting mechanism is generally based on a threading movement of the C-terminus. The latter can be direct, or, instead, involve the population of a conformation with a slipknot. Evidence for the existence of a knotting mechanism based on threading was reported by Jackson for protein YibK, in recent *in vitro* experiments (24). Alternative mechanisms have been proposed in lattice models (spindle mechanism)(91), and for specific proteins (e.g. loop flipping was observed in simulations of MJ0366 (92), and proposed for the protein hypothetical RNA methyltransferase (PDB ID: 1o6d) following *in vitro* experiments (82)).

Although there is some lack of consensus regarding the predominance of direct threading over slipknotting, simulations from different groups, and using different simulation strategies and models, all agree with the timing of the knotting step, i.e., that knotting occurs towards late folding, in structurally consolidated conformations. However, folding of proteins YibK and YibA was observed in a cell-free translating system, mimicking *in vivo* conditions, when one (either the C- or the N-) terminus, or both termini simultaneously were fused to stable (ThiS) domains (24). Because ThiS domains are bulky, the knotting step of fused YibK and YibA proteins should involve a threading movement of the full ThiS domain. The latter can only occur early in the folding process, when the knotting loop is sufficiently loose to allow for the passage of large structures. Because knotting is entropically costly (85) such mechanism may seem unlikely in the beginning of the folding process since there are not yet enough native interactions established to decrease the free energy barrier to knotting. However, and at least for the YibK model system, one possibility to circumvent this challenge is that knotting occurs co-translationally (i.e. during protein synthesis in the ribosome), via a slipknottted conformation (97), as we discuss later in this article.

6. Knot type and knotting mechanism

While most studies addressing the folding of knotted proteins have been based on knotted trefoils, it is clear that a comprehensive understanding of the folding and knotting process should include the study of model systems embedding other knot types. The number of theoretical articles in the literature addressing this problem is scarce. To the best of our knowledge Virnau and co-workers were the first to explore the folding of protein DehI, embedding a deep 6_1 knot (18), and Soler *et al.* (98) were the first to study the folding of a lattice model system with a shallow 5_2 knot (98). More recently, Sulkowska and co-workers, provided the first off-lattice results for protein Ubiquitin C-terminal Hydrolase L1 (UCH-L1) (PDB ID: 3irt) embedding a shallow 5_2 knot in its native structure (99).

DehI is a homodimer, formed by two 130 residues long monomers connected by a shorter linker region to complete the 6_1 knotted topology. The folding route of Dehl was explored via Molecular Dynamics simulations with a coarse-grained C-alpha model combined with the Gō potential (18). Out of 1000 runs, six trajectories folded into the native knotted state with very similar routes, suggesting that DehI folds via a simple mechanism: two large loops are formed early in the simulations by twists of the partially unfolded protein, and the six-fold knot is created later in a single movement when either one of the loops flips over the other. Experimental analysis of the equilibrium unfolding of DehI by chemical denaturation (100) suggests that DehI first unfolds in to a monomeric intermediate state with 42% of the native secondary structure still formed, which then unfolds into another intermediate state with only 21% of the secondary structure content but a similar level of compactification, and finally total unfolding, and potentially unknotting, is achieved, accompanied by a total loss of secondary structure. Thus, in DehI knotting precedes the formation of most of the secondary structure, but this experimental analysis could not ascertain whether the actual knotting mechanism is as proposed by Virnau and co-workers.

Soler *et al.* considered a lattice protein with chain length $N = 52$, designed by hand to have a shallow 5_2 knot in its native structure (98). The folding mechanism of the 5_2 knot shares with that of shallow the 3_1 knot ($N = 41$) the occurrence of a threading movement of the chain terminus that lays closer to the knotted core. However, in sharp contrast to what is observed for the knot 3_1, knotting occurs particularly late during folding for the 5_2 knot (e.g. when $Q \sim 0.7$ the knotting frequency is 0.82 for the lattice trefoil but is only 0.12 for the twisted-three).

The Ubiquitin C-terminal Hydrolase (UCH) is a family of knotted proteins with the 5_2 knot, and with sub-chains that form the 3_1 knot. Experimental studies of UCH-L3(101) show two parallel folding pathways, each associated with the formation of a rapidly populated intermediate state. Using a structure-based C-alpha model, Sulkowska and co-workers (99) did a computational study of three members of the UCH family that is consistent with the main trends of the experimental results. They also found two, topologically distinct pathways. One in which the N-terminus is structured last and the 5_2 knot is formed directly from the unknot, and another in which the C-terminus is structured last and a 3_1 knotted intermediate appears.

7. Folding properties of knotted proteins

Perhaps one of the most remarkable features of the overall folding behavior of knotted proteins regards the persistence of knotted conformations in chemically denatured states, a finding that was originally reported by Mallam and Jackson for proteins YibK and YibA (86). Recent experiments by Capraro and Jennings for a protein of the same family, the hypothetical RNA methyltransferase (PDB ID: 1o6d), whose native structure also embeds a deep trefoil knot, revealed that in order to untie it, it is necessary to use highly denaturing conditions during several weeks (82). Since original experimental studies (79,101,102) were not aware of this resilience to untie upon chemical denaturation, the first kinetic measurements provided folding rates in line with those exhibited by regular proteins of the same size, and were interpreted accordingly, i.e., knotted proteins fold in about the same timescale. However, they were based in folding processes that started from denatured, but knotted, conformations. As shown through molecular simulations, conformations with a small fraction of native contacts ($Q \sim 0.2 - 0.3$) can fold remarkably fast (more than one order of magnitude faster) than unknotted conformations, if they keep the knotting loop (103). This behavior changes drastically when folding trajectories start from unfolded and unknotted conformations. Faísca *et al.* were the first to compare the folding rate of a knotted lattice protein with a control system (i.e. an unknotted counterpart which is obtained by minimally modifying the backbone connectivity of the knotted system that serves as a template)(89), and predicted that knotted proteins fold considerably slower, especially below the folding transition temperature, T_f. This is in line with in vitro results obtained by Yeates for a cleverly engineered protein, known in the literature as 2ouf, which is designed to embed a 3_1 knot. The latter was shown to fold 20-times slower than a protein that was designed to have a similar tertiary structure but to be unknotted (84). In a subsequent study Soler *et al.* also provided compelling evidence that folding rate decreases dramatically as knot depth (103) and knot complexity increases (98). Further results from lattice (98,103) and off-lattice simulations (104,105) confirmed the same trend, i.e., that knotted proteins fold slower. An important definitive proof that knotted proteins are indeed slow folders, came from *in vitro* experiments carried out by Mallam and Jackson on YibK and YibA. They used a cell free expression system (i.e. an experimental set up containing only the elements necessary for protein translation) to synthesize YibK and YibA as a strategy to assure that the folding process started from unfolded, and unknotted, conformations. They found that YibK and YibA fold spontaneously without populating misfolded states. However, the folding process that starts from denatured conformations exhibits a rate constant that is ~3 (YibA) to ~35 (YibK)-fold greater than that exhibited by newly translated conformations, which was taken as an indication that knotting is the rate-limiting step (23). This conclusion applies if newly synthesized proteins are indeed unknotted. This may, however, not be necessarily the case for YibK since there is theoretical work predicting that knotting may occurs co-translationally (97) .

More recently, single molecule experiments accessed the effect of knot formation on the folding rate of protein ubiquitin C-terminal hydrolase isoenzyme L1, UCH-L1, (PDB ID: 3irt) (106). This protein is 223 residues long and features a shallow 5_2 knot in its native structure. It was shown that folding from an unknotted denatured state is one order of magnitude slower than from knotted unfolded states, indicating that for this family of knotted proteins, knotting is also the rate limiting step.

A major reason why knotted proteins are slower folders is related with the phenomenon of backtracking; the term was introduced by Onuchic and co-workers to describe the process of breaking and re-establishing specific native contacts (107). It is believed that backtracking is prevalent in tangled proteins because knotting is a highly ordered process, whereby the polypeptide chain must arrange itself in a succession of specific conformations (e.g. a conformation with the knotting loop must form first that is subsequently protruded by the chain terminus). Failure in achieving the correct order of events leads to malformed knots, or other misfolded conformations, that must be unfolded to allow for productive knotting and folding. In order to probe the importance of backtracking in the folding of knotted proteins, Soler *et al.* introduced the so-called structural mutations (SM)(98). SMs disrupt (or switches-off in the case of the Gō potential) native interactions involving residues located within the knotted core, which otherwise do not play any role in the energetic stabilization (i.e. nucleation) of the folding transition state. An illustrative example is that of a native interaction between a residue located within the threading terminus (TT), with another one located on the knotting loop (KL). SMs are expected to increase the folding rate because they should decrease backtracking by hampering the premature formation of these critical interactions in conformations that still require a substantial amount of structural consolidation. In line with this hypothesis, a significant increase in the folding rate of knotted lattice proteins was observed upon introducing SMs. The observed enhancement was largest (up to 50%) for the 5_2 knot in comparison with the 3_1 knot ($\sim$36%), indicating that backtracking on lattice polymers is more disruptive in knots of higher complexity (98). This enhancement is particularly striking if one considers the fact that only one TT-KL interaction (i.e. 2% of the total native energy) is being switched-off. The simplicity of lattice models may challenge the relevance of this behavior in real proteins. Nevertheless, it would be interesting to test this prediction in experiments *in vitro*. The importance of the contact map for efficient knotting was addressed in (108) in the context of off-lattice simulations of protein MJ0366.

Topological frustration is expected to lead to folding pathways populated by intermediate states. At the microscopic level, stable non-native interactions, can further contribute to stabilize misfolded intermediates. Therefore, knotted proteins are expected to exhibit complex folding landscapes, populated by intermediate states, which translate into multi-phase kinetics (83,84,101,102,109) and non-linear chevron-behavior (83,95). Recently, single-molecule experiments revealed a rich folding landscape for protein UCH-L1, featuring an ensemble of structurally heterogeneous intermediate states, both on- and off-pathway (106). Some of the intermediate states are mechanically stable and long-lived (i.e. stable for several seconds), and presumably coincide with those previously detected through bulk unfolding studies (109). An ensemble of complex folding pathways with off-pathway misfolded intermediates was also reported for a designed tandem repeat of the hypothetical protein HP0242 (PDB ID: 2bo3 and 4u12), a small (92 residue long) engineered protein (84), which features a knot 3_1 in its native structure(110).

The folding behavior of knotted proteins is rather different from that exhibited by small-single domain proteins (~ 100 residues), which is typically well described by a two-state model, and single exponential kinetics (111). Structure-based models, and particularly pure Gō potentials, correctly capture regions of the folding landscape dominated by native interactions. They should be able to predict topologically trapped intermediates, and intermediates states resulting from side-chain packing effects when combined with full atomistic protein representations (49,112,113). Therefore, they are adequate to recapitulate the folding process of model systems with smooth energy landscapes (such as small, single-domain proteins) but will fail to correctly capture regions of the folding landscape dominated by non-native interactions. Given the importance of non-native interactions in the folding of knotted proteins it is important to develop more realistic (though computationally tractable) potentials that will allow to clarify the role of non-native interactions in the energetics, kinetics and thermodynamics of the knotting process.

8. Towards a mechanistic understanding of knotting *in vivo*

In vivo, efficient folding of many newly synthesized proteins is achieved through the utilization of a special class of molecular machines called chaperones, which prevent protein misfolding and aggregation in the crowded environment of the cell (114). The so-called chaperonins comprise an important and universal class of chaperones, which have the unique ability to fold some proteins that cannot be folded by simpler chaperone systems (e.g. the trigger factor Hsp70)(115). Chaperonins are large cylindrical complexes that contain a central cavity that binds to unfolded and misfolded proteins and allows them to fold in isolation **(Fig. 4A-B)**. The most well studied chaperonin is the bacterial GroEL-GroES complex from *E. coli.* (116). During the GroEl-GroES operating cycle, which is driven by ATP, there is a large conformational change that nearly doubles the cavity's volume and is accompanied by a change of the physical properties of its inner walls, which become predominantly hydrophilic and net-negatively charged. The exact mechanism according to which the chaperonin assists folding has not yet been solved, but two main models have been proposed. The passive (or Anfinsen's) cage model predicts that the role of chaperonin is simply that of increasing folding yield by avoiding aggregation. One the other hand, the active cage model envisages an active role for the chaperonin, according to which the latter is able to accelerate folding process by modulating the free energy landscape (116).

A proteome-wide analysis that studied the contribution of the GroEL-GroES complex to the *E. coli.* proteome revealed that the folding process of ~ 85 of its proteins is stringently chaperon-dependent (117). Interestingly, all these proteins embed a knot in their native structures (118) suggesting that a chaperon-assisted folding is necessary to efficiently fold knotted proteins *in vivo.*

Mallam and Jackson were the first to study the effect of GroEL-GroES on the folding process of *E. coli* proteins YibK and YibA (23). In the presence of GroEL-GroES, folding rate becomes at least 20-fold higher, supporting the view

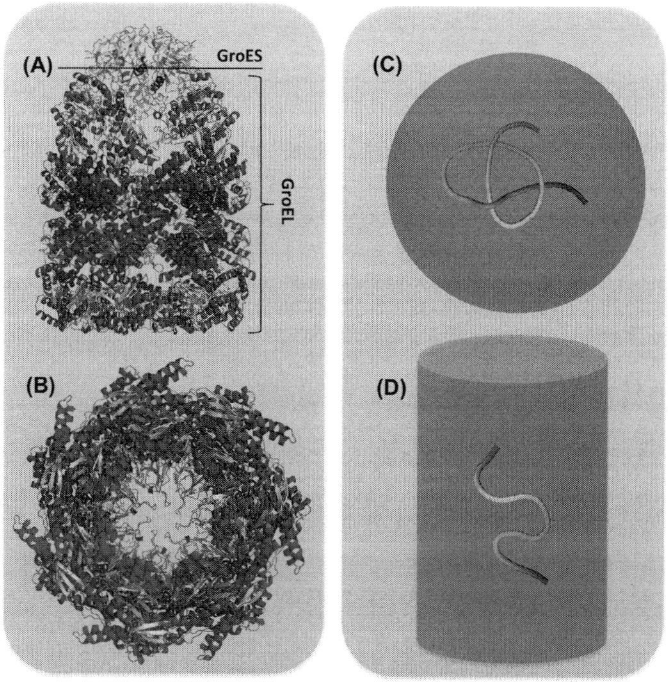

FIGURE 4. Three-dimensional structure of the GroEL-GroES complex (PDB ID: 1aon) (A), and end view of the GroEL-GroES, highlighting the central cavity where folding takes place (B). Simple models of steric confinement used to mimic excluded volume interactions between the protein and the chaperonin chamber. A spherical shape (C) was adopted in (119), while a cylindrical (D) one was adopted in (118).

that this mechanism should indeed operate *in vivo*. It was also found that GroEL-GroES has no effect on the folding rate of denatured (and knotted) conformations. Since none of the proteins populates misfolded species, these results indicate that the chaperonin is likely to play a specific and active role in assisting the knotting step. One possibility put forward by the authors is that it drives the formation of knotted chains similar to those that persist in the denatured ensemble upon chemical denaturation.

Inspired by these results Soler *et al.* were the first to use a Gō model to explore via Monte Carlo simulations the folding and knotting mechanism (of a simple cubic lattice with a trefoil knot and a C-alpha off-lattice representation of protein MJ0366) inside the chaperonin cage (119). The chaperonin was represented by a rigid cavity, i.e., the interactions of the amino acids with the confining walls were strictly limited to excluded volume interactions **(Fig. 4C)**. It was found that steric confinement leads to a stabilization of the transition state relative to the unfolded state at the transition temperature, T_f, resulting into a faster folding kinetics, in line with previous studies that investigated the role of steric confinement in the folding transition of proteins without knots (120,121). However, it was also reported that steric confinement plays a specific role in the folding of knotted proteins by substantially increasing the knotting probability for high degrees of confinement, even above the melting temperature (i.e. in conditions where the native state is destabilized). In particular, the authors reported that native knotting occurs with moderate probability (> 0.5) even before the transition state is crossed, i.e., in conformations with a small degree of structural consolidation. Interestingly, a reduction of backtracking upon steric confinement was also observed, most likely because confinement reduces the frequency of local interactions. Since local interactions contribute to rigidify the protein backbone (i.e. to decrease local flexibility), the authors of (119) performed a systematic study of the effect of local flexibility in the folding process of MJ0366. The conclusion was that when local chain flexibility is increased, MJ0366 samples conformations that are preferably knotted (even slightly above the transition temperature T_f), recapitulating the effects of confinement. These results are in line with theoretical analysis, framed on polymer physics considerations, which predicts that local order is an inhibitor of knotting in proteins (9). The impact of chain flexibility on the structure of the transition state ensemble of MJ0366 was also investigated in (119). An enhanced structural flexibility leads to a structural re-arrangement of the transition state, which becomes devoid of helical content (including the C-terminal helix that is threaded through the knotting loop to tangle the protein). Interestingly, the formation of the beta-hairpin, which is the main structural trait of the intermediate slipknotted state identified in previous, full atomistic simulations (113), starts earlier than for stiffer chains, which facilitates the formation of the knot.

A follow up study by Niewieczerzal and Sulkowska (118), framed on Molecular Dynamics simulations of a simple C-alpha structured-based Gō model parameterized with the SMOG package (122), extended the study of folding under confinement to other small proteins with a trefoil knot: VirC2 (PDB ID: 2rh3), with the same fold as MJ0366, and DndE (PDB ID: 4lrv), with an unclassified fold, whose knots are only slightly deeper (by 2 and 5 amino acids, respectively) than that of MJ0366. Essentially the same conclusion was recapitulated for the three model systems, namely, that steric confinement upon encapsulation within a cylinder **(Fig. 4D)** facilitates knotting at an early stage of folding, while simultaneously reducing backtracking and smoothing the free energy landscape. However, a new and rather interesting effect of steric confinement was observed when the same simulation approach was used to thoroughly investigate the folding process of protein UCH-L1. As previously outlined, this protein exhibits two folding pathways under bulk conditions. A dominant one in which the N-terminus gets structured first and the protein populates an intermediate conformation with a trefoil knot prior to folding into the native one, and a secondary, less populated pathway, in which the C-terminus gets structured first and the formation of the 5_2 knot occurs directly from unknotted conformations. What the authors observed was that confinement does not perturb the frequency of the most dominant bulk pathway but leads to a substantial enhancement (e.g. from 2 to 18.5% at high temperature) of the less dominant pathway (99). Furthermore, they noted a significant number of knots (native and non-native) of different types (3_1, 4_1 and 5_2) in the denatured ensemble of UCH-L1. Surprisingly, the knots are shallow and short lived in contrast with the (less frequent) ones found under bulk conditions.

The results outlined above indicate that steric confinement alone has a direct effect on knotting. By impeding extended conformations, confinement contributes to decrease the entropic cost to knot the chain, facilitating the formation of native knots, even in unfolded conformations. Available data suggests that steric confinement is not able to change the folding pathways observed in the bulk, despite being able to change their frequency. A similar observation holds for the knotting mechanism, which is conserved under steric confinement. In order to broaden and deepen our views on the role of GroEL-GroES on the folding of knotted proteins, more realistic models, encapsulating the physical changes occurring along the chaperonin cycle (123,124) should be applied to knotted proteins. These models may reveal an active role of GroEL-GroES at the level of the folding landscape and knotting mechanism (e.g. by stabilizing key interactions which promote knotting(24)), which add up to the effects of steric confinement to further increase the folding rate and efficiency.

9. Co-translational folding of knotted proteins

Cieplak and Chwastyk were the first to explore the knotting mechanism of protein YibK during protein synthesis (i.e. co-translationally) via Langevin Molecular Dynamics simulations of a simple C-alpha representation, with protein energetics modeled by a Gō potential (97). Protein synthesis occurs vectorially from the N-terminus to the C-terminus, which remains tethered to the peptidyltransferase center on the larger ribosomal unit (**Fig. 5A**). This implies that if knotting of YibK is to occur co-translationally, it cannot be via a direct threading movement of the C-terminus. As in previous lattice simulations (125), Cieplak and Chwastyk used a minimal representation that reduces the ribosome to a steric plane (**Fig. 5B**). They found that when folding occurs co-translationally, successful knotting (representing 3% of the attempted trajectories) occurs exclusively via slipknotting under optimal temperature conditions. In particular, the degree of confinement provided by the wall facilitates the formation of the C-terminal hairpin on the right side of the knotting loop, while simultaneously assisting threading. These results are interesting because they show that not only it is possible to knot YibK without the assistance of non-native interactions, as they also open-up the possibility that in Lim and Jackson's *in vivo* experiments with YibK fused to ThiS (24) the newly translated proteins were already knotted when folding started.

A subsequent account by Dabrowski-Tumanski *et al.* (126) explored the co-translational knotting mechanism of protein Tp0624 (PDB ID: 5jir) with Molecular Dynamics simulations of a C-alpha model. Tp0624 is the deepest knotted protein found to date, having very long knot tails (with chain length exceeding 120 residues). They used a more sophisticated representation of the ribosome, where the exit tunnel is explicitly represented by a purely steric cylinder; the cylinder is capped at one end by a steric plane mimicking the ribosome's surface. However, in this case, they considered the existence of homogeneous attractive intermolecular interactions between selected beads of the N-terminal domain (e.g. arginines, which are positively charged, are likely to interact with the ribosome) and the confining plane. They found that in up to 60% of the successful folding trajectories, knotting is based on what the authors termed *the ribosome mechanism*. According to the latter, a part of the chain protruding from the tunnel gets wrapped around the exit tunnel and sticks onto the surface. Such a loop is then threaded by the nascent chain that is being pushed out of the tunnel with a constant force. While there are

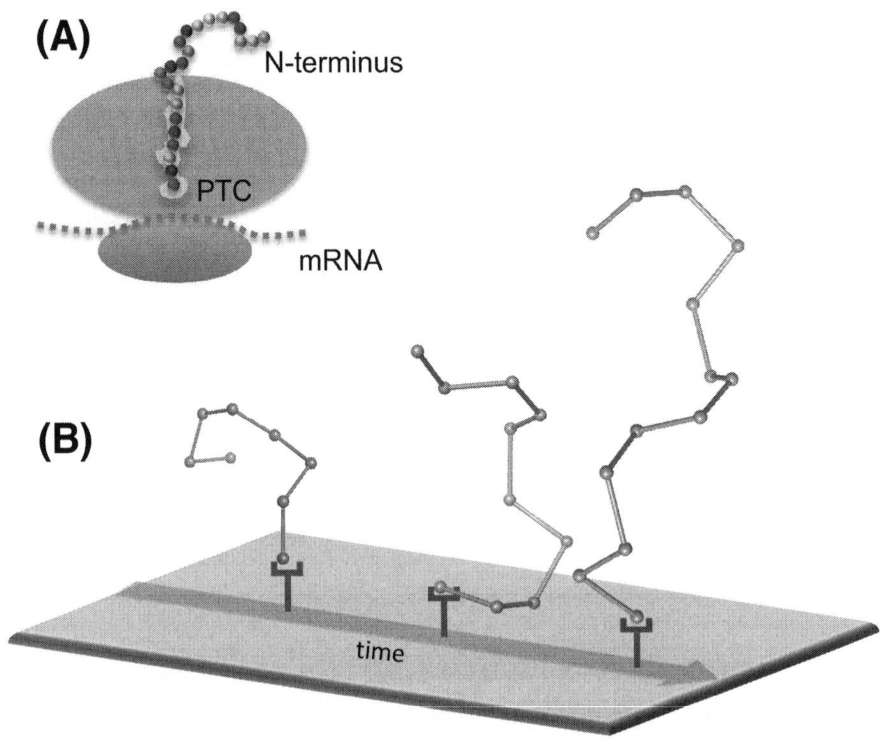

FIGURE 5. Cartoon representation of the ribosome highlighting the exit tunnel (located on the larger ribosomal unit), and the vectorial nature of protein synthesis. The latter proceeds from the N-terminus to the C-terminus that is tethered at the PTC on the larger ribosomal unit (A). Simulation set-up in which the larger ribosomal unit is represented by a steric plane, and the protein (represented by a C-alpha model), is tethered to the plane by the C-terminus (B). Adapted from (125)

not yet *in vitro* results supporting a role for the ribosome on the knotting mechanism, the results predicted by these simulations can, at least partially explain why it is so challenging to knot a protein in molecular simulations.

10. Functional advantages of knots in proteins

Our current understanding of the folding process of knotted proteins shows that these topologically complex molecules are slow folders, which often exhibit relatively complex folding landscapes. In the cell, they are most likely assisted by chaperonins to fold fast and efficiently. This scenario could explain how knotted proteins have withstood evolutionary pressure despite their folding liability (23). However, another possibility is that, to compensate for their slow folding, a knotted native state conveys a functional advantage to the carrier protein. The quest for functional advantages of knots in proteins started with the realization of their existence in the PDB (7), and is also motivated by the observation that knotted motifs are conserved across different families (despite very low sequence similarity)(127).

There is a plethora of studies in the literature, both experimental and computational, which analyzed specific model systems and proposed several biological functions. It has been proposed that knotted proteins have enhanced thermal (20,26), mechanical (26,27,128) and structural (127) stabilities. Another possibility is that knots help shape and stabilize the active site of enzyme (129-131), and in some cases it has even been suggested that knots can control enzymatic (130) and signaling activity (132). A recent study put forward the view that knots can impair protein degradation by ATP-dependent proteases leading to partially degraded products with potentially new functions (133).

By comparing the unfolding rate of 3_1 and 5_2 knotted lattice systems with that of their unknotted counterparts Soler *et al.* found a dramatic enhancement in kinetic stability for the knotted systems (more than one order of magnitude). Furthermore, it was also observed that kinetic stability of lattice proteins increases considerably with the complexity of knot type (98). An enhanced kinetic stability for knotted proteins was also reported in off-lattice Molecular Dynamics simulations (26,134). A reduction of the intrinsic unfolding rate due to the presence of a knot in the native structure was previously reported by Yeates and co-workers through *in vitro* experiments that also compared an engineered knotted model system with its unknotted counterpart (84). Recently, *in vitro* results showed that a truncated (i.e. lacking eleven N terminal residues) variant of 5_2 knotted protein UCHL1, unfolds two-orders of magnitude faster than the full UCHL1(27). Since the truncated variant is unknotted, this finding also supports the view that knots enhance kinetic stability.

Kinetic stability is a property of the native state that is essential to maintain the biological function of the protein during a physiologically relevant timescale (135), and therefore may represent a functional advantage. An enhanced kinetic stability may be particularly advantageous for proteins forming transmembrane channels since they are subjected to mechanical stress. Since knotted patterns appear to be preferentially conserved among transmembrane channels (127) we propose, as Faísca did in (22), that a *systematic* functional role of knots in proteins is precisely that of enhancing their kinetic stability.

11. Conclusions and future prospects

Interest in the field of knotted proteins has been increasing over the last decade, with many important contributions (both theoretical and experimental) from several research groups worldwide. This expansion contributed to significantly advance our understanding of their intricate knotting mechanism, with experiments corroborating previous theoretical predictions for specific knotted proteins. While focus has mostly been placed on proteins embedding knotted trefoils, some recent studies featured model systems with knot types of higher complexity. This is, in our view, a line of research that is worth pursuing as an attempt to establish the different types of knotting mechanisms, how much they depend on knot complexity, how are they conserved across different knotted proteins, and how they affect the overall folding performance.

Microscopically, it is extremely important to use molecular simulations to understand and determine the exact role played by non-native interactions in the kinetics, energetics and thermodynamics of knotting. We also envisage an important role for molecular simulations in what regards the determination of the knotting mechanism inside the "chamber of secrets" (115), for which direct experimental observations are not yet possible. Current contributions focused on the steric effects played by the chaperonin's chamber, but it is crucial to use more realistic models to that take into account the physical changes that take place during the working cycle, and which may play an active role in the physics of knotting. Is the enhanced knotting frequency that occurs under steric confinement conserved when other intermolecular interactions (e.g. hydrophobic and electrostatic) are included? Can the latter modulate the free energy landscape so that productive intermediate states, with enhanced knotting frequency, become populated?

Establishing the role of chaperonin-assisted knotting will not only enlarge our vistas on chaperon Biology, but it will also contribute to clarify our understanding of the evolution of knotted proteins and, consequently, of their functional role.

Acknowledgments

Work supported by UID/MULTI/04046/2019 centre grant from FCT, Portugal (to BioISI). PFNF would like to acknowledge the contribution of the COST Action CA17139.

References

1. Berman, H. M., Westbrook, J., Feng, Z., Gilliland, G., Bhat, T. N., Weissig, H., Shindyalov, I. N., and Bourne, P. E. (2000) The Protein Data Bank. *Nucleic Acids Research* **28**, 235-242

2. Richardson, J. S. (1977) beta-Sheet topology and the relatedness of proteins. *Nature* **268**, 495-500

3. Mansfield, M. L. (1994) Are there knots in proteins? *Nat Struct Mol Biol* **1**, 213-214

4. Adams, C. C. (2004) *The Knot Book: An Elementary Introduction to the Mathematical Theory of Knots,* American Mathematical Society

5. Taylor, W. R. (2000) A deeply knotted protein structure and how it might fold. *Nature* **406**, 916-919

6. Koniaris, K., and Muthukumar, M. (1991) Knottedness in ring polymers. *Physical Review Letters* **66**, 2211-2214

7. Virnau, P., Mirny, L. A., and Kardar, M. (2006) Intricate Knots in Proteins: Function and Evolution. *PLOS Computational Biology* **2**, e122

8. Niemyska, W., Dabrowski-Tumanski, P., Kadlof, M., Haglund, E., Sułkowski, P., and Sulkowska, J. I. (2016) Complex lasso: new entangled motifs in proteins. *Scientific Reports* **6**, 36895

9. Lua, R. C., and Grosberg, A. Y. (2006) Statistics of Knots, Geometry of Conformations, and Evolution of Proteins. *PLOS Computational Biology* **2**, e45

10. Alexander, K., Taylor, A. J., and Dennis, M. R. (2017) Proteins analysed as virtual knots. *Scientific Reports* **7**, 42300

11. Tubiana, L., Polles, G., Orlandini, E., and Micheletti, C. (2018) KymoKnot: A web server and software package to identify and locate knots in trajectories of linear or circular polymers. *The European Physical Journal E* **41**, 72

12. Turaev, V. (2012) Knotoids. *Osaka J. Math.* **49**, 195-223

13. Goundaroulis, D., Dorier, J., Benedetti, F., and Stasiak, A. (2017) Studies of global and local entanglements of individual protein chains using the concept of knotoids. *Scientific Reports* **7**, 6309

14. Goundaroulis, D., Gügümcü, N., Lambropoulou, S., Dorier, J., Stasiak, A., and Kauffman, L. (2017) Topological Models for Open-Knotted Protein Chains Using the Concepts of Knotoids and Bonded Knotoids. *Polymers* **9**, 444

15. Dorier, J., Goundaroulis, D., Benedetti, F., and Stasiak, A. (2018) Knoto-ID: a tool to study the entanglement of open protein chains using the concept of knotoids. *Bioinformatics* **34**, 3402-3404

16. Jamroz, M., Niemyska, W., Rawdon, E. J., Stasiak, A., Millett, K. C., Sułkowski, P., and Sulkowska, J. I. (2015) KnotProt: a database of proteins with knots and slipknots. *Nucleic Acids Research* **43**, D306-D314

17. Dabrowski-Tumanski, P., Sulkowska, J. I., Rubach, P., Stasiak, A., Goundaroulis, D., Dorier, J., Sułkowski, P., Millett, K. C., and Rawdon, E. J. (2018) KnotProt 2.0: a database of proteins with knots and other entangled structures. *Nucleic Acids Research* **47**, D367-D375

18. Bölinger, D., Sułkowska, J. I., Hsu, H.-P., Mirny, L. A., Kardar, M., Onuchic, J. N., and Virnau, P. (2010) A Stevedore's Protein Knot. *PLoS Computational Biology* **6**, e1000731

19. Sułkowska, J. I., Rawdon, E. J., Millett, K. C., Onuchic, J. N., and Stasiak, A. (2012) Conservation of complex knotting and slipknotting patterns in proteins. *Proceedings of the National Academy of Sciences* **109**, E1715-E1723

20. King, N. P., Yeates, E. O., and Yeates, T. O. (2007) Identification of Rare Slipknots in Proteins and Their Implications for Stability and Folding. *Journal of Molecular Biology* **373**, 153-166

21. Mallam, A. L. (2009) How does a knotted protein fold? *The FEBS Journal* **276**, 365-375

22. Faísca, P. F. N. (2015) Knotted proteins: A tangled tale of Structural Biology. *Computational and Structural Biotechnology Journal* **13**, 459-468

23. Mallam, A. L., and Jackson, S. E. (2012) Knot formation in newly translated proteins is spontaneous and accelerated by chaperonins. *Nat Chem Biol* **8**, 147-153

24. Lim, N. C. H., and Jackson, S. E. (2015) Mechanistic Insights into the Folding of Knotted Proteins In Vitro and In Vivo. *Journal of Molecular Biology* **427**, 248-258

25. Micheletti, C., Di Stefano, M., and Orland, H. (2015) Absence of knots in known RNA structures. *Proceedings of the National Academy of Sciences* **112**, 2052-2057

26. Sułkowska, J. I., Sułkowski, P., Szymczak, P., and Cieplak, M. (2008) Stabilizing effect of knots on proteins. *Proceedings of the National Academy of Sciences* **105**, 19714-19719

27. Sriramoju, M. K., Chen, Y., Lee, Y.-T. C., and Hsu, S.-T. D. (2018) Topologically knotted deubiquitinases exhibit unprecedented mechanostability to withstand the proteolysis by an AAA+ protease. *Scientific Reports* **8**, 7076

28. Dabrowski-Tumanski, P., and Sulkowska, J. (2017) To Tie or Not to Tie? That Is the Question. *Polymers* **9**, 454

29. Orlandini, E. (2018) Statics and dynamics of DNA knotting. *Journal of Physics A: Mathematical and Theoretical* **51**, 053001

30. Jackson, S. E., Suma, A., and Micheletti, C. (2017) How to fold intricately: using theory and experiments to unravel the properties of knotted proteins. *Current Opinion in Structural Biology* **42**, 6-14

31. Yeates, T. O., Norcross, T. S., and King, N. P. (2007) Knotted and topologically complex proteins as models for studying folding and stability. *Current opinion in chemical biology* **11**, 595-603

32. Lim, N. C. H., and Jackson, S. E. (2015) Molecular knots in biology and chemistry. *Journal of Physics: Condensed Matter* **27**, 354101

33. Taketomi, H., Ueda, Y., and Gō, N. (1975) Studies on protein folding, unfolding and fluctuations by computer simulation. *International Journal of Peptide and Protein Research* **7**, 445-459

34. Noel, J. K., and Onuchic, J. N. (2012) The Many Faces of Structure-Based Potentials: From Protein Folding Landscapes to Structural Characterization of Complex Biomolecules. in *Computational Modeling of Biological Systems: From Molecules to Pathways* (Dokholyan, N. V. ed.), Springer US, Boston, MA. pp 31-54

35. Plaxco, K. W., Simons, K. T., Ruczinski, I., and Baker, D. (2000) Topology, Stability, Sequence, and Length: Defining the Determinants of Two-State Protein Folding Kinetics. *Biochemistry* **39**, 11177-11183

36. Micheletti, C. (2003) Prediction of folding rates and transition-state placement from native-state geometry. *Proteins: Structure, Function, and Bioinformatics* **51**, 74-84

37. Gromiha, M. M., and Selvaraj, S. (2001) Comparison between long-range interactions and contact order in determining the folding rate of two-state proteins: application of long-range order to folding rate prediction, Edited by P. E. Wright. *Journal of Molecular Biology* **310**, 27-32

38. Krobath, H., Rey, A., and Faisca, P. F. N. (2015) How determinant is N-terminal to C-terminal coupling for protein folding? *Physical Chemistry Chemical Physics* **17**, 3512-3524

39. Onuchic, J. N., and Wolynes, P. G. (2004) Theory of protein folding. *Current Opinion in Structural Biology* **14**, 70-75

40. Leopold, P. E., Montal, M., and Onuchic, J. N. (1992) Protein folding funnels: a kinetic approach to the sequence-structure relationship. *Proceedings of the National Academy of Sciences* **89**, 8721-8725

41. Dill, K. A., and Chan, H. S. (1997) From Levinthal to pathways to funnels. *Nat Struct Mol Biol* **4**, 10-19

42. Lopes, P. E. M., Guvench, O., and MacKerell, A. D. (2015) Current Status of Protein Force Fields for Molecular Dynamics. *Methods in molecular biology* (Clifton, N.J.) **1215**, 47-71

43. Sugita, Y., and Okamoto, Y. (1999) Replica-exchange molecular dynamics method for protein folding. *Chemical Physics Letters* **314**, 141-151

44. Bernardi, R. C., Melo, M. C. R., and Schulten, K. (2015) Enhanced Sampling Techniques in Molecular Dynamics Simulations of Biological Systems. *Biochimica et biophysica acta* **1850**, 872-877

45. Baweja, L., and Roche, J. (2018) Pushing the Limits of Structure-Based Models: Prediction of Nonglobular Protein Folding and Fibrils Formation with Go-Model Simulations. *The Journal of Physical Chemistry B* **122**, 2525-2535

46. Tozzini, V. (2005) Coarse-grained models for proteins. *Current Opinion in Structural Biology* **15**, 144-150

47. Lammert, H., Schug, A., and Onuchic, J. N. (2009) Robustness and generalization of structure-based models for protein folding and function. *Proteins: Structure, Function, and Bioinformatics* **77**, 881-891

48. Li, M. S., Co, N. T., Reddy, G., Hu, C.-K., Straub, J. E., and Thirumalai, D. (2010) Factors Governing Fibrillogenesis of Polypeptide Chains Revealed by Lattice Models. *Physical Review Letters* **105**, 218101

49. Krobath, H., Estácio, S. G., Faísca, P. F. N., and Shakhnovich, E. I. (2012) Identification of a conserved aggregation-prone intermediate state in the folding pathways of Spc-SH3 amyloidogenic variants. *Journal of molecular biology* **422**, 705-722

50. Enciso, M., and Rey, A. (2013) Sketching protein aggregation with a physics-based toy model. *The Journal of Chemical Physics* **139**, 115101

51. Holzgräfe, C., Irbäck, A., and Troein, C. (2011) Mutation-induced fold switching among lattice proteins. *The Journal of Chemical Physics* **135**, 195101

52. Das, S., Eisen, A., Lin, Y.-H., and Chan, H. S. (2018) A Lattice Model of Charge-Pattern-Dependent Polyampholyte Phase Separation. *The Journal of Physical Chemistry B* **122**, 5418-5431

53. Kazmirski, S. L., and Daggett, V. (1998) Non-native interactions in protein folding intermediates: molecular dynamics simulations of hen lysozyme, Edited by B. Honig. *Journal of Molecular Biology* **284**, 793-806

54. Paci, E., Vendruscolo, M., and Karplus, M. (2002) Native and non-native interactions along protein folding and unfolding pathways. *Proteins: Structure, Function, and Bioinformatics* **47**, 379-392

55. Faísca, P. F. N., Nunes, A., Travasso, R. D. M., and Shakhnovich, E. I. (2010) Non-native interactions play an effective role in protein folding dynamics. *Protein Science: A Publication of the Protein Society* **19**, 2196-2209

56. Wallin, S., Zeldovich, K. B., and Shakhnovich, E. I. (2007) The Folding Mechanics of a Knotted Protein. *Journal of Molecular Biology* **368**, 884-893

57. Frisch, H. L., and Wasserman, E. (1961) Chemical Topology1. *Journal of the American Chemical Society* **83**, 3789-3795

58. Sumners, D. W., and Whittington, S. G. (1988) Knots in self-avoiding walks. *Journal of Physics A: Mathematical and General* **21**, 1689

59. Orlandini, E., and Whittington, S. G. (2007) Statistical topology of closed curves: Some applications in polymer physics. *Reviews of Modern Physics* **79**, 611-642

60. Yao, A., Matsuda, H., Tsukahara, H., Shimamura, M. K., and Deguchi, T. (2001) On the dominance of trivial knots among SAPs on a cubic lattice. *Journal of Physics A: Mathematical and General* **34**, 7563-7577

61. Deguchi, T., and Tsurusaki, K. (1997) Universality of random knotting. *Physical Review E* **55**, 6245-6248

62. Shimamura, Miyuki K., and Deguchi, T. (2001) Topological Entropy of a Stiff Ring Polymer and Its Connection to DNA Knots. *Journal of the Physical Society of Japan* **70**, 1523-1536

63. Uehara, E., and Deguchi, T. (2015) Characteristic length of the knotting probability revisited. *Journal of Physics: Condensed Matter* **27**, 354104

64. Lua, R., Borovinskiy, A. L., and Grosberg, A. Y. (2004) Fractal and statistical properties of large compact polymers: a computational study. *Polymer* **45**, 717-731

65. Diao, Y., Ernst, C., and Ziegler, U. (2014) Random walks and polygons in tight confinement. *Journal of Physics: Conference Series* **544**, 012017

66. Diao, Y., Ernst, C., Rawdon, E. J., and Ziegler, U. (2018) Average crossing number and writhe of knotted random polygons in confinement. *Reactive and Functional Polymers* **131**, 430-444

67. Diao, Y., Ernst, C., Rawdon, E. J., and Ziegler, U. (2018) Total curvature and total torsion of knotted random polygons in confinement. *Journal of Physics A: Mathematical and Theoretical* **51**, 154002

68. Virnau, P., Kantor, Y., and Kardar, M. (2005) Knots in Globule and Coil Phases of a Model Polyethylene. *Journal of the American Chemical Society* **127**, 15102-15106

69. Coronel, L., Orlandini, E., and Micheletti, C. (2017) Non-monotonic knotting probability and knot length of semiflexible rings: the competing roles of entropy and bending energy. *Soft Matter* **13**, 4260-4267

70. Rybenkov, V. V., Cozzarelli, N. R., and Vologodskii, A. V. (1993) Probability of DNA knotting and the effective diameter of the DNA double helix. *Proceedings of the National Academy of Sciences* **90**, 5307

71. Micheletti, C., Marenduzzo, D., Orlandini, E., and Sumners, D. W. (2006) Knotting of random ring polymers in confined spaces. *The Journal of Chemical Physics* **124**, 064903

72. Micheletti, C., Marenduzzo, D., Orlandini, E., and Sumners, D. W. (2008) Simulations of Knotting in Confined Circular DNA. *Biophysical Journal* **95**, 3591-3599

73. Marenduzzo, D., Orlandini, E., Stasiak, A., Sumners, D. W., Tubiana, L., and Micheletti, C. (2009) DNA–DNA interactions in bacteriophage capsids are responsible for the observed DNA knotting. *Proceedings of the National Academy of Sciences* **106**, 22269-22274

74. Burton, A. S., Di Stefano, M., Lehman, N., Orland, H., and Micheletti, C. (2016) The elusive quest for RNA knots. *RNA Biology* **13**, 134-139

75. Mallam, A. L., Onuoha, S. C., Grossmann, J. G., and Jackson, S. E. (2008) Knotted Fusion Proteins Reveal Unexpected Possibilities in Protein Folding. *Molecular Cell* **30**, 642-648

76. Potestio, R., Micheletti, C., and Orland, H. (2010) Knotted vs. Unknotted Proteins: Evidence of Knot-Promoting Loops. *PLOS Computational Biology* **6**, e1000864

77. Wüst, T., Reith, D., and Virnau, P. (2015) Sequence Determines Degree of Knottedness in a Coarse-Grained Protein Model. *Physical Review Letters* **114**, 028102

78. Dill, K. A. (1985) Theory for the folding and stability of globular proteins. *Biochemistry* **24**, 1501-1509

79. Mallam, A. L., and Jackson, S. E. (2005) Folding Studies on a Knotted Protein. *Journal of Molecular Biology* **346**, 1409-1421

80. Mallam, A. L., Rogers, J. M., and Jackson, S. E. (2010) Experimental detection of knotted conformations in denatured proteins. *Proceedings of the National Academy of Sciences* **107**, 8189-8194

81. Andersson, F. I., Pina, D. G., Mallam, A. L., Blaser, G., and Jackson, S. E. (2009) Untangling the folding mechanism of the 5_2 knotted protein UCH-L3. *FEBS Journal* **276**, 2625-2635

82. Capraro, D. T., and Jennings, P. A. (2016) Untangling the Influence of a Protein Knot on Folding. *Biophysical Journal* **110**, 1044-1051

83. Zhang, H., and Jackson, S. E. (2016) Characterization of the Folding of a 5_2 Knotted Protein Using Engineered Single-Tryptophan Variants. *Biophysical Journal* **111**, 2587-2599

84. King, N. P., Jacobitz, A. W., Sawaya, M. R., Goldschmidt, L., and Yeates, T. O. (2010) Structure and folding of a designed knotted protein. *Proceedings of the National Academy of Sciences* **107**, 20732-20737

85. Bustamante, A., Sotelo-Campos, J., Guerra, D. G., Floor, M., Wilson, C. A. M., Bustamante, C., and Báez, M. (2017) The energy cost of polypeptide knot formation and its folding consequences. *Nature Communications* **8**, 1581

86. Mallam, A. L., Rogers, J. M., and Jackson, S. E. (2010) Experimental detection of knotted conformations in denatured proteins. *Proceedings of the National Academy of Sciences* **107**, 8189

87. Bottaro, S., and Lindorff-Larsen, K. (2018) Biophysical experiments and biomolecular simulations: A perfect match? *Science* **361**, 355

88. Sułkowska, J. I., Sułkowski, P., and Onuchic, J. (2009) Dodging the crisis of folding proteins with knots. *Proceedings of the National Academy of Sciences* **106**, 3119

89. Faísca, P. F. N., Travasso, R. M. D., Charters, T., Nunes, A., and M., C. (2010) The folding of knotted proteins: insights from lattice simulations. *Physical Biology* **7**, 016009

90. Du, R., Pande, V. S., Grosberg, A. Y., Tanaka, T., and Shakhnovich, E. S. (1998) On the transition coordinate for protein folding. *The Journal of Chemical Physics* **108**, 334-350

91. Soler, M. A., and Faísca, P. F. N. (2012) How Difficult Is It to Fold a Knotted Protein? In Silico Insights from Surface-Tethered Folding Experiments. *PLOS ONE* **7**, e52343

92. Beccara, S., Ŝkrbić, T., Covino, R., Micheletti, C., and Faccioli, P. (2013) Folding Pathways of a Knotted Protein with a Realistic Atomistic Force Field. *PLOS Computational Biology* **9**, e1003002

93. Ŝkrbić, T., Micheletti, C., and Faccioli, P. (2012) The Role of Non-Native Interactions in the Folding of Knotted Proteins. *PLOS Computational Biology* **8**, e1002504

94. Noel, J. K., Sułkowska, J. I., and Onuchic, J. N. (2010) Slipknotting upon native-like loop formation in a trefoil knot protein. *Proceedings of the National Academy of Sciences* **107**, 15403

95. Wang, I., Chen, S.-Y., and Hsu, S.-T. D. (2015) Unraveling the Folding Mechanism of the Smallest Knotted Protein, MJ0366. *The Journal of Physical Chemistry B* **119**, 4359-4370

96. Noel, J. K., Onuchic, J. N., and Sulkowska, J. I. (2013) Knotting a Protein in Explicit Solvent. *The Journal of Physical Chemistry Letters* **4**, 3570-3573

97. Chwastyk, M., and Cieplak, M. (2015) Cotranslational folding of deeply knotted proteins. *Journal of Physics: Condensed Matter* **27**, 354105

98. Soler, M. A., Nunes, A., and Faísca, P. F. N. (2014) Effects of knot type in the folding of topologically complex lattice proteins. *The Journal of Chemical Physics* **141**, 025101

99. Zhao, Y., Dabrowski-Tumanski, P., Niewieczerzal, S., and Sulkowska, J. I. (2018) The exclusive effects of chaperonin on the behavior of proteins with 52 knot. *PLOS Computational Biology* **14**, e1005970

100. Wang, I., Chen, S.-Y., and Hsu, S.-T. D. (2016) Folding analysis of the most complex Stevedore's protein knot. *Scientific Reports* **6**, 31514

101. Andersson, F. I., Pina, D. G., Mallam, A. L., Blaser, G., and Jackson, S. E. (2009) Untangling the folding mechanism of the 52-knotted protein UCH-L3. *The FEBS Journal* **276**, 2625-2635

102. Mallam, A. L., and Jackson, S. E. (2007) A Comparison of the Folding of Two Knotted Proteins: YbeA and YibK. *Journal of Molecular Biology* **366**, 650-665

103. Soler, M. A., and Faísca, P. F. N. (2013) Effects of Knots on Protein Folding Properties. *PLOS ONE* **8**, e74755

104. Sułkowska, J. I., Noel, J. K., and Onuchic, J. N. (2012) Energy landscape of knotted protein folding. *Proceedings of the National Academy of Sciences* **109**, 17783

105. Prentiss, M. C., Wales, D. J., and Wolynes, P. G. (2010) The Energy Landscape, Folding Pathways and the Kinetics of a Knotted Protein. *PLoS Computational Biology* **6**, e1000835

106. Ziegler, F., Lim, N. C. H., Mandal, S. S., Pelz, B., Ng, W.-P., Schlierf, M., Jackson, S. E., and Rief, M. (2016) Knotting and unknotting of a protein in single molecule experiments. *Proceedings of the National Academy of Sciences* **113**, 7533-7538

107. Gosavi, S., Chavez, L. L., Jennings, P. A., and Onuchic, J. N. (2006) Topological Frustration and the Folding of Interleukin-1β. *Journal of Molecular Biology* **357**, 986-996

108. Dabrowski-Tumanski, P., Jarmolinska, A. I., and Sulkowska, J. I. (2015) Prediction of the optimal set of contacts to fold the smallest knotted protein. *Journal of Physics: Condensed Matter* **27**, 354109

109. Lou, S.-C., Wetzel, S., Zhang, H., Crone, E. W., Lee, Y.-T., Jackson, S. E., and Hsu, S.-T. D. (2016) The Knotted Protein UCH-L1 Exhibits Partially Unfolded Forms under Native Conditions that Share Common Structural Features with Its Kinetic Folding Intermediates. *Journal of Molecular Biology* **428**, 2507-2520

110. Wang, L.-W., Liu, Y.-N., Lyu, P.-C., Jackson, S. E., and Hsu, S.-T. D. (2015) Comparative analysis of the folding dynamics and kinetics of an engineered knotted protein and its variants derived from HP0242 of Helicobacter pylori. *Journal of Physics: Condensed Matter* **27**, 354106

111. Jackson, S. E. (1998) How do small single-domain proteins fold? *Folding and Design* **3**, R81-R91

112. Estácio, S. G., Krobath, H., Vila-Viçosa, D., Machuqueiro, M., Shakhnovich, E. I., and Faísca, P. F. N. (2014) A Simulated Intermediate State for Folding and Aggregation Provides Insights into ΔN6 β2-Microglobulin Amyloidogenic Behavior. *PLOS Computational Biology* **10**, e1003606

113. Noel, J. K., Sułkowska, J. I., and Onuchic, J. N. (2010) Slipknotting upon native-like loop formation in a trefoil knot protein. *Proceedings of the National Academy of Sciences* **107**, 15403-15408

114. Hartl, F. U., and Hayer-Hartl, M. (2002) Molecular Chaperones in the Cytosol: from Nascent Chain to Folded Protein. *Science* **295**, 1852

115. Spiess, C., Meyer, A. S., Reissmann, S., and Frydman, J. (2004) Mechanism of the eukaryotic chaperonin: protein folding in the chamber of secrets. *Trends in Cell Biology* **14**, 598-604

116. Hayer-Hartl, M., Bracher, A., and Hartl, F. U. (2016) The GroEL–GroES Chaperonin Machine: A Nano-Cage for Protein Folding. *Trends in Biochemical Sciences* **41**, 62-76

117. Kerner, M. J., Naylor, D. J., Ishihama, Y., Maier, T., Chang, H.-C., Stines, A. P., Georgopoulos, C., Frishman, D., Hayer-Hartl, M., Mann, M., and Hartl, F. U. (2005) Proteome-wide Analysis of Chaperonin-Dependent Protein Folding in Escherichia coli. *Cell* **122**, 209-220

118. Niewieczerzal, S., and Sulkowska, J. I. (2017) Knotting and unknotting proteins in the chaperonin cage: Effects of the excluded volume. *PLOS ONE* **12**, e0176744

119. Soler, M. A., Rey, A., and Faisca, P. F. N. (2016) Steric confinement and enhanced local flexibility assist knotting in simple models of protein folding. *Physical Chemistry Chemical Physics* **18**, 26391-26403

120. Baumketner, A., Jewett, A., and Shea, J. E. (2003) Effects of Confinement in Chaperonin Assisted Protein Folding: Rate Enhancement by Decreasing the Roughness of the Folding Energy Landscape. *Journal of Molecular Biology* **332**, 701-713

121. Mittal, J., and Best, R. B. (2008) Thermodynamics and Kinetics of Protein Folding under Confinement. *Proceedings of the National Academy of Sciences of the United States of America* **105**, 20233-20238

122. Noel, J. K., Levi, M., Raghunathan, M., Lammert, H., Hayes, R. L., Onuchic, J. N., and Whitford, P. C. (2016) SMOG 2: A Versatile Software Package for Generating Structure-Based Models. *PLOS Computational Biology* **12**, e1004794

123. Betancourt, M. R., and Thirumalai, D. (1999) Exploring the kinetic requirements for enhancement of protein folding rates in the GroEL cavity, Edited by A. R. Fersht. *Journal of Molecular Biology* **287**, 627-644

124. Jewett, A. I., Baumketner, A., and Shea, J.-E. (2004) Accelerated folding in the weak hydrophobic environment of a chaperonin cavity: Creation of an alternate fast folding pathway. *Proceedings of the National Academy of Sciences of the United States of America* **101**, 13192-13197

125. Krobath, H., Shakhnovich, E. I., and Faísca, P. F. N. (2013) Structural and energetic determinants of co-translational folding. *The Journal of Chemical Physics* **138**, 215101

126. Dabrowski-Tumanski, P., Piejko, M., Niewieczerzal, S., Stasiak, A., and Sulkowska, J. I. (2018) Protein Knotting by Active Threading of Nascent Polypeptide Chain Exiting from the Ribosome Exit Channel. *The Journal of Physical Chemistry B* **122**, 11616-11625

127. Sułkowska, J. I., Rawdon, E. J., Millett, K. C., Onuchic, J. N., and Stasiak, A. (2012) Conservation of complex knotting and slipknotting patterns in proteins. *Proceedings of the National Academy of Sciences* **109**, E1715

128. Alam, M. T., Yamada, T., Carlsson, U., and Ikai, A. (2002) The importance of being knotted: effects of the C-terminal knot structure on enzymatic and mechanical properties of bovine carbonic anhydrase II 1. *FEBS Letters* **519**, 35-40

129. Nureki, O., Shirouzu, M., Hashimoto, K., Ishitani, R., Terada, T., Tamakoshi, M., Oshima, T., Chijimatsu, M., Takio, K., Vassylyev, D. G., Shibata, T., Inoue, Y., Kuramitsu, S., and Yokoyama, S. (2002) An enzyme with a

deep trefoil knot for the active-site architecture. *Acta Crystallographica Section D* **58**, 1129-1137

130. Nureki, O., Watanabe, K., Fukai, S., Ishii, R., Endo, Y., Hori, H., and Yokoyama, S. (2004) Deep Knot Structure for Construction of Active Site and Cofactor Binding Site of tRNA Modification Enzyme. *Structure* **12**, 593-602

131. Dabrowski-Tumanski, P., Stasiak, A., and Sulkowska, J. I. (2016) In Search of Functional Advantages of Knots in Proteins. *PLOS ONE* **11**, e0165986

132. Haglund, E., Pilko, A., Wollman, R., Jennings, P. A., and Onuchic, J. N. (2017) Pierced Lasso Topology Controls Function in Leptin. *The Journal of Physical Chemistry B* **121**, 706-718

133. San Martín, Á., Rodriguez-Aliaga, P., Molina, J. A., Martin, A., Bustamante, C., and Baez, M. (2017) Knots can impair protein degradation by ATP-dependent proteases. *Proceedings of the National Academy of Sciences* **114**, 9864

134. Xu, Y., Li, S., Yan, Z., Luo, Z., Ren, H., Ge, B., Huang, F., and Yue, T. (2018) Stabilizing Effect of Inherent Knots on Proteins Revealed by Molecular Dynamics Simulations. *Biophysical Journal* **115**, 1681-1689

135. Sanchez-Ruiz, J. M. (2010) Protein kinetic stability. *Biophysical Chemistry* **148**, 1-15

Departamento de Física and BioISI - Biosystems and Integrative Sciences Institute
Email address: pffaisca@fc.ul.pt

Faculdade de Ciências, Universidade de Lisboa, Campo Grande, Ed. C8, Lisboa, Portugal

Contemporary Mathematics
Volume **746**, 2020
https://doi.org/10.1090/conm/746/15008

Knotoids and protein structure

Dimos Goundaroulis, Julien Dorier, and Andrzej Stasiak

Abstract. Many proteins form open knots. To study and classify proteins in terms of their topology, the protein chain needs to be artificially closed in order to analyze it as a knot. The theory of knotoids provides a workaround to this approach that gives a more refined overview of the topology of knotted proteins. In this review we explain how to analyze open protein chains using the theory of knotoids.

1. Introduction

Individual proteins are built of single, linear polypeptide chains composed of amino acids connected by peptide bonds. In living cells, the majority of polypeptide chains are folded into quasi-rigid 3D structures whose forms are dictated by their amino acid sequence. Proteins with knotted polypeptide chains fold at a slower rate than unknotted proteins of a similar length [**1**–**3**]. It is unknown why some proteins maintained their knottedness throughout evolution despite the burden connected with their slower and less efficient folding. Apparently, in some proteins, the knotted character of the polypeptide chains provides advantages that cannot be easily achieved without knotting. It has been postulated [**4**] and then experimentally demonstrated [**5**] that the 5_2-knotted ubiquitin C-terminal hydrolase protein, which assists proteasomes in recycling unneeded proteins, is protected against being sucked into the proteasome by the existence of the knot. In addition, proteins with knots are thermally more stable [**6**]; and in some cases, knotted protein cores contain active sites with internal contacts providing a favourable environment for catalytic sites [**7**].

A topological approach will enable us to better understand the relationship between the structure and function of proteins as well as the complex process of folding knotted proteins. Knots are formally defined as equivalence classes of closed curves in 3-space where the equivalence relation is given by ambient isotopy. The majority of proteins are open curves, and hence are topologically trivial. For this reason, several methods have been introduced to produce a closed curve from an open one in an unbiased way. These methods can be divided into two categories: the single closure techniques and the stochastic closure techniques. An example of a single closure technique is *the direct closure*, where the protein endpoints are connected with a straight segment. However, that segment may pass through the

2010 *Mathematics Subject Classification.* Primary 57M25; Secondary 92D20.
Key words and phrases. Knot theory, knotoids, invariants, protein structure, protein folding.

entangled part of the chain potentially changing the apparent knot type. Nonetheless, this approach is computationally fast. Another single closure method is the *out of the center of mass closure* technique, where the center of mass of the chain is first determined and then two rays are extended in the direction of the lines that connect the center of mass with each of the endpoints. The two arcs are then connected at infinity to form a closed curve. Stochastic methods rely on probabilistic closures of the chain and then consider the probability distribution of the resulting knot types. One example of a stochastic method is *the infinity closure technique* where the open chain is positioned inside of a large ball, and parallel lines are extended from the endpoints of the chain towards points on the boundary of the ball and then joined together at infinity [**8**]. Closing the open curve in all possible directions provides a probability distribution of knot types that characterizes the curve. Even though all of these techniques provide a reasonable approach to determining whether an open curve is knotted, they all have the weakness that they replace an open chain by a closed curve and thus lose information about the actual conformation of the protein.

The theory of knotoids, which was introduced by V. Turaev in [**9**], generalizes the concept of a knot projection to a projection of a knotted arc. Knotoids have been used to study entanglement of open curves in 3-space [**10**], and more specifically to analyze the topology of open protein chains [**11**]. Furthermore, it has been shown that knotoids provide a more refined overview of a protein chain's topology than any technique for closing the open chain to obtain a knot [**12**].

In this review, we present the mathematical machinery required for the analysis of protein structure using knotoids. More precisely, we give the basic definitions of knotoids in S^2 and in the plane, and we present some polynomial invariants for knotoids that can be used in protein topology analysis. We discuss how the entanglement of an open curve can be measured by probability distributions of its knotoids and how this applies to proteins. We also explain how the knotoid approach can be extended to include proteins with intra-chain bonds (e.g. lasso proteins) by introducing the concept of bonded knotoids and, finally, we briefly present Knoto-ID [**13**], the computational tool that we developed for this analysis.

2. Knotoids

Suppose that we started drawing a knot diagram on a piece of paper and at some point we decided to stop and leave the diagram open. Can this diagram be classified as knotted? Or, even better, does this diagram describe an open knot? If we perform planar isotopies and Reidemeister moves that change the position of the endpoints, we will soon find out that we can unknot the diagram. However, if we were biased in our selection of moves and isotopies, and we chose only those that kept the endpoints from sliding over or under the rest of the diagram then we would observe that the knottedness of the diagram would be preserved.

This new family of knot-like objects are called knotoid diagrams (see for example Fig. 1a) and they can formally be defined as a generic immersion of the closed unit interval $[0, 1]$ in the interior of a surface Σ such that only finitely many double points are allowed. The double points carry the information of over/under crossings, while the endpoints of the interval are taken via the generic immersion to distinct points in Σ. These points in Σ are respectively called the *tail* and the *head* of the knotoid diagram, or just *endpoints* of the knotoid diagram, and they

FIGURE 1. (a) Examples of knotoid diagrams. (b) Knotoids to knots: The underpass, the overpass and the virtual closure.

are distinct from the double points [9, 10]. A knotoid diagram is by convention oriented from tail to head.

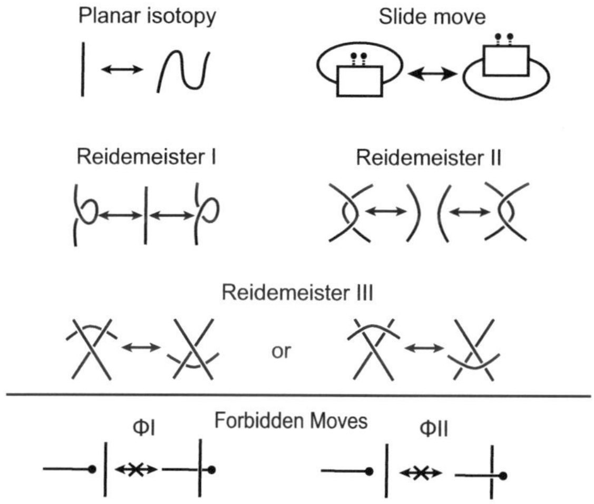

FIGURE 2. The upper part of the figure shows a planar isotopy, a slide move and the Reidemeister moves away from the endpoints of a knotoid diagram. The lower part shows the two forbidden moves. Note that in the slide move the rectangle indicates the entangled part of the diagram. The endpoints are chosen to be placed outside the rectangle in order to simplify the figure however, they can appear in any combination in the diagram.

Knotoid diagrams are usually studied in S^2. Two knotoid diagrams in S^2 are equivalent if they can be continuously deformed to one another by a finite sequence of planar isotopies and Reidemeister moves that are performed away from the endpoints. There are two forbidden moves in this procedure, one where an endpoint slides over an arc of the diagram and one where an endpoint slides under (see Fig. 2). An arc that connects two crossings is allowed to slide around the surface of the sphere and come from the other side of the diagram, as long as it does not cross any of the endpoints. This shall be called the slide move. The equivalence classes of knotoid diagrams under the above moves are called *knotoids.* Knotoids with both endpoints in the same local region of the diagram are called *knot-type* knotoids while those with their end points in different local regions of the diagram are called *proper* knotoids.

The theory of knotoids generalizes knot theory by providing a new diagrammatic approach to knots. Given a knotoid k, one can obtain a knot by connecting the endpoints of k with an embedded arc α that passes always under (resp. over) k.

The arc is called a *shortcut* for k and the closure is the *underpass* (resp. overpass) closure. Given a choice of overpass or underpass closure, the shortcut is unique for every knotoid diagram, up to isotopy. On the other hand, every knot can be represented by a knotoid diagram. This can be easily seen if one removes an underpassing arc from a knot diagram. Of course, two knotoid diagrams represent the same knot if they are related by planar isotopy and the Reidemeister moves away from the endpoints. If all crossings that the shortcut α creates with k are chosen to be virtual, then α is called the virtual closure (See Figure 1b). In [**10**], the extension of knotoids to virtual knot theory is explored. It is still an open question whether every virtual knot is the virtual closure of a knotoid.

Knotoids in $\mathbb{R}^2$ are called *planar* knotoids. Two planar knotoid diagrams are equivalent if one can be deformed to the other using local planar isotopy and Reidemeister moves away from the endpoints. In this case, the slide move is forbidden in addition to the moves ΦI and ΦII (see Fig. 2). The inclusion $\mathbb{R}^2 \hookrightarrow S^2 \cong \mathbb{R}^2 \cup \{\infty\}$ allows every knotoid in $\mathbb{R}^2$ to be seen as a knotoid in S^2. If $\mathcal{K}(\mathbb{R}^2)$ is the set of all knotoids in $\mathbb{R}^2$ and $\mathcal{K}(S^2)$ is the set of all knotoids in S^2, then $\iota : \mathcal{K}(S^2) \to \mathcal{K}(\mathbb{R}^2)$ is an inclusion map. On the other hand, any knotoid in S^2 can represent a knotoid in $\mathbb{R}^2$ if it is pushed past the point at infinity. The map $\rho : \mathcal{K}(S^2) \hookrightarrow \mathcal{K}(\mathbb{R}^2)$ is then well-defined and the composition $\iota \circ \rho = id$, meaning that ι is surjective. There are pairs of non-isotopic planar knotoids that are equivalent as knotoids in S^2 but not as planar knotoids. Note that the slide move is not valid for the planar knotoid equivalence relation. For example, in Figure 1a, the two diagrams are not equivalent in $\mathbb{R}^2$ but they are equivalent in S^2 since they differ by a slide move.

3. Knotoid invariants

Even though the theory of knotoids is relatively new, there is already quite an extensive arsenal of knotoid invariants, many of which are natural extensions of knot invariants. Classical invariants like the crossing number and the knot group as well as polynomial invariants like the Kauffman bracket and the Jones polynomial can be extended to knotoids [**9, 10**]. More recently, invariants for knotoids using biquandle colourings and double-branched covers have been defined in [**14**] and [**15**], respectively. In this review, we focus on polynomial invariants that are regularly used in the analysis of the topology of open protein chains. These invariants are the Jones polynomial [**9, 10**], the Turaev loop bracket [**9, 12**], the arrow polynomial [**10**], and the loop arrow polynomial [**12**].

3.1. The bracket polynomial and the Jones polynomial for knotoids. The Kauffman bracket polynomial for knotoids in any surface Σ can be defined recursively on unoriented knotoid diagrams using the skein relation and axioms depicted in Fig. 3 in an analogous way to its definition for knots.

Applying the skein relation to all crossings of a knotoid diagram k, one obtains a linear sum of disjoint circular components together with a long segment component containing the two endpoints. Each of the circular components contributes a factor of $(-A^2 - A^{-2})$ to the polynomial. There could be cases in which the skein relation on a knotoid diagram yields states where the long segment is nested within a number of circles. This happens for diagrams that have both of their endpoints in one of the inner (bounded) regions of the diagram. Knotoid diagrams in $\mathbb{R}^2$ are said to be *normal* if they have their tail in the outermost region (unbounded region of the plane) of the diagram [**9**]. However, by using slide moves, every knotoid in S^2 can

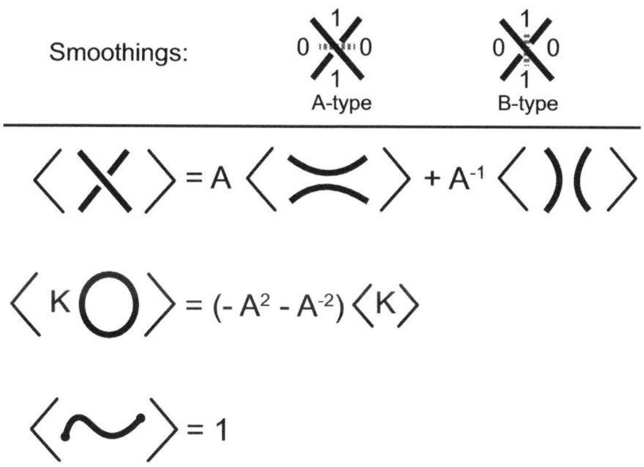

FIGURE 3. In the upper part of the figure, we see the two smoothing types of a crossing. Type 0 (resp. 1) is when the crossing is smoothed in such a way that the two regions labeled as 0 (resp. 1) are merged. The marker indicates the smoothing operation for each case. The lower part of the figure presents the skein relation of the bracket polynomial for knotoids and its axioms.

be represented by a normal knotoid diagram and so for computations in S^2 there is no distinction between nested and not-nested components. The bracket polynomial

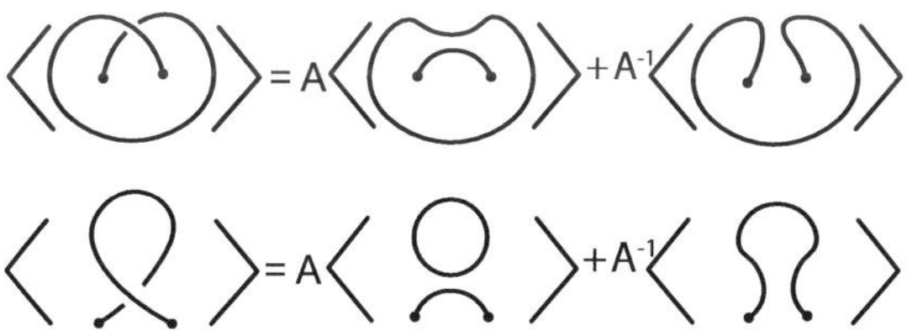

FIGURE 4. The bracket on the two knotoid diagrams is the same in S^2 since they differ only by a slide move. The bracket does not distinguish between circular components that enclose the long segment and those who do not.

for knotoids is a regular isotopy invariant; however, it can be normalized to obtain an ambient isotopy invariant which is the Jones polynomial for knotoids. It is defined via the following normalized state summation formula over all states S of k:

$$f_k(A) = \left(-A^3\right)^{-\operatorname{wr}(k)} \sum_S A^{\sigma(S)}(-A^2 - A^{-2})^{|S|-1} \tag{3.1}$$

where $\operatorname{wr}(k)$ is the writhe of k, $\sigma(S)$ is the number of 0-smoothings minus the number of 1-smoothings of a state S of k and $|S|$ is the number of components of the state S of k.

3.2. The Turaev loop bracket polynomial. For planar knotoids, it is important to distinguish between the states where the long segment is nested in k circular components and the states where the long segment lies outside of all circles. This can be clearly seen in the example of Fig. 4. For this reason, we introduce one extra variable into the axioms of the bracket (see Fig. 5). Using the skein relations and axioms in Fig. 5 one can recursively define the Turaev loop bracket polynomial. The following normalized state summation formula gives the Turaev loop bracket polynomial [**12**]:

$$\widehat{f}_k(A, v) = \left(-A^3\right)^{-\operatorname{wr}(k)} \sum_S A^{\sigma(S)}(-A^2 - A^{-2})^{\alpha(S)} v^{\beta(S)} \tag{3.2}$$

where $\operatorname{wr}(k)$ and $\sigma(S)$ are as in 3.1, $\alpha(S)$ is the number of circular state components that do not enclose the long segment, and $\beta(S)$ is the number of circular components that contain the long segment of the state S of k. For $v = -A^2 - A^{-2}$, we obtain the Jones polynomial for knotoids. Notice that for a state S, we have that $\alpha(S) + \beta(S) = |S| - 1$. Note that 3.2 can also be obtained as a specialization of $[k]_\circ \in \mathbb{Z}[A^{\pm 1}, u^{\pm 1}, v]$, the three-variable extended bracket for knotoids in $\mathbb{R}^2$ [**9**], for $u = 1$.

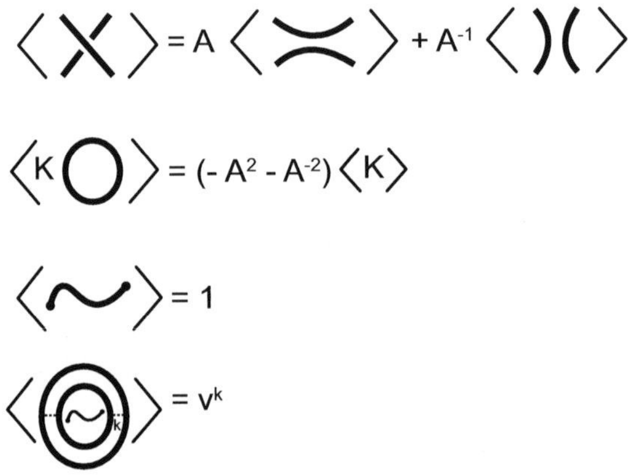

FIGURE 5. The skein relation and the axioms of the Turaev loop bracket polynomial.

3.3. The arrow polynomial. The arrow polynomial is based on the oriented expansion of the bracket polynomial and it was initially defined in [**16**] as an invariant for virtual knots. The extension to classical knotoids in S^2 and virtual knotoids appeared in [**10**]. The skein relation in this case involves smoothings with matching or conflicting orientations (see Fig. 6). Note that each smoothing with a conflicting orientation results in a pair of cusps, and therefore the set of axioms for the recursive definition of the arrow polynomial à la bracket polynomial has to include rules that take into account these additional combinatorial structures. Each state includes a number of circular components and a long segment component, all of which may contain a number of consecutive cusps. A cancellation rule allows the simplification of two consecutive cusps into a straight arc, if the acute angles of both cusps are in the same local region of the diagram (see Fig. 7). The cancellation rule is not applicable to cases where the acute angles of two consecutive cusps are in different local regions of the diagram. The arrow polynomial assigns a new variable to each long segment of a state with a number of surviving cusps. In particular, two consecutive surviving cusps form a zigzag and a long segment with $2i$ surviving cusps is evaluated at m_i. The skein relation together with the axioms discussed above recursively define the arrow polynomial. Equivalently, the arrow polynomial can be defined by the following state summation formula:

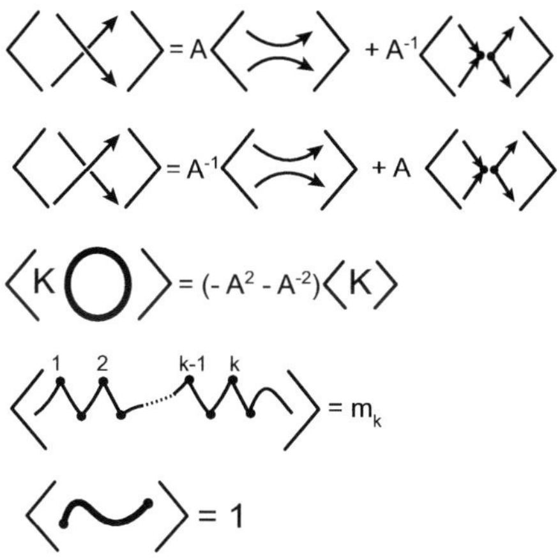

FIGURE 6. The skein relation and the axioms of the arrow polynomial.

(3.3) $$A\,[K] = (-A^3)^{-\operatorname{wr}(k)} \sum_{S} A^{\sigma(S)} m_{\iota(S)} (-A^2 - A^{-2})^{|S|-1},$$

where $\operatorname{wr}(k)$, $\sigma(S)$ and $|S|$ are as in 3.1, and $\iota(S)$ is the number of zigzags of the long segment of a state S. Note that the smoothing types of a crossing are the same, with the addition that the orientations of the arcs are taken into account.

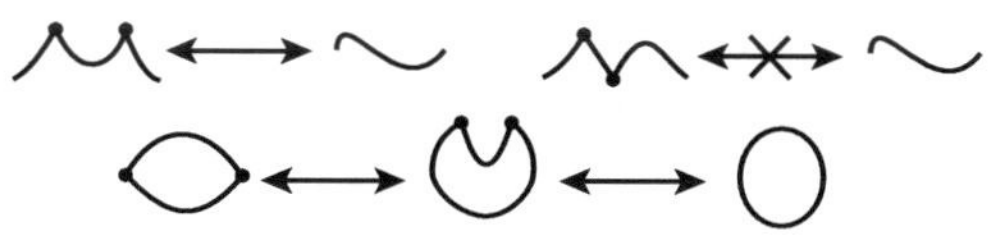

FIGURE 7. The allowed and forbidden cancellation rules.

3.4. The loop arrow polynomial. In analogy to the Turaev loop bracket, the loop arrow polynomial is the planar version of the arrow polynomial and it was mentioned first in [**12**]. The loop arrow distinguishes two different types of zigzags (compare the zigzags in Fig. 6 to those in Fig. 8), and it assigns one of the variables m_i or w_i, depending on the type of zigzag. Furthermore, much like the Turaev loop bracket, the loop arrow polynomial assigns different variables to circular components that enclose the long segment with or without any of the two types of zigzags (see Fig. 8).The loop arrow polynomial is defined by the following state summation formula:

(3.4)
$$\widehat{A}\,[K] = (-A^3)^{-\operatorname{wr}(k)} \sum_{S} A^{\sigma(S)} m_{\iota_1(S)} w_{\iota_2(S)} (-A^2 - A^{-2})^{\alpha(S)} v^{\beta(S)} p^{\gamma(S)}_{\eta_1(S)} q^{\delta(S)}_{\eta_2(S)},$$

where $\operatorname{wr}(k)$, $\sigma(S)$, $\alpha(s)$ and β are as in 3.2, $\iota_1(S)$ and $\iota_2(S)$ are the number of zigzags of the first and second type respectively on a long state segment not enclosed by a circular component,$\gamma(S)$ and $\delta(S)$ are the number of circular state components that enclose a long state component with zigzags of the first or second type respectively, and $\eta_1(S)$ and $\eta_2(S)$ are the number of zigzags of the first and second type, respectively, on a long state segment enclosed by circular components. Note that for a state S, the following holds : $\alpha(S) + \beta(S) + \gamma(S) + \delta(S) = |S| - 1$, where $|S|$ is as in 3.1.

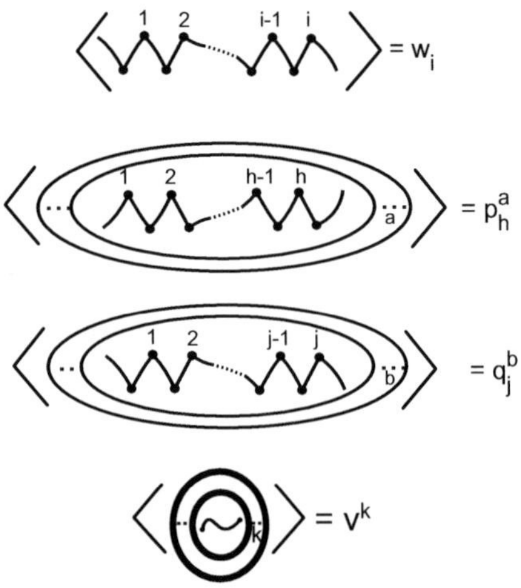

FIGURE 8. The loop arrow polynomial satisfies the skein relation as well as all axioms of the arrow polynomial together with the four additional axioms presented in this figure.

4. Embedded open curves in 3-space and knotoids

Consider a smooth curve embedded in 3-space. We would like to find whether if it is knotted and to detect its knot type. If the curve is closed then all one needs to do is to consider the curve's generic projection on a plane together with the over/under crossing information at double points, and then evaluate the resulting diagram using a knot invariant. Note that all projections of the curve are equivalent up to ambient isotopy. The case of open curves, however, is not so direct. While projecting the curve on a plane may seem a reasonable approach, one has to notice that different choice of projection planes may lead to the placement of the endpoints in different regions of the diagram, changing the resulting knotoid (see Fig. 9). Throughout this section we shall use the same open curve for all of our examples.

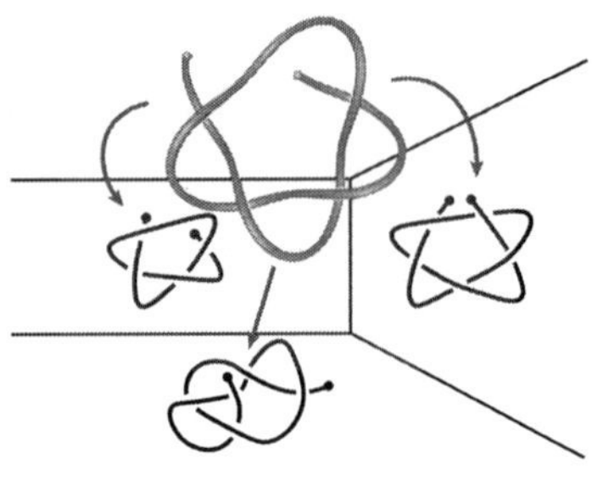

FIGURE 9. Projections of the same open curve to different planes yields different, non-isotopic knotoid diagrams.

By analogy with various methods to form a closed curve from an open one (see e.g. [**8**]), the approach to the characterization of an open curve as a knotoid is a probabilistic one. In this context, the space of all projections of the curve are considered. Sampling of this space gives a probability distribution of knotoid types which characterizes the topology of the open curve, while the knotoid that appears with the highest probability is said to be *dominant*. The knotoids are identified by evaluating each projection with a knotoid invariant. It is important to note that ambient isotopy of an open curve may displace the endpoints of its projections to a chosen plane and thus change the knotoid type of the projections. This means that the forbidden moves for the knotoid diagrams equivalence are not respected, which require the endpoints of the diagrams to be fixed region-wise. For this reason, two infinite lines, each of which passes through one of the endpoints, that are perpendicular to the projection plane are considered, see Fig. 10.

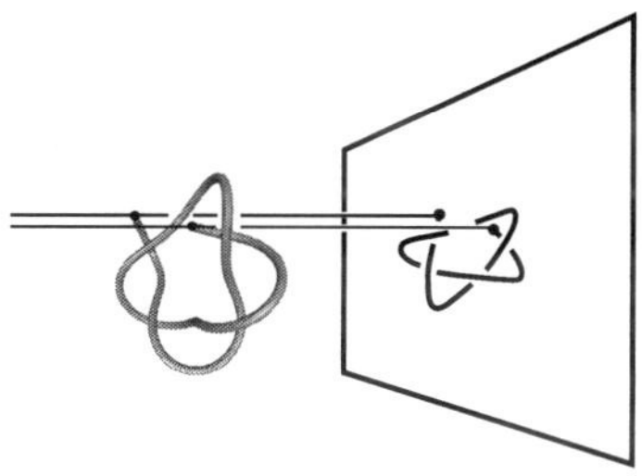

FIGURE 10. An open curve together with the two lines that fix its endpoints with respect to a plane of projection and its corresponding knotoid diagram .

If two curves, each with a pair of such infinite lines perpendicular to the same projection plane, are ambient isotopic in the complement of the lines, by an isotopy taking infinite lines to infinite lines and endpoints to endpoints, then we say the curves are *line-isotopic* [**10**]. Knotoid diagrams that correspond to line-isotopic curves belong to the same equivalence class of knotoids. It is important to note that while studying the topology of an open curve with the probabilistic method discussed above, each choice of a projection plane determines a pair of infinite lines through the endpoints of the curve perpendicular to the chosen plane.

In practice, global topology is analyzed using computational means and, in the case of knotoids, this can be achieved using Knoto-ID [**13**]. The curve is considered inside a large enough ball where each point on the boundary of the ball corresponds to a tangent plane of projection. The visualization of this analysis uses the so-called projection globes where each point is colour coded depending on the knotoid type obtained when it is projected on a tangent plane. Consequently, regions with the same colour correspond to the same knotoid type. In the example of Fig. 11, it is easy to see that the dominant knotoid is the 4_6^m knotoid (where m means that it is the 4_6 knotoid but with all of its crossings inverted). The dominant knotoid is not always easy to detect directly from the projection globes and careful examination of the probability distribution of knotoid types is required. Note also that closing 4_6^m with an overpass closure we obtain the 5_1 knot. The knotoid notation that we use follows the classification of [**17**].

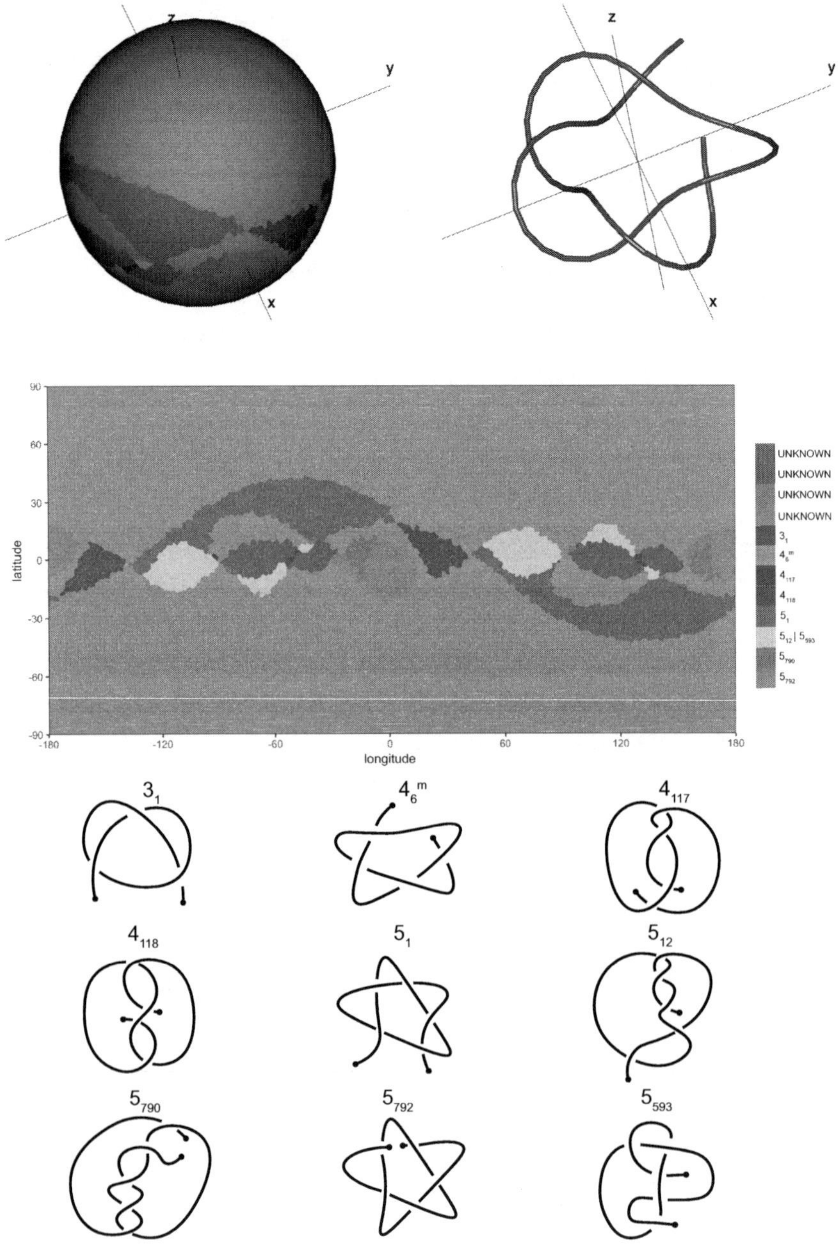

FIGURE 11. The top row shows the projection globe and the analyzed curve, while the middle row shows the corresponding projection map together with the colour map of knotoid types. In this particular example we sampled 10,000 projections of the open curve. Each projection point is associated to a Voronoi cell on the surface of the sphere which is coloured according to the corresponding knotoid type.

5. Application to protein studies

Proteins are long biopolymers whose *backbone* consists of a single long chain of amino acids, which are organic compounds containing carbon, nitrogen and oxygen atoms connected in pairs with covalent peptide bonds. The sequence of amino acids of a protein determines its shape and function. [**18**]. In order to perform its biological function, a protein has to reach first its native folded state. The specifics of the folding mechanism of a protein are not fully understood and especially intriguing is the mechanism by which knotted proteins are able to achieve their native folded state. From protein structures that are deposited in the Protein Data Bank [**19**], we can extract the coordinates of their carbon atoms and reconstruct their backbones as piece-wise linear curves. We can then smooth the curve and apply the methods described in the previous section to analyze the topology using knotoids in S^2 or planar knotoids [**11, 12**]. It is important to mention here that analyzing proteins using virtual knots [**20**] provides equivalent results to knotoids in S^2 [**11**]. However, it has been shown that planar knotoids provide a more refined overview of the topology of an open chain than knotoids in S^2 [**12**].

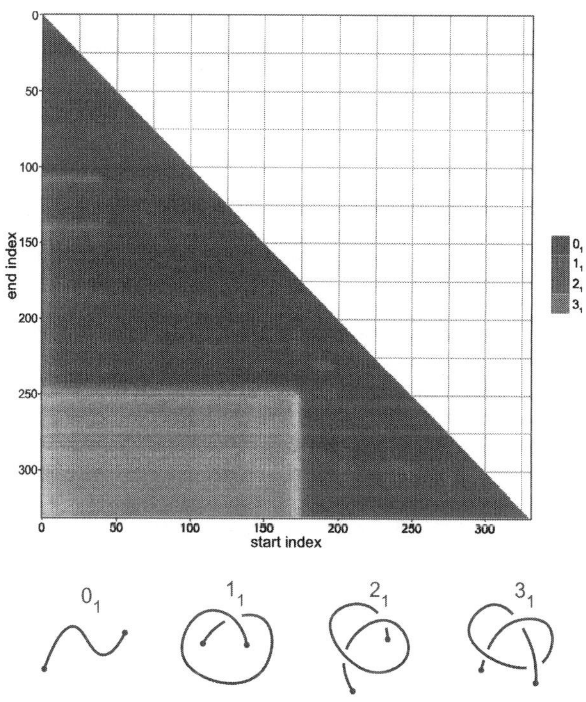

FIGURE 12. The fingerprint matrix of the protein N-acetyl-L-ornithine transcarbamylase [**24**]. Each entry in this matrix corresponds to a subchain of the protein backbone. The lower left corner corresponds to the whole chain. Moving to the right along the x-axis, we trim the chain starting from one endpoint (the N-terminus) while moving up the y-axis we are clipping vertices from the other endpoint (C-terminus) of the chain. This protein contains a deep trefoil knotoid, meaning that we need to remove approximately 170 amino acids from the N-terminus in order to get a trivial subchain and about 60 amino acids from the C-terminus to do the same. We use a colour code to indicate the dominant knotoid types appearing in this fingerprint matrix.

The piecewise linear curve corresponding to the protein backbone is usually simplified with the Koniaris-Muthukumar algorithm [**23**] which preserves the underlying topology of the chain and also decreases computational load. The Koniaris-Muthukumar or *triangle elimination method* checks first whether three consecutive vertices of the curve form a triangle. If the surface of the triangle is not pierced by any other arc of the chain then the middle vertex of the three is removed. This procedure is repeated until no such triangles appear in the chain. In the case of open curves analyzed with knotoids a modified version of this algorithm is used which additionally checks if the triangle surface is pierced by any of the infinite lines [**10, 11**].

We have discussed so far how to determine the global topology of a chain, that is, the knotoid type of the whole backbone. There have been cases however where a protein chain that is globally topologically trivial contains a locally knotted portion [**21**]. Such a conformation is called a *slipknot* and was first observed in proteins in [**21**].

These local non-trivial topologies have biological significance but, unfortunately, the global analysis is blind to them. However, by truncating enough vertices from a slipknotted piece-wise linear curve a topologically non-trivial subchain is obtained and so localized knots can be captured by considering the subchain that contains them. Systematic study of all subchains of a protein reveals the global and local topology of its conformation. This information can be summarized in a triangular matrix called the fingerprint where the dominant knotoid types of all subchains of the chain are presented [**21**] (see Fig. 12). Each entry in the matrix represents the dominant knotoid type of the subchain that starts at the amino acid number indicated along the x axis and ends at the amino acid position indicated along the y axis, while its knotoid type is listed in the colour coded bar.

6. Knotoids and intra-chain protein bonds

Apart from the covalent bonds between carbon, nitrogen and oxygen atoms that make up the protein backbone, there can be other types of intra-chain bonds covalent or non-covalent that exist in protein molecules. The most well known of them are disulphide bonds that tie together two cysteines (one of twenty types of amino acids present in polypeptide chains) that are not adjacent on the polypeptide chain but which are close together in space. Intrachain bonds together with backbone bonds create closed paths that may have complex topology. A mathematical characterization of such paths can lead to a better understanding of a protein structure. In [**12**], the concept of *bonded knotoids*, which is a topological model that takes into account possible intra-chains bonds, was proposed. In this model, each bonding site between two amino acids is represented by a bonding arc that connects them. Each bonding site is considered as a locally rigid planar formation. An isotopy relation for bonded knotoid diagrams is generated by the bonded moves (see Fig. 13) and the corresponding class is a bonded knotoid.

By replacing the bonding sites with tangles (see Fig. 14) and evaluating the resulting knotoids with an invariant like the ones presented in Section 3, one can derive invariants for bonded knotoids.

By including backbone bonds and other intra-chain bonds in our representations of protein structures, we obtain lasso structures and theta curves (see for

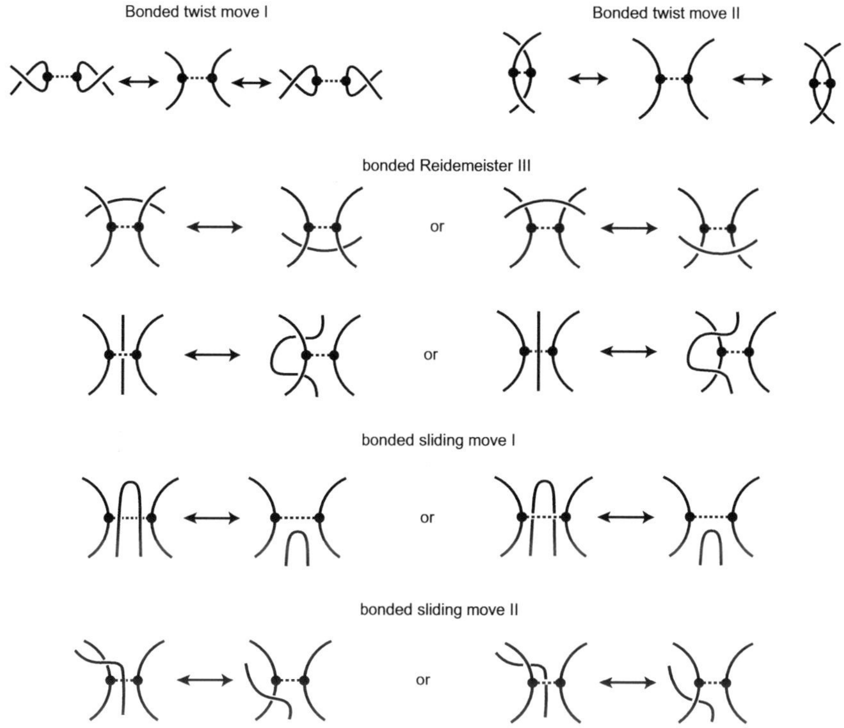

FIGURE 13. The bonded Reidemeister moves.

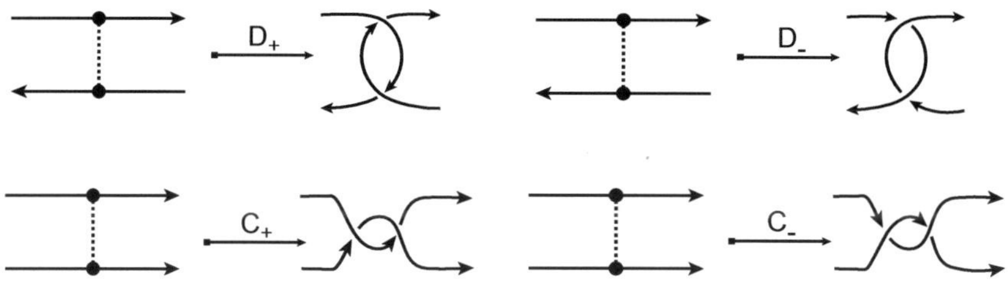

FIGURE 14. Anti-parallel (up) and parallel (down) strands are substituted by a full twist. The substitution preserves the orientation of the overall diagram. The twist can be either positive or negative.

example [**12**, **25**]). Note that the resulting conformation may correspond to a knotoid diagram with multiple circular components. Such a diagram is called a multi-knotoid. The equivalence relation on knotoids as well all invariants discussed above extend naturally to multi-knotoids.

7. Knoto-ID: the first software for the analysis of open ended curves using knotoids.

Studies of protein topology rely heavily on computational tools for the identification of knot types formed by the protein backbone. Until recently, there wasn't any specialized computational tool that could analyze an open protein chain using knotoids. In fact, all the available tools required open chains to be closed in order to form a knot. Knoto-ID [**13**] is the first software that can analyze the topology of any open curve using the methods discussed in Sections 4 and 5. It accepts as input the curve in xyz coordinates while for the knotoid identification, in its current version, it supports the Jones polynomial for knotoids and the Turaev loop bracket. The arrow and loop arrow polynomials are implemented in the development version of Knoto-ID and will appear in a future release. Moreover, it supports input of diagrams in Gauss code as well as in PD code and it can plot knotoid diagrams. Knoto-ID comes bundled with a set of helpful scripts for the visualization of the resulting topological data. Note that, as mentioned above, Knoto-ID can analyze the topology of any open chain in xyz coordinates and so it can be used for the topological analysis not only of proteins but also of chromosomes, random linear polymers, random walks and so on. All analyses included in this work have been achieved using Knoto-ID. Knoto-ID is open source and it can be downloaded from `https://github.com/sib-swiss/Knoto-ID`. A detailed user manual is provided together with the distribution that also includes the basics for protein analysis with Knoto-ID. Finally, we note that Knoto-ID is now part of the Knotprot 2.0 database [**26**].

Acknowledgements

This work has been funded by Leverhulme Trust (Grant RP2013-K-017 to A.S.) and by the Swiss National Science Foundation (Grant 31003A 166684 to A.S.). The authors would like to thank Louis H. Kauffman for many fruitful conversations on the subject.

References

[1] A. L. Mallam and S. E. Jackson, *Knot formation in newly translated proteins is spontaneous and accelerated by chaperonins.* Nat. Chem. Biol. 8 no 2 (2012), 147-53. doi: 10.1038/nchembio.742.

[2] P. Dabrowski-Tumanski, A. I. Jarmolinska, and J. I. Sulkowska JI. *Prediction of the optimal set of contacts to fold the smallest knotted protein.* Journal of Physics-Condensed Matter. 27 no 35 (2015). doi: 10.1088/0953-8984/27/35/354109.

[3] J. I. Sulkowska, J. K. Noel, and J. N. Onuchic, *Energy landscape of knotted protein folding.* Proc. Natl. Acad. Sci. U.S.A. 109 no 44 (2012), 17783-8. doi: 10.1073/pnas.1201804109.

[4] P. Virnau, L. A. Mirny, and M. Kardar, *Intricate knots in proteins: Function and evolution.* PLoS Comput. Biol. 2 (2006), 1074-1079.

[5] M. K. Sriramoju, Y. Chen, Y-T. C. Lee, and S-T. D. Hsu, *Topologically knotted deubiquitinases exhibit unprecedented mechanostability to withstand the proteolysis by an AAA+ protease* Sci. Rep. 8 no 1 (2018), 7076.

[6] J. I. Sulkowska, P.Sułkowski, P. Szymczak, and M. Cieplak *Stabilizing effect of knots on proteins* Proc. Natl. Acad. Sci. U.S.A. 105 no 50 (2008) ,19714-19719. https://doi.org/10.1073/pnas.0805468105

[7] P. Dabrowski-Tumanski, A. Stasiak, and J.I. Sulkowska, *In Search of Functional Advantages of Knots in Proteins.* PLoS ONE 11 (2016), e0165986, doi:10.1371/journal.pone.0165986.

[8] J. I. Sulkowska, Joanna, E. J. Rawdon, K. C. Millett, J. N. Onuchic, Jose N and A. Stasiak, *Conservation of complex knotting and slipknotting patterns in proteins.* Proc. Natl. Acad. Sci. U.S.A. 109 no 26 (2012), E1715-E1723.

[9] Vladimir Turaev, *Knotoids*, Osaka J. Math. **49** (2012), no. 1, 195–223. MR2903260

[10] Neslihan Gügümcü and Louis H. Kauffman, *New invariants of knotoids*, European J. Combin. **65** (2017), 186–229, DOI 10.1016/j.ejc.2017.06.004. MR3679845

[11] D. Goundaroulis, J. Dorier, F. Benedetti, and A. Stasiak, *Studies of global and local entanglements of individual protein chains using the concept of knotoids.* Sci. Rep. 7 no 1 (2017), 6309.

[12] D. Goundaroulis, N. Gügümcü, S. Lambropoulou, J. Dorier, A. Stasiak, and L. Kauffman, *Topological Models for Open-Knotted Protein Chains Using the Concepts of Knotoids and Bonded Knotoids.* Polymers 9 no 9 (2017), 444.

[13] J. Dorier, D. Goundaroulis, F. Benedetti, and A. Stasiak, *Knoto-ID: a tool to study the entanglement of open protein chains using the concept of knotoids.* Bioinformatics, 34 no 19 (2018), 3402-3404, DOI `https://doi.org/10.1093/bioinformatics/bty365`.

[14] Neslihan Gügümcü and Sam Nelson, *Biquandle coloring invariants of knotoids*, J. Knot Theory Ramifications **28** (2019), no. 4, 1950029, 18, DOI 10.1142/S0218216519500299. MR3943696

[15] A. Barbensi, D. Buck, H. A. Harrington, and M. Lackenby, *Double branched covers of knotoids.* arxiv:1811.09121 (2018).

[16] H. A. Dye and Louis H. Kauffman, *Virtual crossing number and the arrow polynomial*, J. Knot Theory Ramifications **18** (2009), no. 10, 1335–1357, DOI 10.1142/S0218216509007166. MR2583800

[17] D. Goundaroulis, J. Dorier, and A. Stasiak, *A systematic classification of knotoids on the plane and on the sphere* arXiv:1902.07277 (2019).

[18] B. Alberts, A. Johnson, J. Lewis, M. Raff, K. Roberts, and P. Walter, *Molecular Biology of the Cell.* Garland Science, New York, 2008.

[19] H. M. Berman, J. Westbrook, Z. Feng, G. Gilliland, T. N. Bhat, H. Weissig, I. N. Shindyalov, and P. E. Bourne, *The protein data bank.* Nucleic Acids Res. 28 no 1 (2000), 235-242.

[20] K. Alexander, A. J. Taylor, and M. R. Dennis *Proteins analysed as virtual knots* Sci. Rep. 7 no 1 (2017) 42300.

[21] N. P. King, E. O. Yeates, and T. O. Yeates, *Identification of rare slipknots in proteins and their implications for stability and folding.* J. Mol. Biol 373 no 1 (2007), 153-166.

[22] Kenneth C. Millett, *Knots, slipknots, and ephemeral knots in random walks and equilateral polygons*, J. Knot Theory Ramifications **19** (2010), no. 5, 601–615, DOI 10.1142/S0218216510008078. MR2646649

[23] K. Koniaris, M. Muthukumar, *Self-entanglement in ring polymers.* J. Chem. Phys. 95 no. 4 (1991), 2873-2881.

[24] D. Shi, X. Yu, L. Roth, H. Morizono, M. Tuchman, and N. M. Allewell, *Structures of N-acetylornithine transcarbamoylase from Xanthomonas campestris complexed with substrates and substrate analogs imply mechanisms for substrate binding and catalysism.* Proteins Struct. Funct. Bioinform. 64 (2006), 532-542.

[25] W. Niemyska, P. Dabrowski-Tumanski, M. Kadlof, E. Haglund, P. Sułkowski, and J. I. Sulkowska, *Complex lasso: new entangled motifs in proteins.* Sci. Rep. 6 (2016), 36895.

[26] P. Dabrowski-Tumanski, P. Rubach, D. Goundaroulis, J. Dorier, K. C. Millett, E. J. Rawdon, P. Sułkowski, A. Stasiak, and J. I. Sulkowska, *KnotProt 2.0: a database of proteins with knots and other entangled structures.* Nucleic Acids Res. 47 D1, (2018) D367-D375. `https://doi.org/10.1093/nar/gky1140`.

Center for Integrative Genomics, University of Lausanne, Lausanne, Switzerland and SIB Swiss Institute of Bioinformatics, Lausanne, Switzerland
Email address: `dimoklis.gkountaroulis@unil.ch`

Vital-IT, SIB Swiss Institute of Bioinformatics, Lausanne, Switzerland and Center for Integrative Genomics, University of Lausanne, Lausanne, Switzerland
Email address: `julien.dorier@sib.swiss`

Center for Integrative Genomics, University of Lausanne, Lausanne, Switzerland and SIB Swiss Institute of Bioinformatics, Lausanne, Switzerland
Email address: `andrzej.stasiak@unil.ch`

Contemporary Mathematics
Volume **746**, 2020
https://doi.org/10.1090/conm/746/15009

Topological linking and entanglement in proteins

Kenneth C. Millett

ABSTRACT. This paper provides a mathematical introduction to the application of the integral formulation of the Gauss linking number to the analysis of entanglement in macromolecules. This application is inspired by knots and slipknots recently found in proteins. Entanglement is understood as the linking or self-linking of segments of a linear macromolecule, e.g. a protein or polymers in a gel. The mathematical features of Gauss linking, as inspired by protein knotting, provide a method to identify and quantify important sites of entanglement. It will be applied in several ways to proteins to locate sites with these structures that are similar to those of knotting and slipknotting. As a vehicle to illuminate this discussion, we will describe the application of these methods to the case of an important unknotted protein, the nitrogenase molybdenum-iron protein from Clostridium pasteruianum, an oxidoreductase identified in the Protein Data Base as 4WES. Expanding upon this linking analysis, we will also describe how the presence of cysteine bonds in proteins gives another structure within which one finds examples of intrinsic linking. This creates distinctly different classes of linked structures and the application to other ways to study linking.

1. Introduction

The spatial structure of macromolecules such as proteins can be quite varied from topological or geometric prespectives. The relationship between their spatial characteristics and their biological function has long been a research objective. Within functional families, one finds important structural features unchanged across organisms and unchanged during evolution. For example, since 2000, it has been known that there are knotted [**T**, **MDS**, **MS**] and, since 1977, slipknotted [**Ri**, **K**] protein structures that are present within an entire functional family and preserved despite millions of years of evolution [**S**]. A systemic analysis of the knots and slipknots in proteins was published in 2012 [**S**, **R**] with results updated weekly in KnotProt 2.0 [**J**, **DTR**] to include newly reported strutures. In addition to the purely topological facets of knotting and slipknotting, there are geometric features that must be present as well. For example, the Fary-Milnor Theorem [**F**, **MJ**], states that if the total curvature of a simple closed curve in 3-space is not larger than 4π the curve must be unknotted. So knotted and slipknotted chains must support a total curvature in excess of 4π. While knotting and slipknotting are a severe form of self-entanglement of a macromolecule, most proteins do not exhibit

2010 *Mathematics Subject Classification.* Primary 57M25, 57M27; Secondary 92C05, 92C40.
Key words and phrases. Gauss linking, entanglement, proteins, open macromolecules.

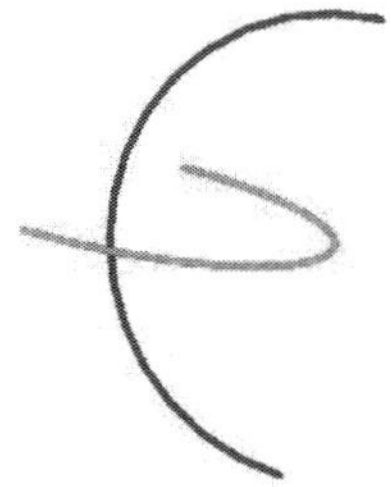

FIGURE 1. The Gauss linking integral, Equation 1.1, gives a real number, in this case approximately 0.565369, quantifying the disposition of two oriented chains.

these structures leading one to seek geometric properties that may be correlated with the biological structure. We find that proteins often consist of multiple unknotted chains, e.g. dimers, trimers, etc., which are entangled through the linking of portions of the chains or, perhaps, contain subchains with substantial self-linking or pairs of subchains that locate spatially critical portions of the macromolecule's functional biological structure.

As protein chains are taken to be oriented from the N terminus to the C terminus thereby giving collections of oriented linear chains. The linking and self-linking in collections of open or closed oriented chains can be determined by the Gauss linking integral [**G**]. The *linking number* between two oriented chains, L_1 and L_2, described by piecewise C^1 parameterizations, $\gamma_1(t)$ and $\gamma_2(s)$, $0 \leq t, s \leq 1$, is defined by the double integral:

$$Ln(L_1, L_2) = \frac{1}{4\pi} \int_{[0,1]} \int_{[0,1]} \frac{(\dot{\gamma}_1(t), \dot{\gamma}_2(s), \gamma_1(t) - \gamma_2(s))}{||\gamma_1(t) - \gamma_2(s)||^3} dtds, \tag{1.1}$$

where $(\dot{\gamma}_1(t), \dot{\gamma}_2(s), \gamma_1(t) - \gamma_2(s))$ is the triple product of the derivatives, $\dot{\gamma}_1(t)$ and $\dot{\gamma}_2(s)$, and of the difference $\gamma_1(t) - \gamma_2(s)$.

The linking number of a pair of closed chains is an integer topological invariant. For open chains, see Figure 1, the Gauss linking number is a real number. For open chains, it is no longer a topological invariant but it still provides a measure of the spatial linking disposition of the two chains.

If L denotes a chain, with a piecewise C^1 parameterization given by $\gamma(t)$ with $0 \leq t \leq 1$, then the *self-linking number* of L is defined by:

$$\begin{aligned} Sl(L) &= \tfrac{1}{4\pi} \int_{[0,1]} \int_{[0,1]} \tfrac{(\dot{\gamma}(t), \dot{\gamma}(s), \gamma(t) - \gamma(s))}{||\gamma(t) - \gamma(s)||^3} dtds \\ &\quad + \tfrac{1}{2\pi} \int_{[0,1]} \tfrac{(\dot{\gamma}(t), \ddot{\gamma}(t), \dddot{\gamma}(t))}{||\dot{\gamma}(t) \times \ddot{\gamma}(t)||^2} dt. \end{aligned} \tag{1.2}$$

where $\dot{\gamma}(t)$, $\ddot{\gamma}(t)$, and $\dddot{\gamma}(t)$ are the first, second, and third derivatives of $\gamma(t)$, respectively, and $(\dot{\gamma}(t), \ddot{\gamma}(t), \dddot{\gamma}(t))$ is their triple product. The self-linking number consists of two terms, the first being the Gauss linking integral and the second being the total torsion of the curve. While one may apply these formula directly to polygonal chains, de facto piecewise C^1 curves, one can use discrete calculation formulae for the Gauss linking and self-linking of polygons [**B**]

We will employ the Gauss linking and self-linking numbers in our analysis of the entanglement present in protein structures. Our first step is to analyse linking features associated with simple mathematical knots and those found in knotted and slipknotted protens identified using KnotProt2.0 [**J**,**DTR**] where one finds a current listing of knotted proteins including the specific location and type of knot present in the protein. The KnotProt2.0 knotting fingerprint, top left Figure 2, shows the knot type of each subchain of a protein segment allowing one to determine the smallest support of the knot within the protein whose cartoon structure is on the right in Figure 2. The same structure appears in the linking fingerprints of other proteins having trefoil knots and slipknots.

This recognition provokes an interest in exploring the presence of linking in protein structures. One can investigate the presence of linking in several ways ranging from the large scale linking between closed loops, between closed and linear chains, and between linear chains. While, in this paper, we will focus on the linking between linear chains, we will also quickly review several other applications focusing on linking. In many instances, such as the nitrogenase molybdenum-iron protein, the protein consists of several chains whose linking and self-linking determines a real symmetric linking matrix whose structure provides a global quantification of the entanglement. Each component of the protein determines a linking fingerprint matrix in which one reports the linking of a subsegment with the initial and final termini. Another facet of linking between two components can be assessed via a fingerprint that reports the linking of a segment of one component with the entirety of the other component and the reversed roles. Furthermore, with the presence of cysteine bonds, the protein can be endowed with the enriched structure of a lasso, a two-component link, or even more complex macromolecular graph structures. These enriched structures provide the foundation for another linking analysis employing minimal oriented spanning surfaces to quantify and characterize another aspect of the linking structures entangling these protens. With respect to this method of analysis, we will describe the corresponding software: LassoProt [**N**,**DT**], LinkProt [**DTS**,**DTJ**], and PyLink [**GN**].

2. Linking entanglement in families of macromolecules

Building upon the Gauss linking number, Equation 1.1, and the Self-linking number, Equation 1.2, we will explore several options to express information that captures the nature of the linking present in the structure. To give an explicit example, we consider the nitrogenase molybdenum-iron protein from Clostridium pasteurianum [**Z**], PDB 4WES, www.rcsb.org [**BW**], consisting of four chains, pairs of two of each of two types, the α chains 4WESA and 4WESC and the β chains 4WESB and 4WESD, Figure 3.

Although having different placements within the protein, chains A and C have essentially the same spatial structure when rotated and translated, Figure 4. Similarly, chains B and D have the same structure. They fit together in a manner analogous to pieces of a puzzle but can be thought of as creating individually separate portions of the protein. As a result, the chains display a weak level of pairwise entanglement that we can measure using the linking numbers as well as assessing the self-entanglement using the self-linking numbers.

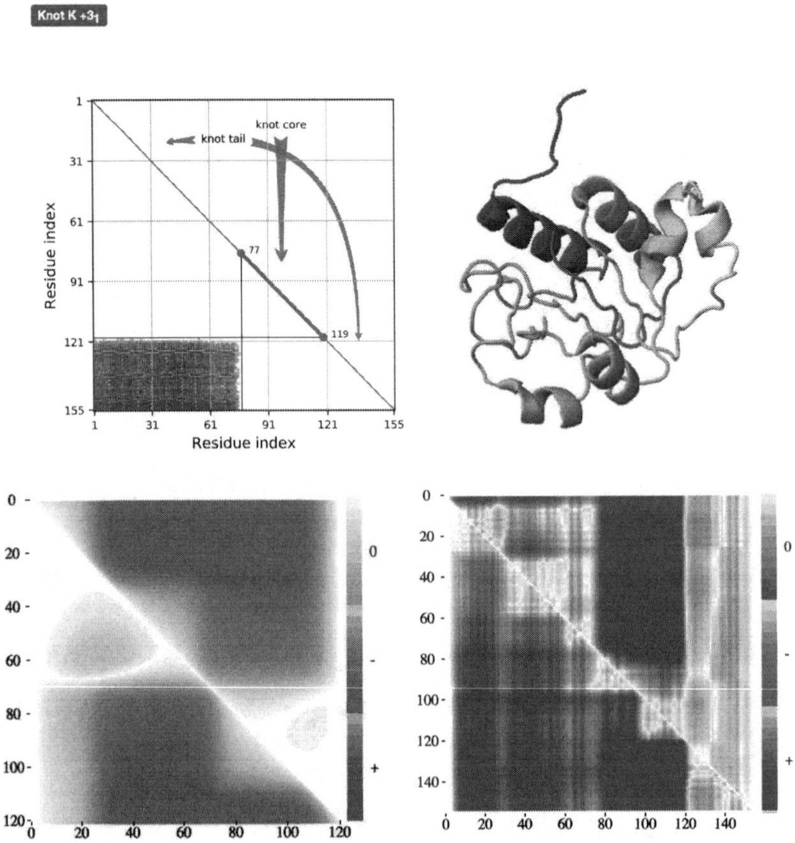

FIGURE 2. The knotting fingerprint of a methyltransferase segment, PDB $1J85_A$, of YIBK from haemophilus influenzae (HI0766) indicates the location of a trefoil knot, the segment: residues 77 through 119 (top left). The correspoinding knot core segment is colored green in the cartoon (top right). Below, on the left, we show the linking fingerprint of a negative trefoil exhibiting a vertical band and a characteristic signature square region along the diagonal of the band that is found, in this case a positive trefoil, in $1J85_A$ with a band from 77 through 119 and the same characteristic signagure at the diagonal. The linking fingerprint is discussed in Subsection 2.2.

2.1. The linking matrix, LN, of 4WES. For a tetramer macromolecular structure such as 4WES, denoted L, consisting of four oriented chains, L_1, L_2, L_3, and L_4, the associated linking matrix is a 4×4 matix whose entries along the diagonal are the self-linking numbers and the off diagonals are the linking numbers of the corresponding pairs of chains. Since the linking number of a pair does not depend on the order, the resulting matrix is a real symmetric matrix. It is, therefore, diagonalizable, having four real eigenvalues and associated eigenvectors that express key characteristics of the global entanglement. It is helpful to think of the linking

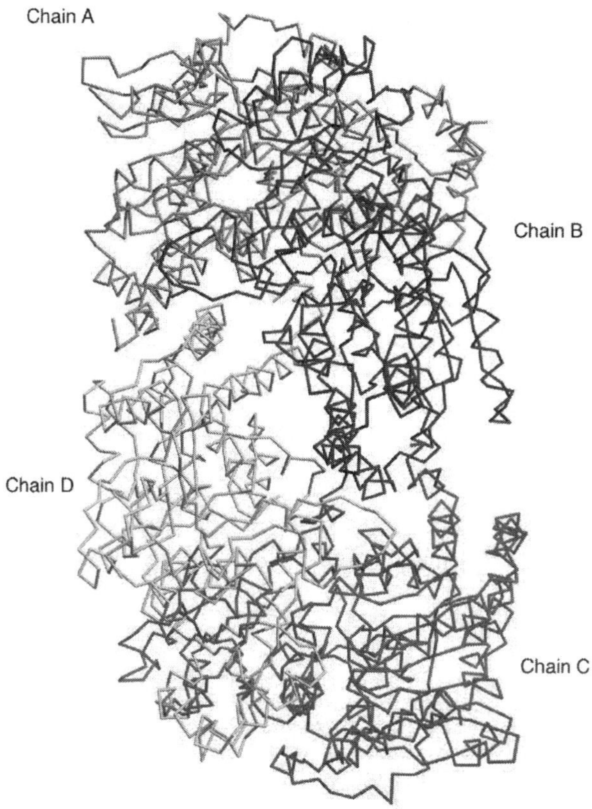

FIGURE 3. The structure of 4WES indicating each of its four chains: Chain A-Green, Chain B-Blue, Chain C-Red, Chain D-Yellow

matrix of a collection of oriented chains as defining a weighted adjacency matrix of a graph representation of a complex system (especially if one has a large number of chains), for which the spectrum of the matrix or of its associated Laplacian matrix provides insight into the global structure of the system, e.g. its Cheeger constant. This perspective inspires the analysis of the linking matrix and its eigenvalues, even for the small structures discussed here.

$$LN(L) = \begin{bmatrix} Sl(L_1) & Ln(L_1, L_2) & Ln(L_1, L_3) & Ln(L_1, L_4) \\ Ln(L_1, L_2) & Sl(L_2) & Ln(L_2, L_3) & Ln(L_2, L_4) \\ Ln(L_1, L_3) & Ln(L_2, L_3) & Sl(L_3) & Ln(L_3, L_4) \\ Ln(L_1, L_4) & Ln(L_2, L_4) & Ln(L_3, L_4) & Sl(L_4) \end{bmatrix}$$

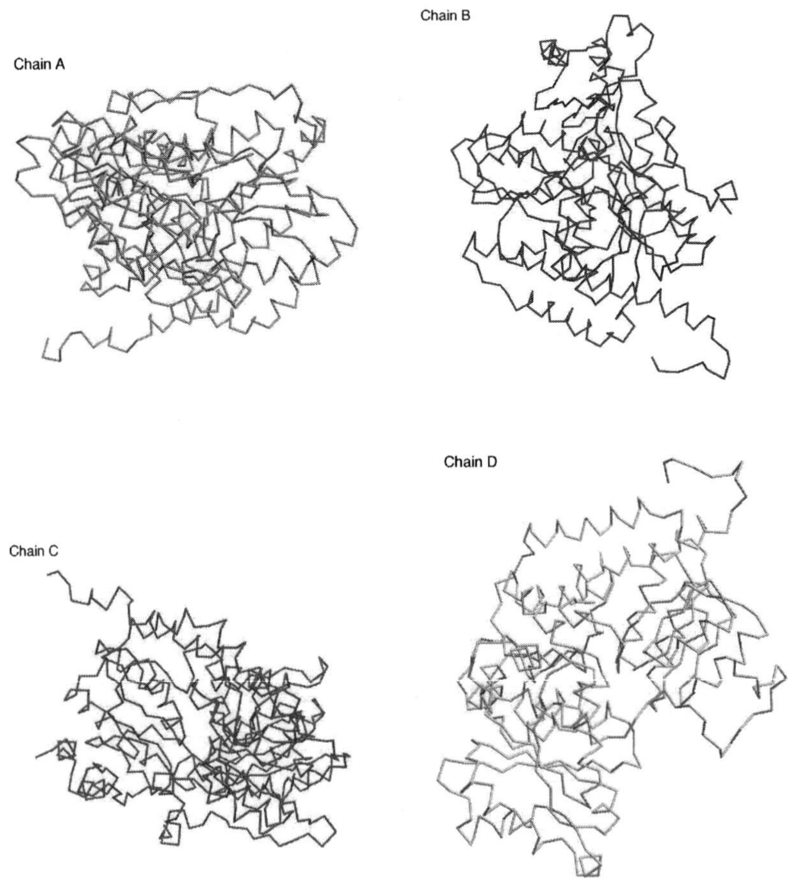

FIGURE 4. The structure of the subchains of 4WES : Chain A-Green, Chain B-Blue, Chain C-Red, Chain D-Yellow

For the case of 4WES, the linking matrix is

$$LN\,(4WES) = \begin{bmatrix} 0.2358710 & 0.0533449 & 0.0428935 & 0.0692668 \\ 0.0533449 & 0.2296200 & 0.0272541 & 0.0284736 \\ 0.0428935 & 0.0272541 & 0.2343980 & 0.0538401 \\ 0.0692668 & 0.0284736 & 0.0538401 & 0.2359280 \end{bmatrix}$$

Observe that the linking numbers are small and the self-linking is significantly stronger than the linking between the pairs of chains. Also, the linking between the pairs of α chains, A and C, and between the pairs of β chains, B and D, is much smaller than the linking of an α and a β chain, for example between chains A and D, reflecting the structure one sees in Figure 3.

2.1.1. *Invariants of the linking matrix of 4WES.* The fundamental measure associated to the linking matrix is its set of eigenvalues and eigenvectors providing a real diagonalization of the linking matrix. For 4WES one has

- 0.374501 $\{0.570846, 0.404258, 0.462702, 0.544625\}$
- 0.21292 $\{-0.200393, -0.768212, 0.497005, 0.350255\}$
- 0.189866 $\{0.386873, -0.364445, -0.709337, 0.462979\}$
- 0.15853 $\{0.691431, -0.335954, 0.167757, -0.617184\}$

The structure of the linking matrix and the associated eigenvalues and eigenvectors suggest that the pair of A and D chains are more strongly entangled. Considering the eigenvectors, one can observe two relationships, that of chains A and D and that of chains B and C. Furthermore, these pairs of relationships are distinct as reflected in the structure of the associated eigenvectors. From the perspective shown in Figure 3, the relative strength of linking between chains A and D, though quite small in absolute size, seems surprising. Furthermore, the interaction between chains B and C is puzzeling. Perhaps these represent interesting biologically significant features of the spatial relationship between these chains.

2.2. The linking fingerprint of a protein chain. The linking fingerprint of an n-atom connected chain, L, in a protein is a visual representation of the *Matrix Fingerprint*, $MLFP$, describing the linking between a subsegment of the chain with its complementary initial and terminal subsegments. Specifically, if $i < j$ then $MLFP_{i,j}$ is the linking number of the subsegment consisting of atoms from i to j with the initial segment consisting of atoms 1 to $i-1$. If $i > j$ then $MLFP_{i,j}$ is the linking number of the subsegment consisting of atoms from i to j with the terminal segment consisting of atoms $j+1$ to n. If $i = j$, we set $MLFP_{i,j} = 0$. The *Linking Fingerprint*, $LFP(L)$, is a visual representation of $MLFP$ defined by coloring the i, jth cell of an $n \times n$ matrix grid by blue of intensity measured by the magnitude of $MLFP_{i,j}$ if positive, red of intensity measured by the magnitude of $MLFP_{i,j}$ if negative, and very light grey if $MLFP_{i,j} = 0$. The intensity of the color is scaled with the unnormalized absolute value of the linking number of the pairs of oriented chains.

2.2.1. *The linking fingerprints of 4WESA.* As the two α chains are spatially similar and the two β chains are spatially similar, one only needs to determine one fingerprint of each class, Figure 5. Despite their apparent differences, for example their different lengths, the two linking fingerprints have strikingly similar structures when one focuses on the structure provided by the blue and the red vertical bands and the presence of similar square regions along the diagonals. At least some part of this similarity can be attributed to the important presence of α helices in both chains.

Inspired by the presence of knots in proteins, one naturally looks for linking structures that resemble those characteristic of the knotted regions. The signature associated to a local trefoil knotted region, shown in Figure 6, is a square region along the diagonal. In Figure 5, one can observe several such potentially interesting square regions such as the one bounded by molecules 206 through 256, within the boxed region in α chain A. While not knotted, one sees some of the features of a knotted region in which the linking captures a specific spatial structure.

Another conspicuous aspect of the fingerprint is the presence of vertical bars, both positive and negative. For example, in chain A in Figure 5 there are red

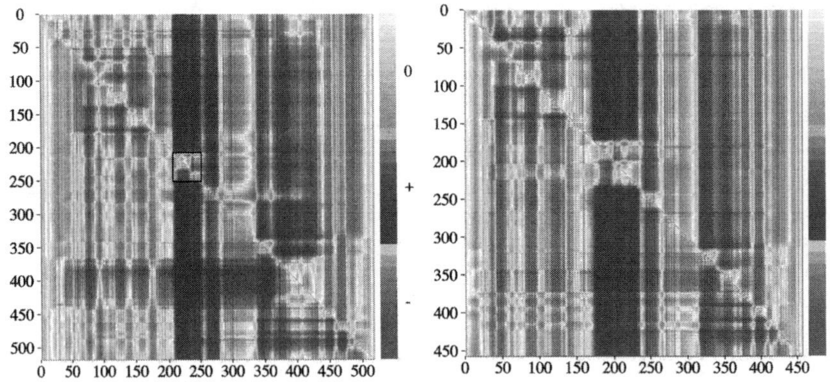

FIGURE 5. The linking fingerprints of the subchains of 4WES: α Chain A (left) and β Chain B (right).

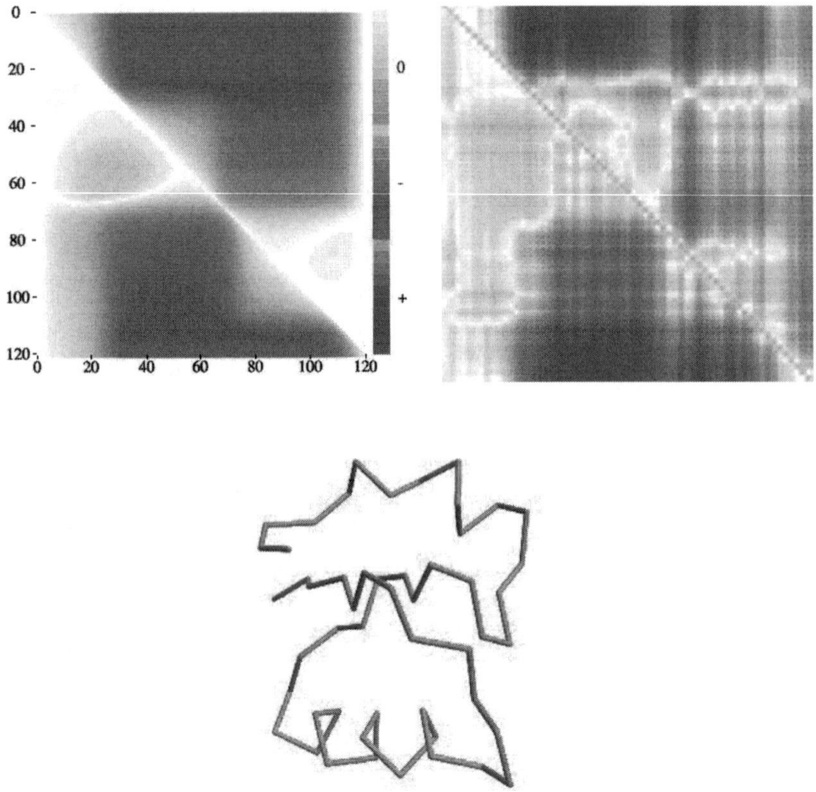

FIGURE 6. The linking fingerprint of the small trefoil knotted segment, and an enlargment of a visually similar region in the fingerprint of A chain 4WES, enclosed in the box in Figure 5, and, below, the subsegment the A chain of 4WES causing this fingerprint.

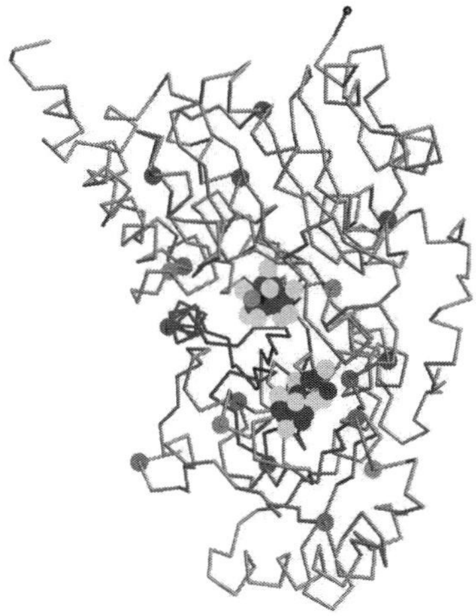

FIGURE 7. The A chain α-carbon backbone structure of 4WES embellished with entanglement and biological information. Looking down along a channel in which one first encounters the iron-sulfur cluster, then the iron-sulfur, molybdenum cluster (whose atoms are colored yellow and purple) and, finally, the large blue entanglement region forming the trefoil stopper in the channel. The trefoil stopper is shown in the box in Figure 5 as well as in Figure 6 . The blue dots identify the centers of the blue bands in the linking fingerprint, Figure 5, and the red dots locate the centers of the red bands. The orange dots are the iron-sulfur binding sites identified in the PDB entry for 4WESA.

bars centered at, roughly, 53, 63, 94, 143, 174 and, 490 and blue bars centered at, roughly, 45, 79, 225, 245, 270, 350, 370, 450 and, 470. The large positive, blue, vertical bar between 206 and 256, in Figure 5, whose diagonal region resembles that of a trefoil knot also has biological importance. The iron-sulfur binding sites and the important iron-sulfur-molybdenum cluster and the iron-sulfur cluster implicated in the functionality of chain A are shown in Figure 7 in which the 206 through 256 segment forming a channel stopper is colored blue. Observe that the centers of the red, negative, bands appear to occur along the inside of the channel whilst the centers of the blue, positive, bands seem occur at the centers of the α-helices on the exterior of the channel of chain A.

One can apply the above analytical procedure to each of the four chains in the tetramer to describe the internal structure of each chain in conjunction with the coarse entanglement measure provided by the linking matrix. Inspired by the linking fingerprint, in the next section we describe a finer method to illuminate the linking between protein chains, the local linking matrix.

2.3. Local linking fingerprint between two chains. Another type of linking fingerprint is defined by the linking of a segment of one chain with the entirety of the other chain. The local linking fingerprint of an n-atom connected subchain, L_1, in one component of a protein and an m-atom connected subchain, L_2, of another component of a protein provides a visual representation of the *Matrix Local Linking Fingerprint*, $MLLFP$, describing the linking between a subsegment of one subchain with another component. Specifically, if $i < j$ then $MLLFP_{i,j}$ is the linking number of the subsegment of L_1 consisting of atoms from i to j with L_2. The *Local Linking Fingerprint*, $LLFP(L_1, L_2)$, is a visual representation of $MLLFP$ defined by coloring the i, jth cell, for $i < j$, of the associated triangular grid by blue of intensity measured by the magnitude of $MLLFP_{i,j}$ if positive, red of intensity measured by the magnitude of $MLFP_{i,j}$ if negative, and light grey if $MLLFP_{i,j} = 0$. One also has the analogous Local Linking Fingerprint, $LLFP(L_2, L_1)$, giving the linking of segments of L_2 with L_1.

2.3.1. *Local linking fingerprint between two chains of 4WES.* As 4WES has four subchains, there will be six local linking fingerprints corresponding to the six distinct pairs of subchains. We will analyze the local linking fingerprint between chains 4WESA and 4WESB, Figure 8. Chain 4WESA has 518 atoms and chain 4WESB has 458 atoms. Thus we compute two triangular linking matrices as shown in Figure 8, showing, first, the linking of segments of 4WESA with 4WESB and, in the second, the linking of segments of 4WESB with 4WESB. As the two chains are topologically unlinked, i.e. there are no pairs of segments with linking as strong as that suggested in Figure 1, the overall linking reflected in the fingerprints is quite weak, reflected by the pale colors: one even observes the existence of regions of both positive and negative linking features. More striking, however, is the presence of distinctive red and blue bands meeting at the diagonal. Recall that the diagonal cells correspond to segments from j to j, i.e. of zero length. Thus, the horizontal bands reflect the linking of the segments whose initial terminus starts at the initial terminus of the chain and gradually approaches j as the cells approach the diagonal. Similarly, the vertical bands reflect the linking of segments whose initial terminus starts at j with terminal end approaching the terminus of the chain.

In the $LLFP(4WESA, 4WESB)$, the top fingerprint in Figure 8, note the blue horizontal band and red vertical band meeting roughly at cell $(45, 45)$. This identifies a positive disposition of the initial 45 atom segment of $4WESA$ with $4WESB$, an "on average" linking turning point near atom 45, and a negative disposition of the terminal segment from atom 45 with the B chain, $4WESB$, see location in Figure 8 identified by arrow. This, roughly, corresponds to the type of transition shown in Figure 9. The corresponding local linking fingerprints, shown in Figure 10, show that while the local linking of the green arc with the blue-red arc is very small due to the cancellation of contributions, the local linking of the blue-red arc with the green arc clearly shows broad regions of positive and negative linking as well as the delineation of their cancellation, when both positive and negative contributions balance each other, along a positive diagonal. This shows the local structure near the diagonal. In the full fingerprint, these linking schemes continue horizontally and vertically due the much weaker contributions of those atoms further away, along the chain, from the atom 45. The cancellation balance visible at the diagonal is obscured by these distal contributions as the length of the chain centered at 45 grows in length and is replaced by the contributions generated from other balance points and their horizontal and vertical bands. In the left local linking fingerprint of A chain segments with the B chain, in addition to that at 45, having similar local linking fingerprint signatures near the diagonal, though with different positive/negative structures. These are found, in Figure 8, at atoms 45, 75, 100, 130, and 470, giving rise to analogous structural relationships, see Figure 11.

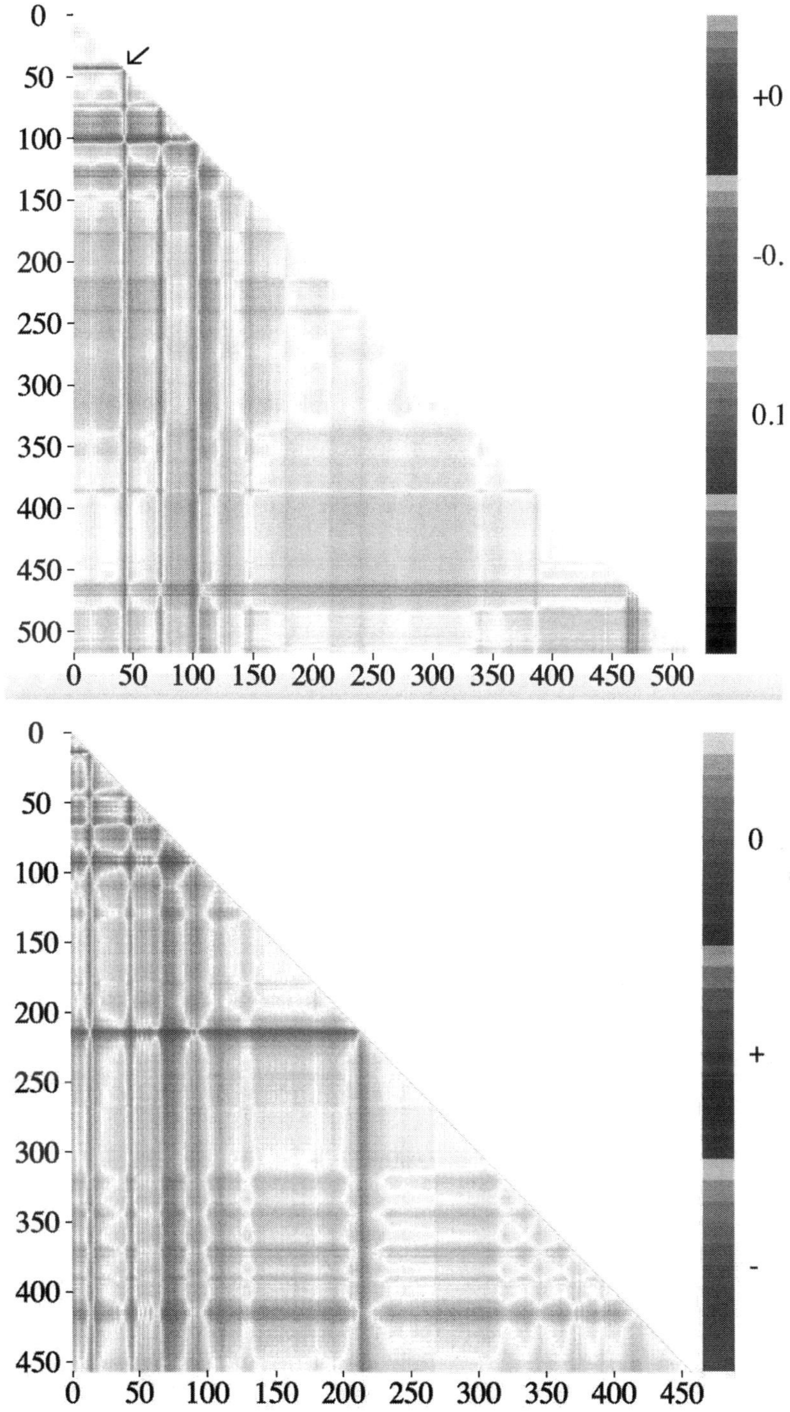

FIGURE 8. The local linking fingerprint of 4WESA and 4WESB. First, one has the local linking of segments of the A chain with the entire B chain. Second, below, one has the local linking fingerprint of the segments of the B chain with the entire A chain.

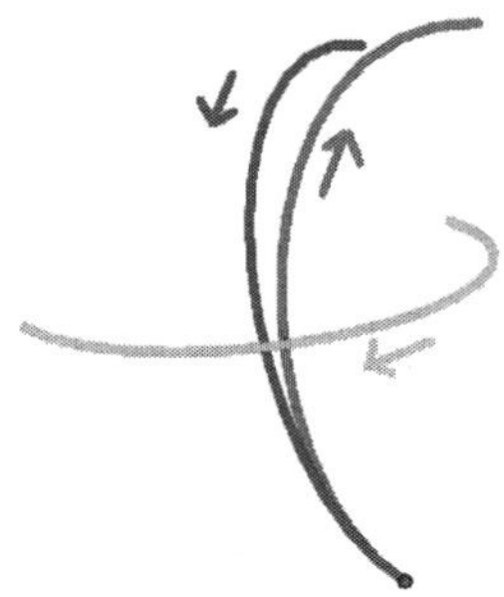

FIGURE 9. A simple illustration of the local linking transition from negative to positive. The orientation of the segments is indicated by the arrows and the blue and red segments have, respectively, positive and negative linking with the green segment as shown in the right most linking fingerprint in Figure 10.

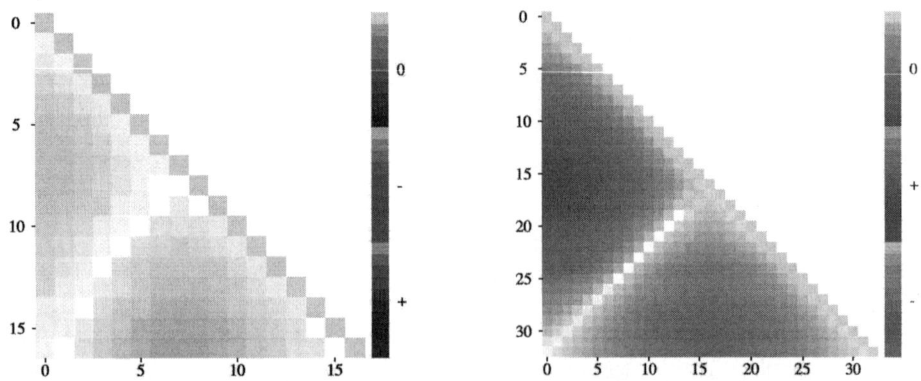

FIGURE 10. The local linking fingerprint of green and blue-red chains in Figure 9. On the left, one has the local linking of segments of the green chain with the entire blue-red chain showing a very weak linking signature. On the right, one has the local linking fingerprint of the segments of the blue-red chain with the entire green chain illustrating the structure that gives rise to the local linking structure example near the diagonal indicated by the arrow in the top fingerprint in Figure 8.

What, perhaps, might be the biological significance of the sites highlighted by this analysis? Could it be that they play a role in the iron-sulpher binding that occurs in three of these locations? What might be the benefit of the other locations this analysis has identified? Questions such as these make this local linking analysis an attractive tool with which to study multi-chain protein complexs such as 4WES.

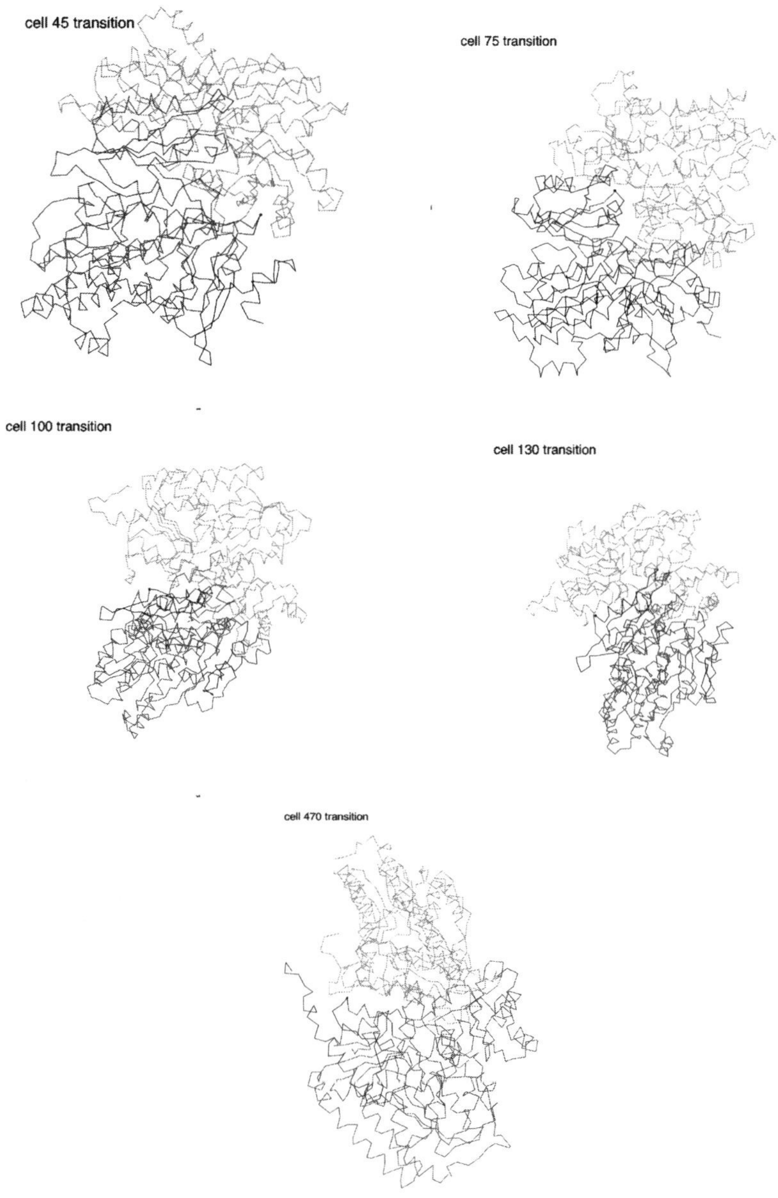

FIGURE 11. The structures generating the transitions of the local linking structure of 4WESA segments beginning at 45, 75, 100, 130, and 470 with 4WESB. For example, in the 45 transition, the horizontal segment in Figure 8 at cell 45 indicated by the arrow, is a positive linking so here the initial segment is colored blue and the terminal segment is colored red are reflected by the red (negative linking) vertical segment at the arrow. A black vertex identifies the transition atom in the figure. The figures for the other linking transitions are colored using the same protocol. So, for the 75 transition, the initial segment is colored red and the terminal segment is colored blue.

3. Lassos and LassoProt: An analysis of linking entanglement in the presence of a single bridge

LassoProt is a server and database, http://lassoprot.cent.uw.edu.pl, that presents the analysis of the linking entanglement structure of proteins and other biopolymers in which a loop is created by the presence of a cysteine, or other, bridge that closes a subchain of the structure thereby creating an oriented lasso loop [**H1**,**H2**,**H3**,**N**,**DT**], see Figure 12, of a variety of types. The complement of the lasso loop segment consists of an initial segment, from the N terminus, and a terminal segment, to the C terminus. LassoProt identifies a minimal area triangulated oriented surface spanning the lasso loop and determines whether and where a terminal segment intersects this surface, see Figure 13. For the A chain of the periplasmic transporter protein, PDB 1KKH, the N chain intersects the surface twice, first in a negative sense and then in a positive manner. As C chain does not intersect the surface, one has a lasso of type L_2.

This analysis does not, however, give the complete picture of the extent to which the N and C chains link the lasso loop. As with multi-chain proteins, one can use the Gauss linking integral, Equation 1.1, to quantify the extent of linking of the loop with these complementary initial and terminal chains using the linking fingerprint method. This can provie a richer picture of the overall entanglement present. Neimyska and I have explored the application of the linking fingerprint method to this end. The protein $1KKH_A$, Figure 14, provides a good example of the analysis. The linking fingerprints showing the linking of the N and the C termini with the lasso loop, Figure 18 show the significant impact of the positive and negative linking segments of the N terminal chain as well the resulting slightly negative overall linking with the lasso loop (observed in the lower left corner of the N chain linking fingerprint in Figure 18). Furthermore, this linking fingerprint provides an independent occurance of the linking transition observed earlier for protein chains in $4WES$, Figures 9 and 10. While the C terminal chain does not intersect the lasso loop minimal surface, it does have a substantial negative with the lasso loop as shown in the right linking fingerprint in Figure 18. As above, one can observe measures of the lasso linking within the local linking fingerprint of the protein, see Figure 16. The N termal linking of -0.012 with the lasso loop is shown at cell (112,286) and the C terminal linking of -0.598 with the lasso loop is shown at cell (286,112).

For the periplasmic transporter protein, $PDB\ 2DVZ_A$, see Figure 17, the N chain intersects the surface once in a positive manner and the C chain intersects the surface once in a negative manner suggesting that the N and C terminal chains link the oriented lasso loop, consisting of residues from 93 through 152, in a positive and negative manner, respectively. The N terminal chain, formed by residues 15 - 92 has a 0.346176 linking with the loop formed by residues from 93 through 152 shown at cell (93,152) in Figure 18. The C terminal chain, formed by residues 153 - 314, has a -1.42215 linking number with the lasso loop shown in Figure 18 at cell (152,93). The linking fingerprint identifies the presence of a negative entanglement supported by the chain of residues from 110 through 245 as well as positive N and C tail structures that appears to be essentially unrelated to the lasso loop structure. This demonstrates that the lasso loop linking and the local linking with initial and terminal chains may identify complementary facets of entanglement within some proteins.

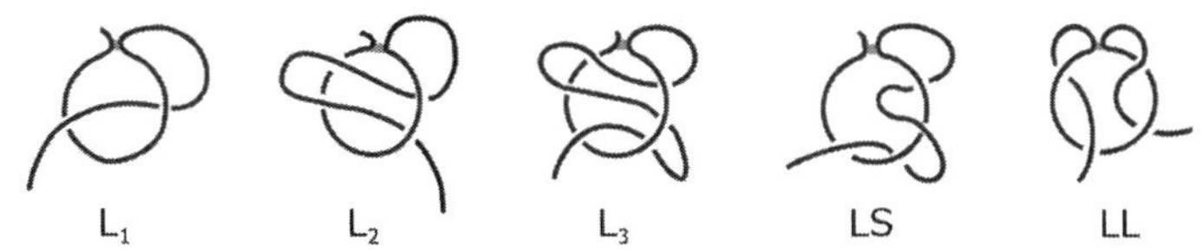

FIGURE 12. Examples of non-trivial lasso configurations [**DT**]

In some cases biological structures have several bridges and, therefore, several lasso loops create structures whose linking relationships one wishes to explore. Such is the case with PDB $6F74A$, a VAO-type flavoprotein MTVAO713 from Myceliophthora Thermophila C1. From LassoProt, one is presented with an analysis of five distinct lasso structures arising from the addition of such bridges. In one case, there is a bridge connection from atom 194 to atom 278 thereby creating an unknotted loop. In the LassoProt analysis, this loop is spanned by a triangulated disc surface of minimal area and the intersections of this minimal disk with the initial and terminal segments exterior to the lasso loop are indicated. In this specific case, one sees that the initial C-terminal tail intersects the disc surface at atom 301 in the negative direction with respect to the right-handed orientation of the surface induced by the oriented lasso loop. While the N-terminal chain does not intersect the disc surface, the Gauss linking integral can detect the extent of linking relationship. Thus, in this case, one has an example of a L_1 lasso, see Figure 12. We note that LassoProt reports that none of the other four lasso loops have any minimal disc intersections. Although the Gauss linking integral can be used to quantify the linking between lasso loops and with two complenentary regions, dimensional limitations seem to prevent an accessible visible quantative presentation of the more complex linking relationships.

4. LinkProt: An analysis of linking entanglement in the presence of multiple bridges

LinkProt is a server and database, http://linkprot.cent.uw.edu.pl, that presents the analysis of the topologically non-trivial linking entanglement structure of proteins and other biopolymers, Figure 19, in which loops are created by the deterministic presence of a cysteine, or other, bridges that close the subchain of the structure thereby creating a special type of graph structure that supports multiple closed loops that may or may not be linked [**DTJ**]. In addition, it presents structures that are evaluated stochastically, see Figure 20, by random independent closure "to infinity" of segments in two chains having pairs of residues suggesting the presence of an "almost loop". Each of the pairs of stochastic closures are evaluated for linking thereby generating data whose spectrum is shown at the right in Figure 20 for the A and C chains of the protein PDB 5L7Q. Note that one can also employ the Gauss linking integral to quantitify the linking between such pairs of chains as discussed earlier in section 2.

First found in 1989, [**V**], many more complex linking structures are now known [**BC**]. LinkProt gives hundreds of examples now reported in the Protein Data Bank. LinkProt provides new tools for studying the non-trivial spatial topology and provides a broad spectrum of biological information not available in other databases.

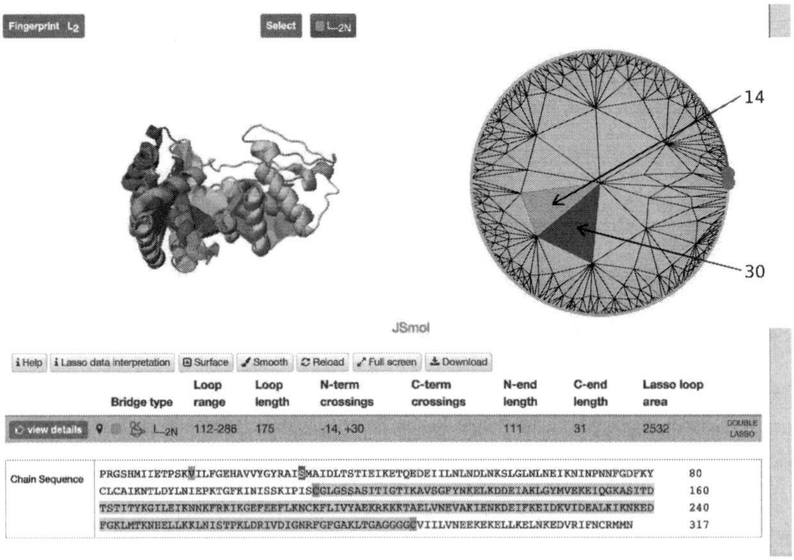

FIGURE 13. The minimal surface intersection data for the protein $PDB\ 1KKH_A$, a lasso of type L_2, showing the intersection cells and the tail lengths for both the C and N terminal chains.

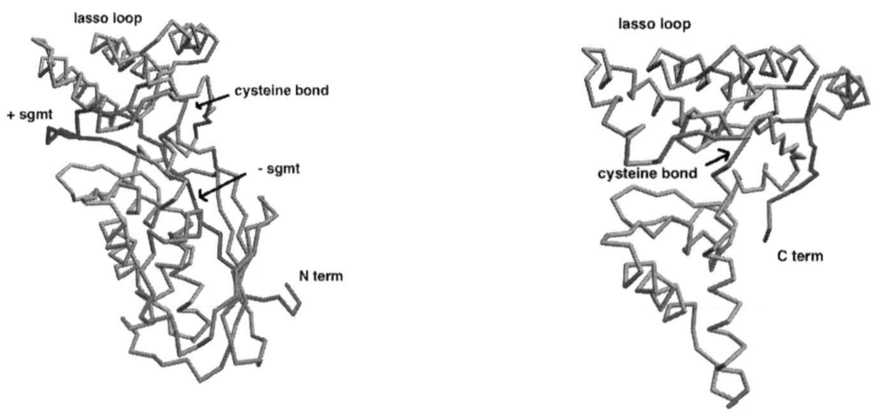

FIGURE 14. On the left the backbone structure of protein $PDB\ 1KKH_A$, a lasso of type L_2, showing the cysteine bond (orange) defining the lasso loop (green). For the complementary N terminal segment, we show the negative intersecting segment (red) and the positive intersecting segment (blue), as well as the remaining essentially neutral segment (pink). On the right, the non-intersecting C terminal segment is also shown (pink).

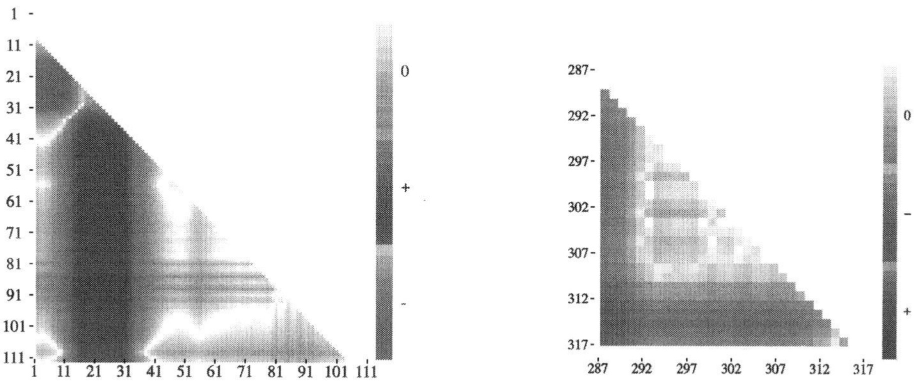

FIGURE 15. Linking fingerprints of N and C terminal segments of the lasso loop in $PDB\ 1KKH_A$. In the N segment, one sees the initial negative linking segment (min linking number -1.07) followed by the positive segment (max linking number 1.24) and the remaining neutral segment. For the C terminal segment one has a minimal negative linking of -0.77 without a minimal surface intersection.

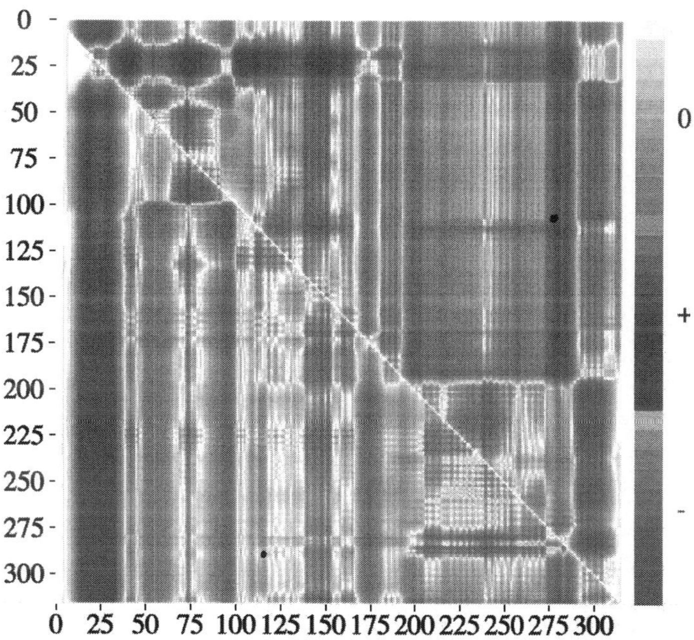

FIGURE 16. Linking fingerprint of the A chain of $1KKH_A$ with the initial N segment and terminal C tail linking with the lasso loop shown at the indicated cells (112,286) and (286,112), respectively.

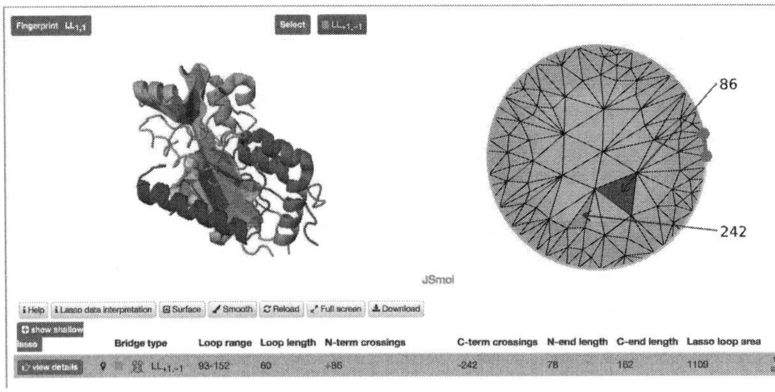

FIGURE 17. The minimal surface intersection data for the A chain of protein $PDB\ 2DVZ_A$ showing the intersection cells and the tail lengths for both the C and N terminal chains.

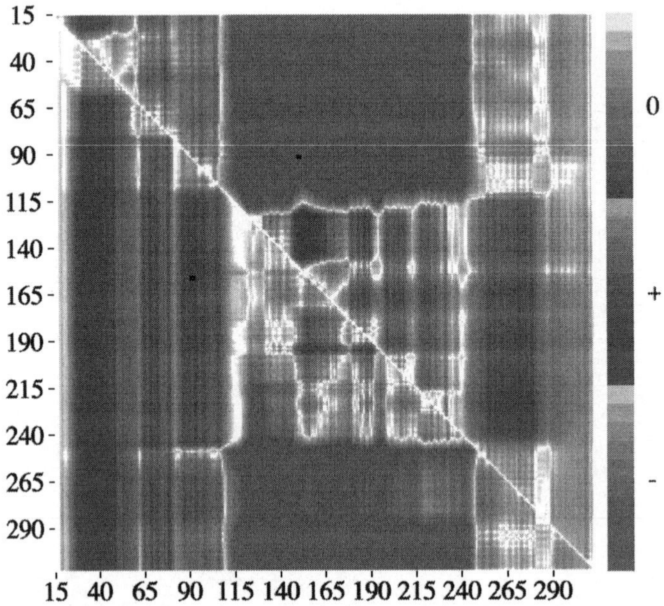

FIGURE 18. Linking fingerprint of the A chain of $2DVZ_A$ with the initial and terminal tail linking with the lasso loop shown at the indicated cells (93,152) and (152,93), respectively.

5. PyLasso & PyLink: PyMol tools to study linking entanglement due to the presence of one or more bridges in proteins

PyLasso [**GD**], https://pylasso.cent.uw.edu.pl, and **PyLink** [**GN**], https://pylink.cent.uw.edu.pl, are multiplatform, user-friendly plugins for PyMol giving us powerful tools to visualize the linking entanglement present in protein structures. PyLasso [**GD**] provides access to the analysis of all types of lassos

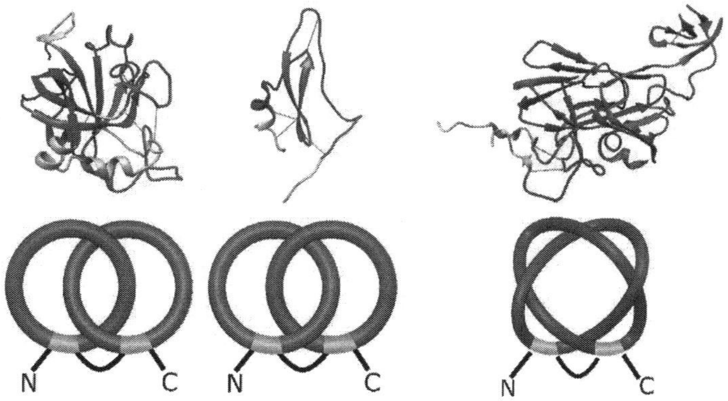

FIGURE 19. Examples of non-trivial link configurations [**DTJ**]

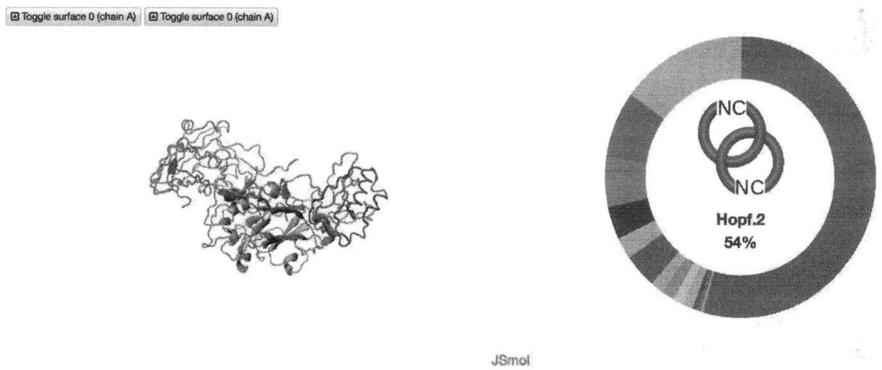

FIGURE 20. Stochastic linking of the A and C chains of a deformed wing virus (a honey bee pathogen) [**DTJ**]

using the methods of LassoProt and that the user can employ to explore potential new loops or can smooth structures to simplify visual analysis as well as minimal surfaces, Figure 21. Based upon both LassoProt and LinkProt, PyLink [**GN**] is further able to identify deterministic links created by the linking entanglement formed by covalent loops, for example, by the inclusion of disulfide bridges. It also supports the probabilistic linking formed by random closures to "infinity" and the macromolecular entanglement formed by two or more chains, e.g. dimers, trimers, etc. It includes the capacity to study spanning minimal surfaces and how other components pierce them, as these locations are likely to have biologial significance, Figure 22, and the capacity to employ a Gauss linking number analysis that has been used to characterize the relationship between entanglement and viscoelastic properties of materials [**P**].

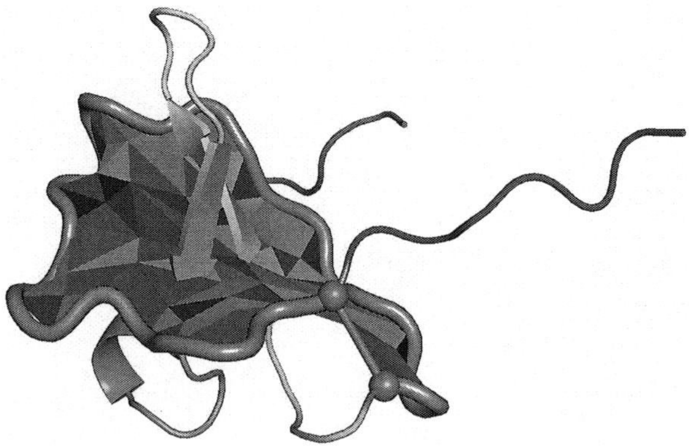

FIGURE 21. A PyLasso example showing the viral macrophage inflammatory protein II, PDB code 1CMG, chain A, exhibiting a lasso spanned by the closed loop between residues 14 and 48, *https://pylasso.cent.uw.edu.pl*

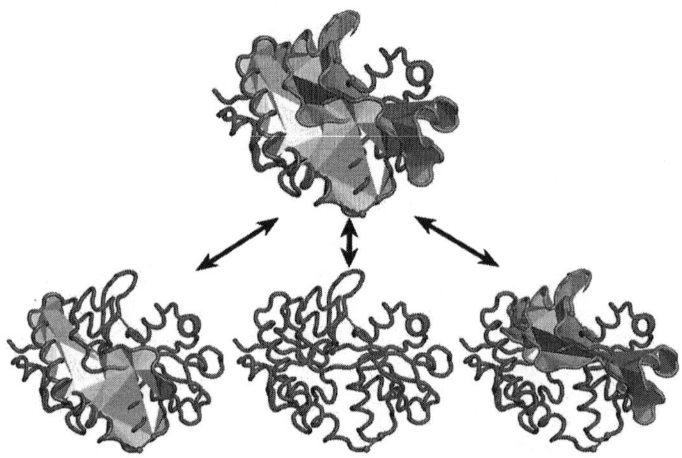

FIGURE 22. A PyLink example showing the transport protein, PDG 3T93 chain F, with the linking of two segments of the chain, residues 63-140 and 206-261, with pierced triangles indicated. Depending upon which surface is selected, the associated surfaces are displayed or hidden, *https://pylink.cent.uw.edu.pl*

6. Conclusion

The identification of knots and slipknots deeply embedded in proteins in 2000 was the catalyst for the development of new methods of identifying and localizing these structures in open molecular chains. The creation of the knotting fingerprint provided a new tool whose systematic application to all structures in the Protein Data Bank showed that while there was a substantial population of knotted and/or slipknotted structures, most structures were unknotted. As a consequence, one is led to search for structures within proteins that, while unknotted, exhibit many aspects of those that are knotted. The Gauss linking integral, Equation 1.1, can be applied to quantify the linking between two open chains as well as the self-linking present in a chain. Analyzing the local linking present within knotted segments in typical proteins provides a target fingerprint for which to search in unknotted proteins. This also provides a tool that can be applied to the quantification of entanglement present in the spatial conformations of one or more macromolecules. Thus, in this discussion, we have given a mathematical description of several strategies that exploit aspects of linking as a vehicle to illuminate topological features of protein structures. These have included the *linking matrix*, the *linking fingerprint* of a macromolecular chain, and the *local linking fingerprint* between two macromolecular chains. We have applied these methods to interesting protein structures in order to illustrate how one can employ them to identify important features to relate to available biological information.

To enable researchers to more widely explore linking structures, we have also described those that arise due to the presence of cysteine bonds and close interactions within structures. As a consequence, we have cited the servers : **LassoProt** and **LinkProt** and the new **PyMol** tools: **PyLasso** and **PyLink** that have followed in the footsteps of **KnotProt2.0**.

Overall our goal has been to describe ways one can study the occurence of linking entanglement in biomolecules arising from the backbone structures or from the additional consideration of disulphide bridges. Here, we have presented the mathematical ideas and associated methods so that those who may be interested in exploring their application to the study of specific structures such as proteins or RNA will have access to these powerful tools.

References

[B] T. Banchoff, *Self linking numbers of space polygons*, Indiana Univ. Math. J. **25** (1976), no. 12, 1171–1188, DOI 10.1512/iumj.1976.25.25093. MR0431226

[BC] D.R. Boutz, D. Cascio, J. Whitelegge, L.J. Perry, T.O. Yeates, *Discovery of a thermophilic protein complex stabilized by topologically interlinked chains* J. Mol. Bio. **368** (2007), 1332–1344.

[BW] H.M. Berman, J. Westbrook, Z. Feng, G. Gilliland, T.N. Bhat, H. Weissig, I.N. Shindyalov, P.E. Bourne. *The Protein Data Bank*, Nucleic Acids Research, **28**, (2000), 235-242.

[F] István Fáry, *Sur la courbure totale d'une courbe gauche faisant un nœud* (French), Bull. Soc. Math. France **77** (1949), 128–138. MR0033118

[G] K.F. Gauss, *Zur mathematischen theorie der electrodynamischen wirkungen*, Werke vol 5 Konigl. Ges. Wiss. Gottingen, (1877), page 605.

[GD] A.M. Gierut, W. Niemyska, P. Dabrowski-Tumanski, P. Sulkowski, J.I. Sulkowksa, *PyLasso: a PyMOL plugin to identify lassos*, Bioinformatics **33** 23 (2017), 3819–3821.

[GN] A.M. Gierut, P. Dabrowski-Tumanski, W. Niemyska, K.C. Millett, J.I. Sulkowksa, *PyLink: a PyMOL plugin to identify links*, Bioinformatics (2019)

[DT] P. Dabrowski-Tumanski, W. Niemyska, P. Pasznik, P. Sulkowski, J.I. Sulkowksa, *LassoProt: server to analyze biopolymers with lassos*, Nucleic Acids Res. **44** (2016), W383–W389.

[DTJ] P. Dabrowski-Tumanski, A. Jarmolinksa, W. Niemyska, E. Rawdon, K.C. Millett, J.I. Sulkowksa, *LinkProt: a database collecting information about biological links*, Nucleic Acids Res. **45** (2017), D233–D249.

[DTR] P. Dabrowski-Tumanski, P. Rubach, D. Goundaroulis, J. Dorier, P. Sułkowski, K. C. Millett, E. J. Rawdon, A. Stasiak, J. I. Sulkowska, *KnotProt 2.0: a database of proteins with knots and other entangled structures*, Nucleic Acids Research (2018), gky1140, https://doi.org/10.1093/nar/gky1140

[DTS] P. Dabrowski-Tumanski, J.I. Sulkowksa, *Topological knots and links in proteins*, Proc. Natl. Acad. Sci. USA **114** (2016), 3415–3420.

[H1] E. Haglund, J.I. Sulkowska, Z. He, G.S. Feng, P.A. Jennings, J.N. Onuchic, *The unique cysteine knot regulates the pleotropic hormone leptin*, PloS ONE (2012), 7:e45654.

[H2] E. Haglund, J.I. Sulkowska, J.K. Noel, H. Lammert, J.N. Onuchic, P.A. Jennings, *Pierced Lasso Bundles Are a New Class of Knot-like Motifs*, PloS Comput. Biol. (2014), 10:e1003613.

[H3] E. Haglund, *Engineering covalent loops in proteins can serve as an on/off switch to regulate threaded topologies*, J. Phys. Condens. Matter (2015), 27:354107

[J] M. Jamroz, W. Niemyska, E.J. Rawdon, A. Stasiak, K.C. Millett, P. Sulkowski, J.I. Sulkowksa, *KnotProt: a database of proteins with knots and slipknots*, Nucleic Acids Res. **43** (2015), D306–D314.

[K] N.P. King, E.O. Yeates, T.O. Yeates, *Identification of Rare Slipknots in Proteins and Their Implications for Stability and Folding*, J. Mol. Biol **373** (2007), 153–166.

[MDS] K.C. Millett, A. Dobay, A. Stasiak, *Linear Random Knots and Their Scaling Behavior*, Macromolecules **38** (2005), 601–606.

[MJ] J. W. Milnor, *On the total curvature of knots*, Ann. of Math. (2) **52** (1950), 248–257, DOI 10.2307/1969467. MR0037509

[MS] Kenneth C. Millett and Benjamin M. Sheldon, *Tying down open knots: a statistical method for identifying open knots with applications to proteins*, Physical and numerical models in knot theory, Ser. Knots Everything, vol. 36, World Sci. Publ., Singapore, 2005, pp. 203–217, DOI 10.1142/9789812703460_0011. MR2197941

[N] W. Niemyska, P. Dabrowski-Tumanski, M. Kadlof, E. Haglund, P. Sułkowski, J.I. Sulkowska, *Complex lasso: new entangled motifs in proteins*, Scientific reports, **6**, (2016), 36895.

[P] E. Panagiotou, K.C. Millett, P. Atzberger, *Topologial methods for polymeric materials characterizing the relationship between polymer entanglement and viscoelasticity*, Polymers (2019) 11(3), 437–457.

[R] E.J. Rawdon, K.C. Millett, J.I. Sulkowska, A. Stasiak, *Knot localization in proteins*, Biochem. Soc. Trans. **41** No. 2, (2013), 538–541.

[Ri] J. S. Richardson, *β-sheet topology and the relatedness of proteins*, Nature **268**, (1977), 495–500.

[S] J.I. Sulkowska, E.J. Rawdon, K.C. Millett, J.N. Onuchic, A. Stasiak, *Conservation of complex knotting and slipknotting petterns in proteins*, Proc. Natl. Acad. Sci. USA **109** No. 26, (2012), E1715-1723.

[T] W.R. Taylor, *A deeply knotted protein and how it might fold*, Nature **406** (2000), 916–919.

[V] B.N. Violand, M. Takano, D.F. Curran, L.A. Bentle, *A novel concatenated dimer of recombinant bovine somotoropin*, J. Protein Chem., **8**, 5, (1989), 619-628.

[Z] L.M. Zhang, C.N. Morrison, J.T. Kaiser, D.C. Rees, *Nitrogenase MoFe protein from Clostridium pasteurianum at 1.08 angstrom resolution: comparison with the Azotobacter vinelandii MoFe protein*, Acta Crystallorgr. Sect. D **71** (2015), 274–281.

Department of Mathematics, University of California, Santa Barbara, Santa Barbara, CA 93108

Email address: `millett@math.ucsb.edu`

Contemporary Mathematics
Volume **746**, 2020
https://doi.org/10.1090/conm/746/15010

A topological study of protein folding kinetics

Eleni Panagiotou and Kevin W. Plaxco

ABSTRACT. Focusing on a small set of proteins that i) fold in a concerted, "all-or-none" (two-state) fashion and ii) do not contain knots or slipknots, we show that the Gauss linking integral, the torsion and the number of sequence-distant contacts provide information regarding the folding rate. Our results suggest that the global topology/geometry of the proteins shifts from right-handed to left-handed with decreasing folding rate, and that this topological change is associated with an increase in the number of more sequence-distant contacts.

1. Introduction

Proteins, a diverse set of macromolecules each comprised by a unique sequence of amino acids, are a fundamental building block of living organisms [**1**]. In order for proteins to function they must attain conformations that belong in a small subset of the possible 3-dimensional conformations a macromolecule can attain, called the *native state. Protein folding* is the process by which a protein molecule folds into this unique, three-dimensional conformation. Understanding protein folding is important in predicting of the native structure of a protein from its amino acid sequence which could lead to a better understanding of protein function as well as to the engineering of biopolymers with desired functions. In this study we use topological tools to understand how the topology of the protein relates to its folding rate, the rate with which it spontaneously folds from a random state to the native state.

The 3-dimensional conformation (also known as *tertiary structure*) is determined by the protein sequence (called *primary structure*) [**10**]. Some patterns of 3-dimensional structure that appear often in proteins are α-helices and β-strands. The sequence of such patterns is called the *secondary structure* of the protein. Structure is a very important clue to understanding and manipulating biological function. However, protein folding is very complex and it has been very difficult to provide a model for it. An interesting fact is that the rate at which proteins fold varies significantly from one protein to the other. At first, this might seem intuitive, since the length of the proteins also varies and one might expect a longer macromolecule to require a longer time to organize itself into a particular 3-dimensional structure. However, studies have shown that the protein length correlates only weakly with the protein folding rate [**24**]. This lead to the study of the topological

---

2010 *Mathematics Subject Classification.* Primary 57M25, 92B05.

complexity of the native state[**5**, **11**, **13**, **14**, **23**, **24**]. In [**23**], a measure of topological complexity, the *relative contact order*, was introduced and showed a very strong correlation with protein folding rates. Moreover, a simpler measure, the *number of sequence distant contacts*, which depicts the sequence distant amino acids that are close in 3-space, provided a similarly strong correlation with the protein folding rate. The number of sequence distant contacts is a global measure of structure which captures features beyond those captured in secondary structure motifs. Its strongly statistically significant correlation with experimental folding rates highlights the extent to which the topology of the native state influences the folding process. The number of contacts was also successfully used to provide a model for protein folding [**11**]. The underlying idea is that a high number of sequence distant contacts corresponds to few local interactions, suggesting that the route from the unfolded conformation to the native state is slow due to the need to overcome the barriers arising from these spatial restraints. However, it is unknown if the number of sequence distant contacts is a proxy for some more physically meaningful topological measure, which could help us improve our understanding of protein folding kinetics.

In this study we explore the ability of various parameters more directly associated with topology to predict protein folding rates. To do so we have looked at the sequence of CA atoms (all amino acids possess a central, or α, carbon) of the proteins as vertices connected by edges, such that we can represent the protein as a polygonal curve. To explore the complexity of such curves we invoke knot theory. Specifically, a measure of entanglement that can be directly applied to open chains and is a continuous function of the chain coordinates is the Gauss linking integral. This measure has been successfully applied to polymers in order to study polymer entanglement and it has been shown that it correlates with the physical properties of polymeric material [**19**–**22**]. The Gauss linking integral has likewise been used to classify protein conformations successfully [**3**, **25**] and it was shown that the maximum linking between any two parts of a protein correlates with its folding rate [**4**]. In this study we further use the Gauss linking integral to reveal new characteristics of protein entanglement relevant to protein folding kinetics that can be useful in future models of protein folding. More precisely, we study the correlation between folding rate and the geometry/topology of the entire protein and also that of the secondary structures that compose it.

The paper is organized as follows: in Section 2 we define the measures of geometrical/topological complexity we use to characterize protein structure and Section 3 presents our results.

2. Characterization of protein structure

2.1. The number of contacts. A way in which a straight conformation of a chain differs from a more complex, curved conformation is that in the straight conformation, the 3-space distance of the i-th and j-th monomer of the molecule is $|i-j|$ times the bond length. This is not the case for any other conformation, where chains whose sequential distance is $|i-j|$ may have been brought closer in 3-space. Therefore, a way to measure the topological complexity of the native state of a protein is to measure how much this property deviates from that of the straight configuration, by accounting for the number of sequence distant contacts [**23**].

We say that two monomers form a contact if their CA-CA distance is less than 6 Å (a value typical for amino acids in close physical contact in the native structure) and their sequence distance greater than 12 amino acids. Closer contacting pairs were excluded because contacts of closer amino acids are trivial as these are enforced by the connectivity of the chain. We denote by Q_D the number of sequence distant contacts in a protein. It has been shown that the number of contacts and measures related to that (the absolute and relative contact order) correlate well with the protein folding rate [**5, 23**]. Moreover, in [**11**] the number of sequence-distant contacts was used to provide a model for protein folding. In this paper we will not provide a model for protein folding, but we will examine the correlations of several topological parameters with folding rates and their relation to the number of contacts.

2.2. Gauss linking number and the total torsion. A measure of the degree to which polymer chains interwind and attain complex configurations is given by the Gauss linking integral:

DEFINITION 2.1. (Gauss Linking Number). The Gauss *Linking Number* of two disjoint (closed or open) oriented curves l_1 and l_2, whose arc-length parametrizations are $\gamma_1(t), \gamma_2(s)$ respectively, is defined as the following double integral over l_1 and l_2 [**9**]:

$$L(l_1, l_2) = \frac{1}{4\pi} \int_{[0,1]} \int_{[0,1]} \frac{(\dot{\gamma}_1(t), \dot{\gamma}_2(s), \gamma_1(t) - \gamma_2(s))}{||\gamma_1(t) - \gamma_2(s)||^3} dtds, \tag{2.1}$$

where $(\dot{\gamma}_1(t), \dot{\gamma}_2(s), \gamma_1(t) - \gamma_2(s))$ is the *scalar triple product* of $\dot{\gamma}_1(t), \dot{\gamma}_2(s)$ and $\gamma_1(t) - \gamma_2(s)$.

The Gauss Linking Number is a topological invariant for closed chains and a continuous function of the chain coordinates for open chains.

We also define a one chain measure for the degree of intertwining of the chain around itself:

DEFINITION 2.2. (Writhe). For a curve ℓ with arc-length parameterization $\gamma(t)$ is the double integral over l:

$$Wr(l) = \frac{1}{4\pi} \int_{[0,1]} \int_{[0,1]} \frac{(\dot{\gamma}(t), \dot{\gamma}(s), \gamma(t) - \gamma(s))}{||\gamma(t) - \gamma(s)||^3} dtds. \tag{2.2}$$

The Writhe is a continuous function of the chain coordinates for both open and closed chains. The Average Crossing Number (ACN) is obtained when we consider the absolute value of the integrand in the Writhe. ACN measures the expected number of crossings of a chain with itself in a random projection, while the writhe gives the expected number of signed crossings, which accounts for which strands come over and under. Similarly, the linking number of two chains, measures the expected number of signed inter-crossings of two arcs with each other accounting for which comes over and under in a random projection, divided by two.

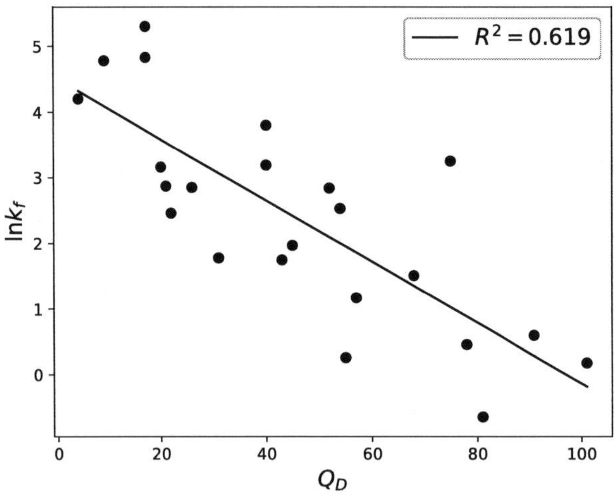

FIGURE 1. The folding rate is decreasing with increasing number of sequence distant contacts.

The total torsion of a chain, describes how much it deviates from being planar and is defined as:

DEFINITION 2.3. The *torsion* of a curve ℓ with arc-length parameterization $\gamma(t)$ is the double integral over l:

$$T(l) = \frac{1}{2\pi}\int_{[0,1]} \frac{(\gamma'(t) \times \gamma''(t)) \cdot \gamma'''(t)}{||\gamma'(t) \times \gamma''(t)||^2} dt. \tag{2.3}$$

Notice that it is possible to construct a random walk with high writhe or torsion that does not contain a knot, however, the mean absolute value of the writhe and torsion increases with knot complexity [**2, 6, 7, 15, 22**]. In the same way, one can construct a configuration of a non-flat, polygonal chain with zero writhe. However, the writhe of a random walk is non-zero with probability one.

3. Results

We analyze a set of simple, single domain, non-disulfide-bonded proteins that have been reported to fold in a concerted, "all-or-none", two-state fashion. Despite their simplicity compared to other proteins, the slowest folding rate of the proteins in the set is a million times slower than the fastest folding rate in the same set. Therefore, this provides a dataset where geometrical and topological features important in protein folding may become aparent. The set of proteins and their corresponding extrapolated folding rates in water can be found in [**24**].

3.1. The number of contacts. Figure 1 shows the folding rate as a function of the number of sequence-distant contacts, Q_D, in a protein. We see that the number of contacts increases with decreasing folding rate. This suggests that the more global interactions are required to form the native state, the slower folding is. The correlation coefficient for the relationship between $\log k_f$ and Q_D is $R^2 = 0.619$.

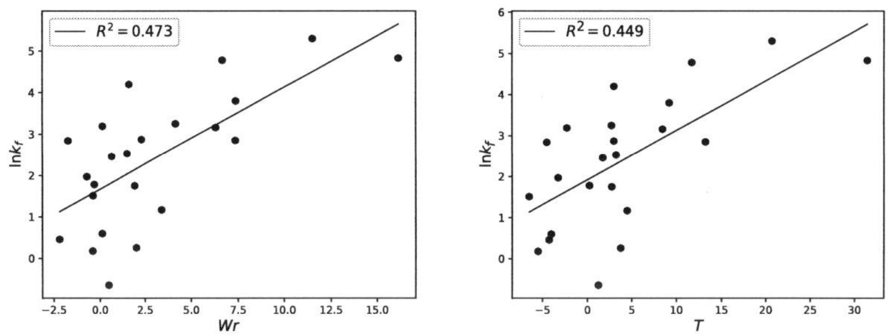

FIGURE 2. The writhe and torsion of the proteins. It is known that the folding rate is influenced by the number of helices. A decreasing number of helices may lead to a decrease of the writhe and torsion of the proteins.

3.2. The writhe and total torsion of the proteins. The set of proteins we analyze in this study do not contain knots or slipknots and their length is in the range of 50-150 amino acids. The mean absolute writhe of unknotted random coils of lengths in the range of the proteins under study is small and one might expect the writhe of the proteins to be small as well [**7, 8, 19, 22**]. Figure 2 shows the folding rate as a function of the writhe and torsion of the proteins. The values of the writhe of the proteins significantly exceed the expected values of comparable length random walks. The writhe and the torsion take mostly positive values and are decreasing with decreasing folding rate with correlation coefficients $R^2 = 0.473$ and $R^2 = 0.449$, respectively.

At first the fact that the folding rate decreases with decreasing writhe and torsion of the proteins might seem counterintuitive, since one might expect folding rates to decrease as topological complexity increases. However, we notice that the writhe is influenced by the presence of secondary structures. The helices found in proteins are almost invariably right handed due to steric (physical excluded volume) interactions between individual amino acids (which are themselves chiral) and the chirality of the helix. This causes them to contribute a large, positive amount of writhe. β-strands have particular configurations that have a small in absolute value, but negative writhe (see Section 3.3). It is also known that the folding rate decreases with decreasing number of helices, while the number of β-strands increases [**12**]. Therefore, the observed behavior of the writhe and torsion of the proteins may be reflecting exactly this change in secondary structure. Notice that this change is local and may be hiding the underlying global topology/geometry of the protein. In order to extract information about the global geometry of a protein we propose to study the conformation of its *primitive path.* We define the primitive path (PP) of the protein (inspired by the tube model for polymer melts [**26**]) to be the axis of the thinnest tube that surrounds the chain, with no self-intersections and whose diameter (approximately 6 Å) is that of a helix.

To compute the writhe of the PP in practice, we obtain a semi-analytical formula for the writhe of the PP of the protein using the Gauss linking integral only, without constructing the tube.

First, we notice that the writhe of the protein can be expressed as:

$$(3.1)\quad Wr(\text{protein}) = \sum_{i\in S} Wr(s_i) + \sum_{j\in \text{ coils}} Wr(\text{coil } j) + \sum_{i,j\in S} Lk(s_i, s_j) + \sum_{i\in S} \sum_{j\in \text{coils}} Lk(s_i, \text{coil } j) + \sum_{i,j\in \text{ coils}} Lk(\text{coil } i, \text{coil} j)$$

where S denotes the set of secondary structure elements of the protein.

The protein PP is formed by replacing the protein secondary structure elements by their axis, giving PP$= \{e_i | e_i$ axis of secondary element $i\} \cup \{\text{coils}\}$. Notice that $Wr(e_i) = 0$ for all axes e_i. Thus,

$$(3.2)\quad Wr(\text{PP}) = \sum_{i,j\in S'} Lk(e_i, e_j) + \sum_{i,j\in S'} Lk(e_i, \text{coil j}) + \sum_{j\in \text{ coils}} Wr(\text{coil } j) + \sum_{i,j\in \text{ coils}} Lk(\text{coil } i, \text{coil} j)$$

where $S' = \{e_i | e_i$ axis of secondary element $i\}$.

We notice that, since the secondary structure elements lie in disjoint cells (disjoint convex hulls) from coils and from each other and due to sign cancellations, we can make the following approximations:

$$(3.3)\quad Lk(e_i, \text{coil } j) \approx Lk(s_i, \text{coil j})$$

and

$$(3.4)\quad Lk(e_i, e_j) \approx Lk(s_i, s_j)$$

Using Eq. 3.3 and 3.4 in Eq. 3.2 and comparing it with Eq. 3.1, the writhe of the PP can be obtained by the writhe of the protein and its secondary structure elements as:

$$(3.5)\quad Wr(\text{PP}) \approx Wr(\text{protein}) - \sum_{\alpha-\text{helices}} Wr_{\alpha-\text{helix}} - \sum_{\beta-\text{strands}} Wr_{\beta-\text{strand}}$$

Similarly, we define the ACN and the torsion of the PP to be:

$$(3.6)\quad ACN(\text{PP}) = ACN(\text{protein}) - \sum_{\alpha-\text{helices}} ACN_{\alpha-\text{helix}} - \sum_{\beta-\text{strands}} ACN_{\beta-\text{strand}}$$

and

$$(3.7)\quad T(\text{PP}) = T(\text{protein}) - \sum_{\alpha-\text{helices}} T_{\alpha-\text{helix}} - \sum_{\beta-\text{strands}} T_{\beta-\text{strand}}$$

The folding rate as a function of the writhe and the torsion of the PP are shown in Figure 3. The range of values of the writhe and the torsion is now comparable to those of random walks of similar lengths. We see that the folding rate is decreasing with decreasing writhe and torsion of the PP which start from positive values and attain negative values with correlation coefficients $R^2 = 0.483$ and $R^2 = 0.456$, respectively. This reveals a significant change in the global conformation of proteins which affects their folding rate: the proteins fold more slowly when they need to attain a configuration with negative global writhe and torsion. This may suggest a role of handedness of proteins not only in local organization but also in global structure.

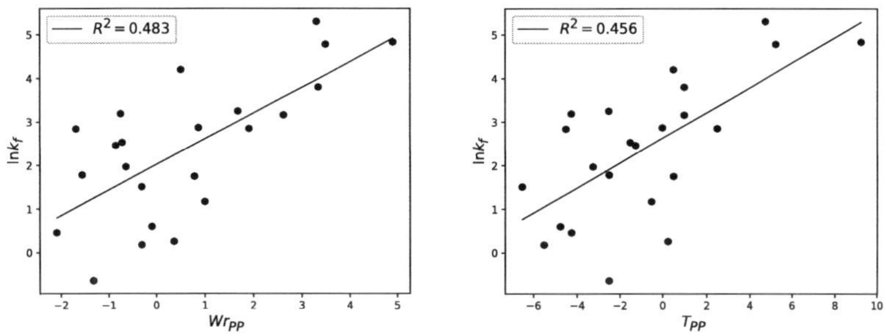

FIGURE 3. Writhe and torsion of the primitive path. The folding rate is decreasing with the writhe and the torsion of the PP shifting from positive to negative values.

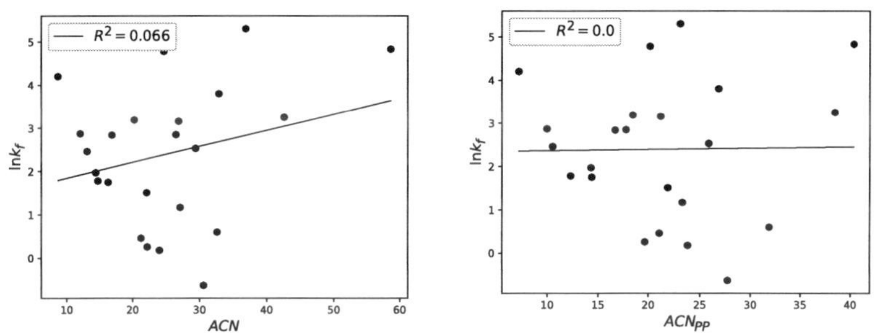

FIGURE 4. The ACN of the protein and its PP. The ACN of the PP shows that the complexity is not decreasing even when the writhe is zero, confirming that a change in global handedness of the protein may be the reason for decreasing folding rates.

The folding rate as a function of the ACN of the proteins and their PP is shown in Figure 4. The ACN shows a small correlation with folding rate and its values are larger than those for similar length random walks [**8**]. This is expected due to the high ACN values of helices. As the folding rate decreases, the ACN of the PP remains large, even when the writhe is decreasing, confirming that the decrease of the writhe is not reflecting a decrease in complexity but a preference for left-handed global conformations. This may suggest that some left-handed confromations have low energy but they are more difficult to attain [**18**].

3.3. The topology and geometry of secondary structures. It has been shown that proteins whose native state has only α-helices fold more rapidly than proteins that also contain β-strands, while the proteins that only contain β-strands fold even more slowly [**16, 17**]. Our results so far suggest that the folding rate is related to the global rearrangement of the secondary structure elements in space and not on the local structure of α-helices and β-strands. In order to understand the origins of this global rearrangement which results in negative global writhe, in this section, we study how the different secondary structure elements (α-helices, β-strands and coils) link with each other and how this correlates with the folding rate. More precisely, we examine the linking between α-helices, coils and β-strands.

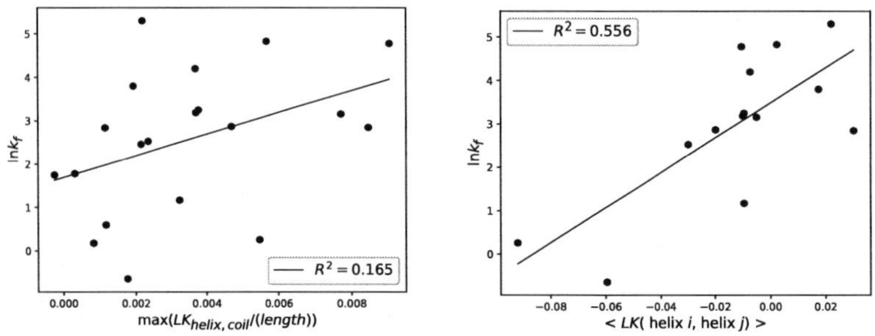

FIGURE 5. Left: The folding rate as a function of the maximum absolute linking number between helices with neighboring coils. Right: The folding rate as a function of the average linking number of helices in a protein. The folding rate is decreasing with decreasing linking number, suggesting that the proteins fold slowly when a left-handed conformation of helices is required.

(In the following we report only the results that showed the strongest correlations for this dataset.)

When examining the linking of α-helices with coils in a protein we notice that the linking number is rather random in sign and we find only a small correlation with folding rate, $R^2 = 0.077$. However, we find that the linking of a helix with the neighboring (in secondary structure) coils is mostly positive and the protein folding rate tends to decrease with decreasing linking. The maximum absolute linking number relative to the length of the coil and helix involved is shown in Figure 5. We see that the folding rate is decreasing with decreasing maximum linking number of a helix with a neighboring coil, with $R^2 = 0.165$. The fact that it is positive shows a right-handed preference for the conformation helix-neighboring coil. The tendency of the protein folding rate to decrease with decreasing maximum linking between helices and coils may indicate an increasing difficulty of the protein to keep right-handed conformations, presumably due (again) to interactions between the chirality of the chain's fold and the chirality of the individual amino acids. Figure 5 also shows the average linking number between helices in each protein of the dataset that contains at least two helices. The folding rate decreases as the average linking number between helices becomes more and more negative, with $R^2 = 0.556$, confirming our results in the previous section which suggest a change in the preferred handedness of the global linking of the proteins.

Our data do not show a correlation of the folding rate with the absolute value of the linking between secondary structures, suggesting that the folding rate may depend only on the sign of the linking between secondary structures. To further explore this, we examine if the folding rate correlates with the number of occurrences of pairs of secondary structure elements which have negative linking. Figure 6 shows the folding rate as a function of the number of pairs of helices or pairs of helices-coils with negative linking number over the total number of pairs. Figure 6 also shows the number of sequence distant contacts as a function of the number of pairs of helices and helices coils with negative linking. The results suggest that conformations of negative linking between helices and conformations of negative linking between a helix and a coil, require many contacts, which causes the folding

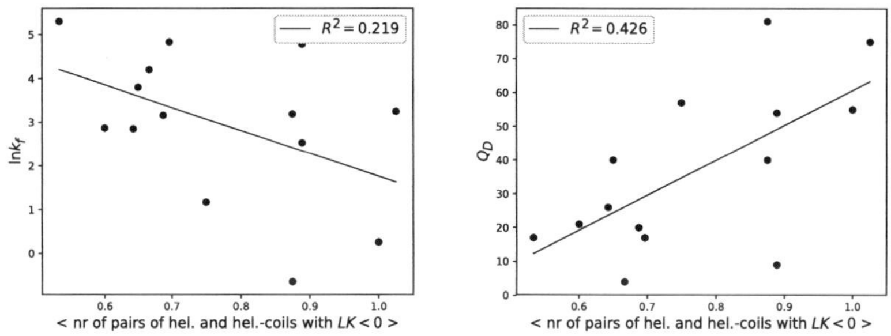

FIGURE 6. The folding rate and the number of sequence distant contacts as a function of the relative number of pairs of α-helices or pairs of α-helices and coils with negative linking number.

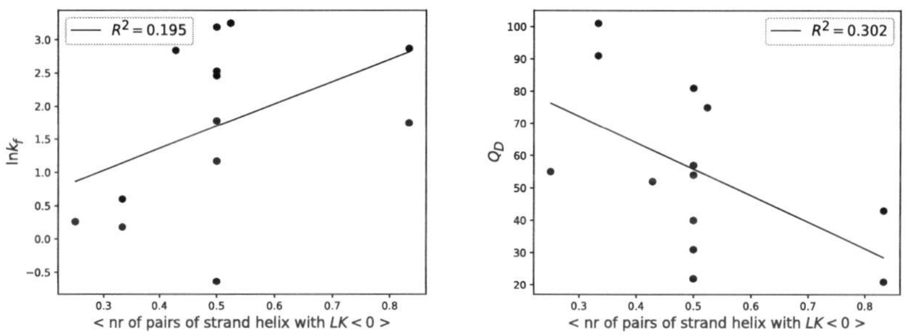

FIGURE 7. The folding rate and the number of sequence distant contacts as a function of the relative number of pairs of α-helices and β-strands with negative linking number. The 0.5 ratios indicate the presence of antiparallel β-strands which contribute a negative and positive linking number.

rate to decrease. Conversely, it may be true that negative linking is difficult for the chain to achieve (perhaps due to interactions between the "handedness" (all amino acids in proteins are of the "L" chirality) of the chain and the handedness of negative linking interactions, and that the number of sequence distant contacts is a proxy for this property.

The sign of the linking number between α-helices and β-strands is rather random ($R^2 = 0.089$). However, the folding rate is decreasing with decreasing number of pairs of α-helices and β-strands with negative linking over the total number of pairs, with $R^2 = 0.195$ (see Figure 7). We notice a large set of proteins with exactly 0.5 ratio of pairs of helices-strands with negative linking. We believe that this is because of the particular conformations of sheets, with antiparallel β-strands, which contribute an opposite sign to the average.

These results suggest a dependence of the folding rate simply on the number of negative linking occurrences involving secondary structure elements, and not particularly on the values of these linking occurrences. Since the secondary structure elements follow an almost straight axis, the source of change in their relative orientation stems from the coils in-between. Indeed, the coils can attain any random configuration. Our results show that the writhe of the coils is on average positive, but negative writhe coils also exist and our results suggest that the folding rate

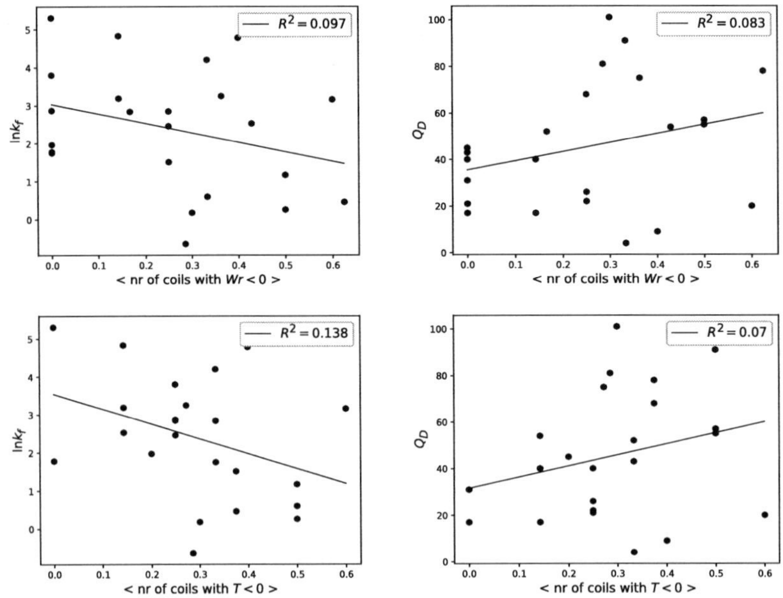

FIGURE 8. The folding rate and the number of sequence distant contacts as a function of the relative number of coils with negative writhe (above) and torsion (below). The folding rate is related to the number of coils with negative torsion, even if their presence does not correlate with the number of sequence distant contacts.

decreases as the number of coils with negative writhe is increasing. In Figure 8 we see that folding rate is decreasing with increasing number of coils with negative writhe and also with number of coils with negative torsion over the total number of coils in a protein with $R^2 = 0.097$ and $R^2 = 0.138$.

Figure 8 shows also the number of sequence distant contacts as a function of the number of coils with negative writhe and as a function of coils with negative torsion. The results suggest that more sequence distant contacts are required for increasing number of negative writhe and negative torsion coils. However, the results show that the folding rate is more sensitive to the number of negative writhe/torsion coils than the number of sequence distant contacts. Therefore, this may be a feature that is only weakly captured by the number of sequence distant contacts but influences the folding rate.

4. Conclusions

Studies have shown that the folding rates of proteins that fold in a concerted all-or-none fashion correlate very well with the number of sequentially distant contacts [**11**]. By studying a set of proteins which fold in a concerted, all-or-none fashion and do not contain knots, we showed that their folding rates and the number of sequence-distant contacts correlate with their writhe, ACN and torsion. By examining the global entanglement of the proteins, ignoring the local secondary structure, we showed that the folding rate decreases as the global writhe and torsion of the proteins becomes more and more negative. We find that the folding rate is related to the relative orientation of helices and strands in space, with negative linking conformations being associated with more sequence distant contacts and, in turn, slower folding. We also see that coils with negative writhe or torsion slow down the folding process even if they do not require many contacts. Our results confirm that the highly organized native structure is too complex to be captured by a single parameter, and suggest that the combination of the topological and geometrical tools, such as the number of sequence distant contacts and the Gauss linking integral, could provide complementary information useful in the search of a model for protein folding kinetics.

References

[1] B Alberts, A. Johnson, J. Lewis, M. Raff, K. Roberts, and P. Walter. *Molecular Biology of the Cell.* New York: Garland Science, 2002.

[2] J. Arsuaga, T. Blackstone, Y. Diao, E. Karadayi, and M. Saito, *Linking of uniform random polygons in confined spaces*, J. Phys. A **40** (2007), no. 9, 1925–1936, DOI 10.1088/1751-8113/40/9/001. MR2316305

[3] G. A. Arteca and O. Tapia. Relative measure of geometrical entanglement to study folding-unfolding transitions. *Int. J. Quantum Chem.*, 80:848–855, 2000.

[4] M. Baiesi, E. Orlandini, F. Seno, and A. Trovato. Exploring the correlation between the folding rates of proteins and the entanglement of their native state. *J. Phys. A: Math. Theor.*, 50:504001, 2017.

[5] A. Broom, S. Gosavi, and E. A. Meiering. Protein unfolding rates correlate as strongly as folding rates with native structure. *Prot. Sci.*, 24:580–7, 2015.

[6] Y. Diao, A. Dobay, and A. Stasiak, *The average inter-crossing number of equilateral random walks and polygons*, J. Phys. A **38** (2005), no. 35, 7601–7616, DOI 10.1088/0305-4470/38/35/001. MR2169479

[7] Y. Diao, C. Ernst, K. Hinson, and U. Ziegler, *The mean squared writhe of alternating random knot diagrams*, J. Phys. A **43** (2010), no. 49, 495202, 21, DOI 10.1088/1751-8113/43/49/495202. MR2740363

[8] Y. Diao, A. Dobay, R. B. Kusner, K. Millett, and A. Stasiak, *The average crossing number of equilateral random polygons*, J. Phys. A **36** (2003), no. 46, 11561–11574, DOI 10.1088/0305-4470/36/46/002. MR2025860

[9] K. F. Gauss. *Werke.* Kgl. Gesellsch. Wiss. Göttingen, 1877.

[10] C. Lawrence, J. Kuge, K. Ahmad, and K. W. Plaxco. Investigation of an anomalously accelerating substitution in the folding of a prototypical two-state protein. *J. Mol. Biol.*, 403:446–458, 2010.

[11] D. E. Makarov and K. W. Plaxco. The topomer search model: asimple, quantitative theory of two-state protein folding kinetics. *Protein Science*, 12:17–26, 2003.

[12] S. Malik, T. Ray, and S. Kundu. Transiently disordered tails accelerate folding of globular proteins. *FEBS Lett*, 591:2180–2191, 2017.

[13] K. L. Maxwell, D. Wildes, A. Zarrine-Afsar, M. A. De Los Rios, A. G. Brown, C. T. Friel, L. Hedberg, J. Horng, D. Bona, E. J. Miller, A. Vallée-Bélise, E. R. G. Main, F. Bemporad, L. Qiu, K. Teilum, N. Vu, A. M. Edwards, I. Ruczinski, F. M. Poulsen, B. B. Kragelung, S. W. Michnick, F. Chiti, Y. Bai, S. J. Hagen, L. Serrano, M. Oliveberg, D. P. Raleigh, P. Wittung-Stafshede, S. E. Radford, S. Jackson, T. R. Sosnick, S. Marqusee, A. R. Davidson, and K. Plaxco. Protein folding: defining a "standard" set of experimental conditions and a prelimiray kinetic data set of two-state proteins. *Protein Science*, 14:602–616, 2005.

[14] C. Micheletti. Prediction of folding rates and transition-state placement from native-state geometry. *Proteins*, 51:74–84, 2003.

[15] K. C. Millett, A. Dobay, and A. Stasiak. Linear random knots and their scaling behavior. *Macromolecules*, 38:601–606, 2005.

[16] V. Munoz and M. Cerminara. When fast is better: protein folding fundamentals and mechanisms from ultrafast approaches. *Biochem. J.*, 473:2545–2559, 2016.

[17] H. Nguyen, M. Jäger, J. W. Kelly, and M. Gruebele. Engineering a β-sheet toward the folding speed limit. *J. Phys. Chem. B Lett*, 109:15182–15186, 2005.

[18] J. N. Onuchic, P. G. Wolynes, Z. Luthey-Schulten, and N. D. Socci. Toward an outline of the topography of a realistic protein-folding funnel. *PNAS*, 92:3626–3630, 1995.

[19] E. Panagiotou and M. Kröger. Pulling-force-induced elongation and alignment effects on entanglement and knotting characteristics of linear polymers in a melt. *Phys. Rev. E*, 90:042602, 2014.

[20] E. Panagiotou, M. Kröger, and K. C. Millett. Writhe and mutual entanglement combine to give the entanglement length. *Phys. Rev. E*, 88:062604, 2013.

[21] E. Panagiotou, K. C. Millett, and P. Atzberger. Topological methods for polymeric materials: Characterizing the relationship between polymer entanglement and viscoelasticity. *Polymers*, 11:437, 2019.

[22] E. Panagiotou, K. C. Millett, and S. Lambropoulou, *The linking number and the writhe of uniform random walks and polygons in confined spaces*, J. Phys. A **43** (2010), no. 4, 045208, 28, DOI 10.1088/1751-8113/43/4/045208. MR2578732

[23] K. W. Plaxco, K. T. Simons, and D. Baker. Contact order, transition state placement and the refolding rates of single domain proteins. *J. Mol. Biol.*, 277:985–994, 1998.

[24] K. W. Plaxco, K. T. Simons, I. Ruczinski, and D. Baker. Topology, stability, sequence and length: defining the determinants of two-state protein folding kinetics. *Biochemistry*, 39:11177–11183, 2000.

[25] P. Rogen and B. Fain. Automatic classification of protein structure by using gauss integrals. *Proc. Natl Acad. Sci*, 100:119–24, 2003.

[26] M. Rubinstein and R. Colby. *Polymer Physics.* Oxford University Press, 2003.

Department of Mathematics and SimCenter, University of Tennessee at Chattanooga, TN 37403

Email address: `eleni-panagiotou@utc.edu`

Department of Chemistry and Biochemistry and Center for Bioengineering, University of California Santa Barbara, CA 93106-3080

Email address: `kwp@chem.ucsb.edu`

Selected Published Titles in This Series